Earth and Environmental Sciences Library

Earth and Environmental Sciences Library (EESL) is a multidisciplinary book series focusing on innovative approaches and solid reviews to strengthen the role of the Earth and Environmental Sciences communities, while also providing sound guidance for stakeholders, decision-makers, policymakers, international organizations, and NGOs.

Topics of interest include oceanography, the marine environment, atmospheric sciences, hydrology and soil sciences, geophysics and geology, agriculture, environmental pollution, remote sensing, climate change, water resources, and natural resources management. In pursuit of these topics, the Earth Sciences and Environmental Sciences communities are invited to share their knowledge and expertise in the form of edited books, monographs, and conference proceedings.

Nitish Kumar
Editor

Heavy Metal Remediation

Sustainable Nexus Approach

Springer

Editor
Nitish Kumar
Department of Biotechnology
Central University of South Bihar
Gaya, Bihar, India

ISSN 2730-6674 ISSN 2730-6682 (electronic)
Earth and Environmental Sciences Library
ISBN 978-3-031-53687-8 ISBN 978-3-031-53688-5 (eBook)
https://doi.org/10.1007/978-3-031-53688-5

This Springer imprint is published by the registered company Springer Nature Switzerland AG
The registered company address is: Gewerbestrasse 11, 6330 Cham, Switzerland

Preface

In recent years, contamination of the environment by heavy metals has become a major concern.

Their multiple industrial, domestic, agricultural, medical, and technological applications have led to their wide distribution in the environment, raising concerns over their potential effects on human health and the environment. Owing to their toxic, non-degradable and bio-accumulative nature, the health burden on the population has increased significantly. Heavy metals such as arsenic, lead, mercury, cadmium and uranium don't play a significant role in metabolism in the human body and are thus toxic. Their exposure in high concentration can cause acute toxicity resulting in acute health conditions, which is easy to observe and regulate while similar is not visible for immediate action when their exposure is in trace amounts over the years. Heavy metals enter in the food chain through consumption of plant material. A high concentration of heavy metals has been found to be harmful to vegetation. As the heavy metals concentration in plants increases, it adversely affects several biological parameters and eventually renders the soil barren.

This edited volume brings together a diverse group of environmental science, sustainability and health researchers to address the challenges posed by global mass poisoning caused by heavy metals contamination of soil and plants. The book sheds light on this global environmental issue and proposes solutions to contamination through multi-disciplinary approaches and case studies from different parts of the world. This book is a valuable resource to students, academicians, researchers and environmental professionals who are doing field work on heavy metals contamination throughout the world.

Gaya, Bihar, India Nitish Kumar

Acknowledgment

First and foremost, I would like to praise and thank God, the almighty, who has granted countless blessing, knowledge and opportunity to accomplish the book project work. Thanks to all the authors of the various chapters for their contributions. It had been a bit of a long process from the initial outlines to developing the full chapters and then revising them in the light of reviewer's comments. I sincerely acknowledge the author's willingness to go through this process. I also acknowledge the work and knowledge of the members of our review panels, many of which had to be done at short notice. Thanks to all the people at Springer Nature, especially Dr. Qiao Shu and Ms. Saranyaa Vasuki Balasubramanian with whom I corresponded for their advice and facilitation in the production of this book. I am grateful to my family members Mrs. Kiran (wife), Miss Kartika Sharma and Laavanya Sharma (daughters) and parents for their incredible and selfless support all the time.

Gaya, Bihar, India Nitish Kumar

Contents

The Source and Distribution of Heavy Metals in the Atmosphere Across Southeast Asia 1
Ly Sy Phu Nguyen, Thi Thuy Linh Do, Truong Gia Han Vo, Quoc Hau Le, and To Thi Hien

Heavy Metals in the Ecosystem; Sources and Their Effects 27
Abderrahim Maftouh, Omkaltoume El Fatni, Asma Ben Moussa, Faissal Boukir, and Waliha Noor us Subha

Management of New Methods Used in the Reclamation of Soils Contaminated with Heavy Metals for Sustainable Agriculture: The Case of Biochar 45
Ahmet Çelik, Korkmaz Bellitürk, and Mehmet Fırat Baran

Management of Soil and Water Contaminated with Heavy Metals with Sustainable Green Technologies: The Case of Vermicomposting 65
Ahmet Çelik, Korkmaz Bellitürk, and Mehmet Fırat Baran

Phytoremediation Potential of Bioenergy Plants for Heavy Metals: Supportive Method for Environmental Management 95
Daka Gopamma, Kolli Suresh Kumar, Parida Tamanna, and Namuduri Srinivas

Recent Advances Towards Improved Microbial Bioremediation of Heavy Metal Pollution 115
Farzaneh Dianatdar and Zahra Etemadifar

Biotechnological Strategies for Effective Remediation of Heavy Metals 139
Muibat Omotola Fashola, Olajide Solomon Anagun, Mautin Lawrence Ogun, Prince Emeka Ndimele, and Olubukola Oluranti Babalola

Role of Biosurfactants in Remediation of Heavy Metals 165
Muibat Omotola Fashola, Olajide Solomon Anagun, Ahmeed Olalekan Ashade, and Olubukola Oluranti Babalola

Role of Rhizobacteria in Phytoremediation of Heavy Metal 183
Zain Mushtaq, Ariba Muzammil, Korkmaz Bellitürk, Waheed Anwar, Adnan Akhter, Hafiz Azhar Ali Khan, Amir Parnian, and Shafeeq Ur Rahman

Bioremediation of Heavy Metals—Its Pros and Cons 213
Seragadam Padma, Samaroha Ghosh, Badri Srinivas, Kartik Chandra Ghanta, and Susmita Dutta

Adsorptive Removal of Heavy Metals from Wastewater Using Low-Cost Adsorbents Derived from Agro-based Materials 237
Jonas Bayuo, Mwemezi J. Rwiza, Kelvin Mark Mtei, and Joon Weon Choi

An Introduction to Heavy Metals and Oxidative Stress in Ageing and Cancer: A Role for Detoxification Through Diet 273
Chanda Siddoo-Atwal

Propolis as a Bioindicator of Contamination with Toxic Metals 289
Ana Paola Balderrama-Carmona, Víctor Ramos-García, Liliana Ruiz-López, and Ximena Felipe-Ortega-Fonseca

Ecotechnology Approach for Remediation of Heavy Metals 319
Shaik Riyazuddin, Daka Gopamma, Kolli Suresh Kumar, Nyayapathi Priyanka Priyadarshini, and Namuduri Srinivas

The Source and Distribution of Heavy Metals in the Atmosphere Across Southeast Asia

Ly Sy Phu Nguyen, Thi Thuy Linh Do, Truong Gia Han Vo, Quoc Hau Le, and To Thi Hien

Abstract Atmospheric heavy metal (HMs) pollution that impacts the environment and human health is one of the increasingly concerning problems in developing countries. Southeast Asia (SEA) is now described as a dynamic and rapidly developing region and is facing problems of air pollution, including HMs in the atmosphere. According to the International Cancer Research Institute (IARC), some metals (e.g., Pb, As, Ni, Cd, Hg, Cd, Cr) are particularly harmful to human health. In this chapter, we focus on the knowledge of atmospheric HMs in the SEA region over the last 15 years, in which, the potential sources and spatio temporal distribution of atmospheric HMs in SEA were discussed in detail. Research on atmospheric HMs is unevenly distributed in the SEA region and most of the studies concentrated in countries such as Thailand, Malaysia, and Vietnam. By employing multivariable models, including PCA and PMF, studies show that both anthropogenic sources (e.g., transportation, biomass burning, and industrial processes) and natural sources (e.g., volcanic eruption and dust storms) can contribute to elevated atmospheric HMs. HMs concentrations in industrial areas are often higher than in urban/background locations. On a seasonal basis, HMs concentrations in the dry season are often higher than in the rainy season. Certain knowledge gaps pertaining to atmospheric HMs in the SEA region necessitate comprehensive investigation. Specifically, there is a need for rigorous research focused on elucidating the deposition of atmospheric HMs as well as research regarding atmospheric mercury (Hg). This chapter presents scientific information on atmospheric HMs pollution in SEA from both regulatory and research perspectives. It aims to enhance the understanding of HMs contaminations in the SEA region by providing updated and rigorous scientific analysis.

L. S. P. Nguyen (✉) · T. G. H. Vo · Q. H. Le · T. T. Hien
Faculty of Environment, University of Science, Ho Chi Minh City, Vietnam
e-mail: nlsphu@hcmus.edu.vn

Vietnam National University, Ho Chi Minh City, Vietnam

T. T. L. Do
Institute for Environment and Resources (IER), Ho Chi Minh City, Vietnam

Department of Science and Technology, Vietnam National University, Ho Chi Minh City, Vietnam

N. Kumar (ed.), *Heavy Metal Remediation*, Earth and Environmental Sciences Library,
https://doi.org/10.1007/978-3-031-53688-5_1

Keywords Southeast Asia · Heavy metals · Atmosphere · Source · Spatiotemporal distribution · Deposition

1 Introduction

Heavy metals (HMs) constitute a cluster of metallic elements (Pb, As, Ni, Cd, Cu, Mn, V, Hg...) that stand out by virtue of their relatively substantial atomic masses and densities. HMs are well-known environmental pollutants due to their inherent toxicity to various life forms, notably including humans, particularly when they accumulate within ecosystems or living organisms [115, 57, 33]. Inhalation and ingestion of HMs-contaminated air, water, and food can lead to a range of health issues, including respiratory problems, neurological disorders, organ damage, and even cancer [53, 68]. According to the International Agency for Research on Cancer (IARC), HMs can be classified into class I carcinogens (As, Ni, Cd, Cr), class II(B) carcinogens (Pb), and non-carcinogenic (Cu, Zn, and Mn). Among HMs, mercury (Hg) is a ubiquitous toxic HMs of global concern due to its persistence and bioaccumulation in the environment and organisms. Hg is widely accepted as a global pollutant because it can be dispersed worldwide via long-range atmospheric transport [102]. The impacts of Hg on aquatic and terrestrial systems have been well-documented by many studies, in which the most toxic exposure is through the consumption of methylmercury in fish [23, 27]. A number of health problems, including central nervous system defects, arrhythmias, cardiomyopathy, and kidney damage, are suspected to be linked with Hg exposure [35]. Given the global concern about Hg pollutants, the Minamata Convention on Hg entered into force in August 2017, with the participation of 128 countries and territories to take action to protect human health and the environment from anthropogenic Hg emissions and releases of Hg and Hg compounds.

Most HMs are intrinsic constituents of the Earth's ecosystem, occurring naturally in soil, water, air, and living organisms [61]. The substantial presence of HMs in the atmosphere poses a significant threat to both human health and the environment due to their harmful characteristics and persistent nature [79, 55, 6]. These metallic elements can originate from a range of sources, including natural processes and human-driven activities [74, 101, 62]. For Hg, anthropogenic sources emissions encompass activities like artisanal and small-scale gold mining (ASGM), coal combustion, the manufacturing of cement, and the primary production of non-ferrous metals [69]. Data compiled from various research endeavors demonstrate that global anthropogenic Hg emissions have either remained consistent or risen [102]. Geographically, while anthropogenic Hg emissions are declining in Europe and North America, they are on the rise in East and Southeast Asia (SEA), which is the major emission source region contributing about 40% of the global emissions [69, 102]. The atmosphere is the major pathway for the Hg biogeochemical cycle in the environment. In the atmosphere, Hg exists in three primary forms: gaseous elemental mercury (GEM), gaseous oxidized mercury (GOM), and particulate-bound mercury (PBM) [23, 102]. In contrast to other metals that predominantly occur in the atmosphere as compounds bonded to particulate matter (PM), Hg primarily exists in the atmosphere as GEM, constituting over 95% of the overall Hg content in the atmosphere [58]. GEM has a relatively high vapor pressure (0.18 Pa at 20 °C) and is only slightly water-soluble (Henry's law coefficient of 729 Pa m^3 mol^{-1} at 20 °C) [85]. The typical lifespan of GEM on a global scale is approximately 0.5–1 year, which is notably longer than the lifespans of GOM and PBM. GOM and PBM, in contrast, have considerably higher rates of deposition and solubility in water, causing them to deposit more quickly in areas downwind of emission sources through both wet and dry deposition processes [58, 55, 102]. Presently, there is a significant environmental concern surrounding PBM, mainly due to their reactivity and their role as intermediates between elemental and organic forms of Hg compounds. In general, an in-depth comprehension of the origins and dispersion patterns of atmospheric HMs is fundamental for the formulation and implementation of rigorous regulatory protocols. This understanding is vital for safeguarding human welfare and fostering effective environmental protection strategies.

Numerous investigations of global scope have been undertaken to examine various HMs within the atmospheric milieu [21, 51, 112]. Scholars and scientists systematically amass and scrutinize data from a multitude of sources to elucidate the intricate interplay of distribution patterns, origins, and the conduct of these contaminants [34, 21, 49, 101, 56, 59] . Furthermore, scientific research efforts play a substantial role in elucidating the fate and transport mechanisms of HMs in the atmosphere [101]. Subsequent to their release into the atmosphere, HMs possess the capability to traverse extensive distances and settle within diverse environmental compartments such as terrestrial soil, aquatic bodies, and vegetative matter [5, 52]. In addition, through systematic investigations involving exposed populations and controlled laboratory experiments, scientists can elucidate dose–response correlations, pinpoint vulnerable demographics, and formulate guidelines aimed at safeguarding public health [21, 77, 11]. This information is then used to assess the risks associated with HMs contamination and develop appropriate management strategies.

SEA stands out as a highly dynamic and rapidly advancing global region. The escalating concern of air pollution within this area has garnered significant attention [106]. Studies on air pollution have been carried out in this area, including studies on atmospheric HMs [4, 7, 13, 77, 101]. Presently, research on atmospheric HMs in SEA primarily centers around countries such as Thailand, Malaysia, Vietnam, and Philippines (Fig. 1). Nonetheless, a scarcity of published scientific research on atmospheric HMs is evident in other SEA countries, indicating a significant gap in the scientific knowledge pertaining to this region (Fig. 1). The primary objectives of this chapter are to conduct a comprehensive assessment of the atmospheric HMs within the SEA region. The study focuses on two main aspects: firstly, an in-depth investigation into the sources of HMs, with specific emphasis on Thailand, Vietnam, and Malaysia; and secondly, a thorough examination of the distribution patterns of HMs, considering both spatial and temporal dimensions, across the SEA region. This analysis extends from local to regional scales and involves comparative assessments with other global regions.

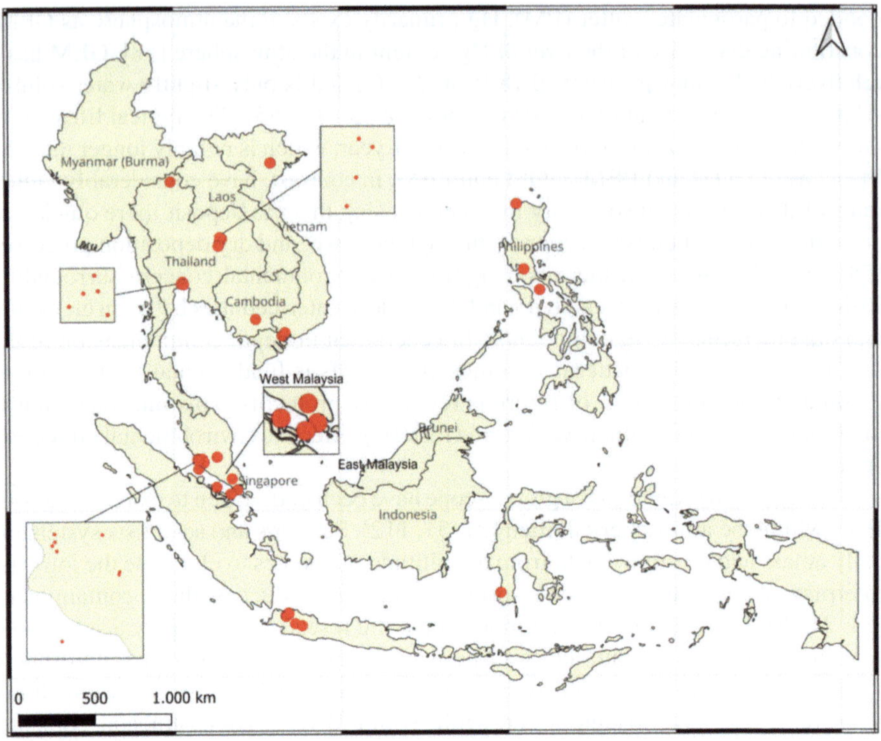

Fig. 1 Distribution of atmospheric HMs studies in SEA

2 HMs in the Atmosphere and Its Deposition

We first discuss on the method to determine HMs concentrations in the atmosphere. There are several methods available for assessing HMs in the atmosphere and the most common method is inductively coupled plasma mass spectrometry (ICP-MS). The analysis of most airborne HMs, except for Hg, typically involves analysis of suspended particulates (i.e., particulate matter). The selection of the method and any necessary amendments depend on the specific requirements of the study or regulations applicable to different regions [2]. Here, we outline a three-step for analyzing HMs from PM samples, which has been widely applied in numerous studies.

- *Step 1: Sample collection and preparation: PM sample collection using specialized samplers (e.g., TSP, PM_{10} or $PM_{2.5}$,* etc.*). Different types of filters can be used depending on the sampling equipment).*
- *Step 2: HMs extraction from filter: The mixtures of strong acids (e.g., HCl + HNO_3) are often used to dissolve all HMs in the sample into the solution. Once the HMs are in the solution, various analytical techniques can be employed to quantify their concentrations.*

- *Step 3: HMs analysis: The analysis method will be selected according to laboratory conditions and the estimated HMs content in the sample. It should be noted that QA/QC steps need to be strictly implemented to ensure reliability, and SRM sample analysis is essential to demonstrate the accuracy of the method.*

In the above steps, choosing the appropriate measurement method is very important. HMs identification is often favored by researchers through various methods, including Energy Dispersive X-ray Fluorescence (EDXRF) [82], Flame Atomic Absorption Spectrometry (FAAS) [25, 110], Graphite Furnace Atomic Absorption Spectrometry (GF-AAS) [111], and Inductively Coupled Plasma Atomic Emission Spectrometry (ICP-AES) [10, 99], Inductively Coupled Plasma Optical Emission Spectrometry (ICP-OES) [40, 94], Inductively Coupled Plasma Mass Spectrometry (ICP-MS) [13, 54, 98, 101]. Among various analytical methods, the ICP-MS method is widely used because of its high sensitivity, precision, and wide dynamic range which can facilitate to analysis of many HMs in the same measurement process [15]. For extraction method (i.e. step 2). Ahmed et al. [3] introduced a novel approach known as convection microwave-assisted digestion. This method is implemented as a preliminary step prior to ICP-MS analysis which could enhance the efficiency and accuracy of sample preparation for HMs analysis.

To determine atmospheric Hg concentrations, including both gas and particle phases, studies often use the cold vacuum atomic fluorescence spectrometer (CV-AFS) method, which is specifically used to quantify volatile HMs (e.g., Hg). On the other hand, the direct thermal decomposition-gold amalgamation method by utilizing direct mercury analyzers (DMA) is widely used to analyze the total amount of Hg collected by glass or quartz fiber filters (i.e. PM samples). Compared to conventional methods (acid digestion + CV-AFS), direct thermal analysis methods are fast, simple, and do not require any additional chemicals, which is beneficial to prevent sample contamination during digestion [63].

In the atmosphere, HMs predominantly reside in the particulate phase [107, 58], except for Hg, which exists mostly in the gas phase (more than 95%, [103]). Due to their link with PM, HMs can deposit through dry and wet deposition processes, exerting significant impacts on both the environment and human populations [58]. Furthermore, they can travel over long distances, causing widespread effects within a large geographical area [41, 58]. Therefore, it is crucial to determine the origin as well as the characteristics of atmospheric HMs.

Bulk deposition is a method in SEA for monitoring the deposition of pollutants in the atmosphere. However, studies on evaluating the deposition process of HMs in air have not been comprehensive [80, 78]. In situations where adequate instruments are available, it is fully possible to determine the amount of wet and dry atmospheric deposits individually [58]. Wet deposition removes pollutants from the atmosphere through rain and snow, while dry deposition occurs directly on surfaces such as leaves and soil [58]. The number of atmospheric particles deposited depends on particle size and origin, natural or anthropogenic. Smaller particles can travel long distances, covering hundreds or thousands of kilometers. The deposition velocities vary depending on the source and location of the particles. For example, deposition on water surfaces increases, especially through seawater sprays and waves that break down, effectively eliminating particles and gases [58].

Currently, in SEA, there is a paucity of scientific investigations concerning the atmospheric deposition of hazardous HMs, despite the presence of comprehensive studies focusing on air pollution. A study led by Samontha et al. [78], delved into the examination of atmospheric HMs deposition in Saraburi, a province located approximately 110 km north of Bangkok, Thailand. The investigation focused on determining the total deposition (dry + wet) of Pb, Cd, and Zn at two specific locations including a lead smelter area and a heavy traffic zone. The research designed a passive air particle collector—a relatively simple method to monitor dust particles deposited in both dry and wet forms. The results of this study show significant differences in Pb deposition between the lead smelter and the heavy-duty traffic area, demonstrating the role of point sources in the atmospheric deposition of HMs. During the dry season, the Pb deposition in the melting zone was 27.6 times higher than in heavy traffic areas. Similarly, during the rainy season, Pb deposits in the melting area were 5 times higher than in areas with heavy traffic [78]. On the other hand, greater atmospheric deposition of Zn in both the dry and wet seasons was obtained in the heavy traffic area as compared lead smelter area, suggesting that Zn emissions are primarily attributed to tire wear [78]. Overall, studies of air pollution have been carried out in SEA, but studies on atmospheric HMs depositions remain limited. The serious impact of HMs exposure on human health and ecosystems makes it imperative to address this research gap. Efforts to understand the sources, concentrations, and impacts of HMs deposition can guide effective policy-making and interventions to mitigate the harmful effects of air pollution in the region.

3 Studies on Atmospheric HMs in SEA

SEA, an expansive subregion of Asia, is situated approximately east of the Indian subcontinent, south of China, and northwest of Australia. Geographically, it is bounded by the western Bay of Bengal to the west, the southern Indian Ocean to the south, the central South China Sea to the north, the eastern Philippine Sea to the northeast, and the vast expanse of the Pacific Ocean to the east [47]. The Association of Southeast Asian Nations (ASEAN) is a regional body established to promote cooperation between countries in many fields such as economics, politics, science, and security. ASEAN includes 10 members: Indonesia, Thailand, Singapore, Vietnam, Malaysia, Philippines, Myanmar, Cambodia, Brunei, Laos, East Timor [1]. The Gross Domestic Product (GDP) of countries ranges from 15.72 billion USD (Laos) to 1.32 trillion USD (Indonesia) [108]. The ASEAN region is characterized by significant economic growth, a large and rapidly growing population, and high levels of biological and cultural diversity. Nevertheless, a prevailing concern shared among the nations in this region centers on the escalating issue of air pollution including HMs contaminations. The surge in industrial activities associated with plating industries, coupled with the extensive utilization [22], and the swift pace of urbanization, collectively pose significant threats to both human health and the environment. Thus, atmosphere pollution in general and HMs pollution in particular are gradually becoming one of the problems that need to be overcome among SEA countries [31, 11].

In this chapter, we endeavor to evaluate and analyze the characteristics of HMs present in the atmosphere within the countries of SEA. This assessment is based on a comprehensive review of scientific papers published between 2006 and 2023, with a primary emphasis on countries including Thailand, Vietnam, Malaysia, Indonesia, Philippines, and Singapore. Our focus lies in delineating the unique attributes of these HMs and exploring their spatial and temporal distributions within the designated research area. Studies have conclusively demonstrated the presence of atmospheric HMs across diverse land-use domains. These encompass industrial sectors [9, 13, 82], urban with densely populated areas [7, 100, 42, 12, 72, 17, 101], and rural areas [50]. The HMs concentrations were recorded on hourly/daily basis [87, 101] and seasonal basis [4, 9, 10, 28, 97]. Moreover, measurements were conducted during diverse pollution events, encompassing occurrences of coal ash combustion, biomass combustion, and hazy days [4, 42, 97, 87]. Among SEA countries, the largest number of studies on atmospheric HMs was found in Malaysia (13 studies from 2006 to 2023). In contrast, Indonesia is recognized as the most economically developed country in ASEAN but exhibits limited engagement in the assessment of HMs, as indicated by the relatively small number of studies that have been conducted [80, 75]. In contrast to that of other HMs, research into atmospheric Hg has generated significant scientific attention in recent decades [102]. However, there has been a constrained focus on comprehending the spatial and temporal distribution of atmospheric Hg specifically within the SEA locale. To our understanding, extremely poor understanding of atmospheric Hg in SEA while only a few studies have been recently conducted in Vietnam and some short-term monitoring data regarding atmospheric Hg pollution in SEA [63, 65, 90, 92, 57]. Most developing nations in the SEA region have yet to make any effort to assess atmospheric Hg despite the high level of Hg emission in this region. Consequently, given the swift expansion of their economies, there is a significant requirement to establish a Hg monitoring network in this area.

4 Sources of Atmospheric HMs Across SEA Region

In the realm of atmospheric science, various methodologies are employed to ascertain the origins of HMs in the atmosphere. These methods encompass the source-receptor model, exemplified by tools such as Hybrid Single Particle Lagrangian Integrated Trajectory (HYSPLIT) and Potential Source Contribution function (PSCF), as well as advanced multivariate statistical techniques like PCA (Principal Component Analysis) and PMF (Positive Matrix Factorization). These analytical approaches are pivotal in discerning the diverse sources contributing to heavy metal presence in the atmosphere. HYSPLIT model conducts source region estimations through back trajectory analysis. HYSPLIT operates as a Lagrangian model, accompanying the air mass and forecasting its trajectory over time [7, 100, 97]. Besides, the PSCF model is derived from the HYSPLIT model to identify source areas for pollutants with a relatively long lifetime [65]. The atmosphere predominantly comprises three primary elements: emissions released into air masses, the movement of these air masses, and the physicochemical processes occurring during their transportation. While HYSPLIT can simulate emission sources and forecast the trajectories of affected air masses, it does not definitively identify a particular source. To accurately ascertain the origins and effects of emissions, a comprehensive model encompassing emissions, chemical reactions, transport phenomena, and transformations is imperative [88, 95, 96]. In addition, based on the dataset of HMs concentrations, the potential sources can also be identified through multivariate statistical methods (e.g., PCA, PMF) [7, 19, 30, 43, 97, 101, 113]. The PCA method is widely used to explore data relationships and source types for atmospheric HMs [7, 30, 113]. PCA is a multivariate statistical method, which reduces a multidimensional system containing many correlated variables into a simpler system of uncorrelated variables, called principal components factors [14]. On the other hand, PMF method represents a receptor model employing a mathematical approach to quantify source contributions to samples and has found applications in the assessment of HMs in several studies [19, 43, 97, 101]. PMF serves as a multivariate factor analysis tool, decomposing the matrix of individual HMs into two distinct matrices: a factor profile (f) and a factor contribution (g) [70].

In the SEA region, atmospheric HMs primarily originate from two main sources: anthropogenic sources (e.g., transportation, biomass burning, and industrial processes) and natural sources (e.g., volcanic eruptions and dust storms). In addition, numerous studies have investigated the influences of biomass burning sources on atmospheric HMs [4, 20, 12, 87, 86]. Carcinogenic HMs such as As, Pb and Cr predominantly originate from industrial processes, whereas elements such as Fe and Mn are naturally occurring constituents of the Earth's crust [74, 62]. Table 1 summarizes the sources of HMs in the region. Specifically, the main sources of HMs in Thailand and Malaysia are burning biomass and traffic activities. In Vietnam and Philippines, in addition to traffic and biomass burning, industrial activities have also played an important source of atmospheric HMs.

The use of a multivariable model to illustrate the potential source of PBM was also implemented at SEA. The first groundwork study on PBM has been carried out in Ho Chi Minh City (HCMC), a tropical megacity in southern Vietnam [63]. By utilizing data from 8 metals and meteorological data in PCA, Nguyen et al. [63] revealed the contributions from both primary and secondary sources to PBM, in which most of the primary source's input are from traffic, combustion and industrial origins. Besides, backward trajectory analysis and PSCF analysis were employed to allocate potential Hg sources that impact on TGM levels at an urban site in HCMC [65, 67]. This study highlighted the importance of East Asian outflow export towards elevated Hg concentration at HCMC as well as implying broader SEA impacts. Moreover, a recent study by Soe et al. [92] summarized TGM data from sites in ASGM areas of SEA, including Indonesia, Myanmar, and Philippines. The research emphasizes how significant the ASGM source's input, i.e., where activities involving the burning of

Table 1 Source of HMs in the atmosphere across SEA region

Site	Country	Potential sources	References
Khon Kaen	Thailand	Power plant, traffic activities, biomass and waste open burning	Sakunkoo [77]
Chiang Mai Chiang Dao	Thailand	Biomass burning, traffic activites and transboundary pollution	Chansuebsri [16]
Phitsanulok	Thailand	Street dust, soil dust	Srithawirat [93]
Chiang Mai	Thailand	Road traffic	Janta [40]
Chiang Rai Bangkok	Thailand	*Chiang Rai:* Crustal and biomass burning, industrial activities *Bangkok:* Industrial activities, and coal combustion, traffic activities	Kayee [42]
Kuala Lumpur	Malaysia	Waste burning, coal burning	Sara [84]
Klang Valley Kuala Lumpur	Malaysia	Industry emissions.	Rahman [72]
Klang Valley	Malaysia	Traffic activities, biomass burning, and soil dust	Amil [9]
Kampar	Malaysia	Traffic activities	Ahmed [4]
Johor	Malaysia	Vehicular exhaust, coal-fired power plants, refineries and industrial activities	Alias [7]
Lembang	Indonesia	Vehicular emissions and metal industry activities	Lestiani [50]
Bandung	Indonesia	Traffic activities, biomass burning, industrial activities, associated with oil-burning	Santoso [80]
Dong Nai	Vietnam	Combustion, and metal production, biomass burning, and waste treatment	[13]
HCMC	Vietnam	Traffic activities, combustion processes, re-suspension of particles, and waste incineration	Truong [101]
HCMC, Hanoi	Vietnam	Industrial activities (non-ferrous metal production, coal power plants)	Hien [100]
HCMC, Hanoi	Vietnam	Source of incineration, Road traffic	Chifflet et al. [17]
Manila	Philippines	Industrial activities	Pabroa [71]
Bulacan	Philippines	Jewelry manufacturing and gold smelting	Galvez [30]

gold amalgamation take place, is to the levels of Hg in the atmosphere. Particularly, TGM peaked at 314 $\mu g\ m^{-3}$ in the gold mining area of Philippines, surpassing the WHO guideline limit by 314-fold [92]. ASGM stands as the rapidly expanding origin of Hg globally, capable of emitting Hg into the air and other environments and the utilization of Hg in ASGM sectors is prevalent across SEA countries [102]. Therefore, it is advisable to explore Hg recovery techniques for the ASGM sector.

The process of rapid urbanization in Thailand has given rise to significant environmental challenges, necessitating comprehensive scientific investigations to address the issue of atmospheric HMs pollution. Extensive research endeavors have been undertaken in the country to comprehend and mitigate the complexities associated with atmospheric HMs pollution resulting from urbanization [16, 40, 42, 77, 93]. Studies have revealed that anthropogenic activities are the primary sources of atmospheric HMs, as demonstrated in the study by Kayee et al. [42] in Chiang Rai, Thailand, where the HMs (Pb, As, Cu, Cr, Mn, Ni, and V) were predominantly traced back to biomass burning. Urban areas predominantly witness HMs emissions from traffic activities, while rural areas are primarily impacted by biomass burning, as elucidated by Chansuebsri et al. [16]. Similarly, the study by Sakunkoo et al. [77] identified additional sources of HMs in $PM_{2.5}$, such as nearby power plants and waste incineration. Furthermore, apart from the presence of HMs in PM samples, it is noteworthy that HMs found in street dust also exert significant influence due to their potential contribution to atmospheric HMs. For instance, Janta et al. [40] in Rural Chiang Mai, Thailand, indicates that Cr, Cu, and Pb in road dust originate from vehicle exhaust. Notably, research efforts in Thailand have primarily focused on HMs' anthropogenic sources in the air, with limited exploration of natural sources in the existing scientific literature.

As for Malaysia, the country is ranked among the three most polluted countries in SEA [89]. The main contributions to atmospheric HMs primarily result from anthropogenic sources, particularly activities such as vehicular emissions and industrial processes. Research by Rahman et al. [72] identified the presence of Pb in the Klang Valley region due to diverse industries including iron and steel, paint, aluminum fabrication, and pharmaceuticals. In southern Malaysia, (i.e. Johor), the concentrations of HMs (Pb, As, Ni, Cu, and V) content in $PM_{2.5}$ indicates that 41% of these emissions are derived from vehicle exhaust, 34% from coal-fired power plants, 4% from refineries and industrial emissions, and 22% from other sources [7]. Amil et al. [9] analyzed potential sources of HMs (Pb, As, Cd, Mn, Ni, and V) in $PM_{2.5}$ samples collected in the Klang Valley. Traffic activities, biomass burning, and soil dust are potential sources of HMs in this area [9]. A study by Ahmed et al. [4] in Kampar, Malaysia, found that sources of HMs (V, Fe, Pb, and Ni) arise from traffic activities, such as fuel and oil combussion, and abrasive products from tires. Sara et al. [84] indicated that sources of As in Kuala Lumpur were from waste burning, coal burning.

In Indonesia, air quality monitoring primarily focuses on total suspended particulate matter (TSP) or PM_{10}, with limited attention given to the concentration and composition of finer PM (i.e. $PM_{2.5}$) [46]. Sources of atmospheric HMs are vehicle exhaust, biomass burning, industrial activities such as oil burning [50, 80]. A study conducted by Santoso et al. [80] in Bandung and the suburbs of Lembang revealed notably high Pb levels in coarse mode of PM (i.e. $PM_{2.5-10}$) samples. The sources of HMs were determined in which Mn was found to be linked to traffic activities and biomass burning, Cu, As, and Pb were identified as originating from industrial activities, and V was associated with oil-burning [80]. Additionally, a significant portion of Pb in PM originated from vehicle exhaust emitting fuels containing tetraethyl lead [37]. Another study in the suburban area of Lembang identified HMs in both $PM_{2.5}$ and PM_{10}, including V, Cr, Mn, Ni, Cu, As, and Pb. The sources of HMs in $PM_{2.5}$ were primarily attributed to vehicular emissions and metal industry activities [50].

In Vietnam, the contamination of HMs poses a significant environmental issue, especially in densely populated urban areas like HCMC, Hanoi, and Dong Nai. The contamination levels are intricately connected to regional emissions, which encompass the burning of fossil fuels, emissions from transportation as well and the long-range transport of polluted air masses (regional sources). Truong et al. [101] applied PMF modeling to pinpoint key sources of emissions in HCMC. Their analysis identified predominant contributors such as transport emissions, combustion processes, re-suspension of particles, and waste incineration. Notably, the research demonstrated a substantial rise in As, Sb, and Pb attributable to incineration activities. Furthermore, Hien et al. [100] indicated that non-ferrous metal production emerged as a significant source of atmospheric soluble HMs in Hanoi. Besides, coal power plants and non-ferrous metal production are anticipated to be the primary sources of trace metals in Hanoi. Chifflet et al. [17] found that the ratio of V/Ni and V/Sb highlight the impact of traffic activities while Cd/Pb ratio indicated contribution from incineration sources. Furthermore, industrial processes such as combustion, metal production, biomass combustion, and waste treatment also play an important role in inducing HMs levels in Bien Hoa [13]. In general, unlike Thailand and Indonesia, biomass burning is not the biggest contribution source of atmospheric HMs in Vietnam.

For Philippines, studies indicated that the sources of atmospheric HMs are mostly attributed to industrial activities [30, 71]. A case study conducted in Manila by Pabroa et al. [71] revealed the detection of Pb originating from the industry. Additionally, research by Galvez et al. [30] identified Pb originating specifically from industrial sources, particularly jewelry manufacturing and gold smelting.

5 Spatial Distribution of HMs in the Atmosphere Across SEA Region.

Toxic elements, such as As, Pb, Ni, Cd, Mn, Cu, V and Cr are important trace elements in $PM_{2.5}$ and PM_{10}, whose concentrations are significantly higher in urban areas than in rural areas due to anthropogenic activities [4, 50, 73]. Atmospheric HMs concentration in industrial sites in Malaysia (i.e. Makassar, 514.2 ng m^{-3},[75] is about 5 times higher than in urban sites (Jakarta, 98.4 ng m^{-3}, [81])). HMs often have low concentrations in urban areas such as V (1.1–5.5 ng m^{-3}) [82, 83] and Cd (0.5–1.8 ng m^{-3}) [44, 17] are also significantly lower 14–117 times than the industrial areas. The concentrations of several HMs and their potential hazards in the atmosphere species are well documented and guidelines level of these are provided World Health Organization.

Figure 2 illustrates the spatial distribution of HMs concentrations (As, Cu, Cr, Mn, Ni, Pb, and V) in $PM_{2.5}$ over the SEA region (i.e. Malaysia, Thailand, Vietnam, Indonesia, Philippines). According to Fig. 2, HMs concentrations are commonly reported in countries such as Vietnam, Thailand, Malaysia, Philippines, and Indonesia [10, 11, 12, 42, 44, 61, 71, 75, 81, 101]. However, it is essential to note that currently, only a singular study is underway, focusing on research conducted in Singapore and Cambodia. No studies have been published in other countries (i.e. Myanmar, Laos, Brunei, and East Timor). Among HMs being considered in Fig. 2, Pb is a common metal, which appears in most of the studies and has a relatively high concentration in SEA followed by Mn, Cu, Cr, Ni, As, V, and Cd.

Table 2 summarizes the HMs concentrations in SEA countries and elsewhere worldwide. The HMs concentrations in countries within the SEA region are typically elevated compared to developing nations like the United States, Canada, and European countries (Table 2). However, it is noteworthy that HMs concentrations in SEA are still notably lower, ranging from 2 to 6 times less, than those observed in East and South Asia [24, 32]. In more detail, the concentration of As in SEA urban areas is about 3.7 ng m^{-3}, ranging from 0.5 ng m^{-3} in Chiang Rai (Thailand) to 9 ng m^{-3} in Singapore, significantly higher than that observed in Alberta (Canada) (0.2 ng m^{-3}), and Bogota (Colombia) (0.5 ng m^{-3}) but lower than those reported in China (19 ng m^{-3}) and India (25 ng m^{-3}). The average concentration of Ni in SEA is about 8.7 ng m^{-3} ranging from 0.6 ng m^{-3} in Hanoi, Vietnam to 49.7 ng m^{-3} in Singapore which is considerably higher than those reported in Portland (2.1 ng m^{-3}), Taiwan (5.2 ng m^{-3}). Furthermore, the average concentration of Cd for the urban

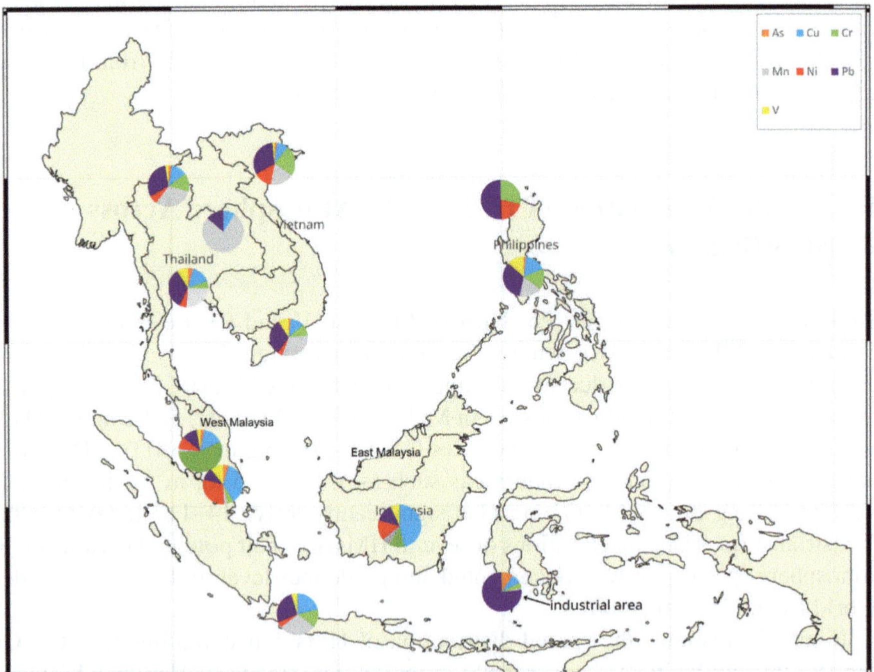

Fig. 2 Spatial distribution of HMs concentrations in $PM_{2.5}$ within SEA region

SEA is 6.2 ng m^{-3}, which ranged from 0.3 ng m^{-3} in Singapore to 19.5 ng m^{-3} in Thailand, which is higher than those of Elefsis (Greek) (0.7 ng m^{-3}) but comparable to that obtained in China (5.8 ng m^{-3}). Nevertheless, these studies are relatively outdated, conducted over 10 years ago, and only a few studies have been conducted recently.

Studies on HMs in SEA have predominantly been conducted in Malaysia, particularly in the Peninsula region (i.e. Kuala Lumpur, Klang Valley) [7, 44, 72, 83]. Peninsular Malaysia, a constituent region of Malaysia, adjoins Thailand to the north and is geographically linked to the island country of Singapore in the south. Studies indicate elevated concentrations of HMs in the Klang Valley [44] surpassing levels observed in Kuala Lumpur for most metals except V [7, 44, 72]. Compared to the Peninsula area, As, Ni and Pb concentrations in the Klang Valley were about 1.5, 6.9, and 2.8 times greater, while V is 3.8 times lower [7]. The THR values (total health risk) calculated for the carcinogenic metals (i.e. Pb, Ni, As, and Cd) in Malaysia did not exceed the guidelines of USEPA cancer risk [44].

In Thailand, studies have been carried out in several regions including Chiang Rai provinces in the northern [42], Khon Kaen province in the central northeastern [77], and Bangkok in the Southern [42, 18]. Land use is an important driving factor responsible for HMs variation at the sampling sites. For example, [77] reported concentrations of HMs in $PM_{2.5}$ in Khon Kaen, a province in the center of northeastern Thailand. This study focused on 4 sampling categories including the university area, residential area, industrial area, and agricultural area. The HMs concentrations observed in industrial, urban, and university areas exhibited a similar pattern, in which the HMs concentrations decreased in the following order: Mn > Pb > Cu > Cd. Industrial areas showed the highest concentrations (2296.0 ng m^{-3} > 272.9 ng m^{-3} > 185.7 ng m^{-3} > 101.1 ng m^{-3}), followed by residential areas (427.9 ng m^{-3} > 209.0 ng m^{-3} > 57.3 ng m^{-3} > 19.5 ng m^{-3}) and the university area (284.7 ng m^{-3} > 75.4 ng m^{-3} > 33.8 ng m^{-3} > 11.1 ng m^{-3}). On the other hand, agricultural area (rural site) showed slightly different pattern (Mn (247.7 ng m^{-3}) > Cu (103.6 ng m^{-3}) > Pb (23.9 ng m^{-3}) > Cd (3.5 ng m^{-3})). The main contributors to such spatial distribution of HMs in the area were differences in potential HMs sources such as power plants, vehicles, biomass burning, and open waste burning [77].

In Vietnam, several studies on atmospheric HMs have been conducted in both the northern areas (e.g., Hanoi) and the southern areas (e.g., Bien Hoa City and HCMC) [13, 100, 17, 101]. The study conducted by Chifflet et al. [17] revealed significant disparities in HMs (As, Cu, and Pb) concentrations in Hanoi (3.9 ng m^{-3}, 56.8 ng m^{-3}, and 65.7 ng m^{-3}) were significantly higher than those observed in HCMC (2.8 ng m^{-3}, 16.2 ng m^{-3}, and 44.1 ng m^{-3}) (Table 2). The major reason to explain such a pattern is mostly attributed to the difference in potential emission sources between the northern and southern Vietnam [17, 101]. Besides, the density of industrial parks and potential HMs sources in the northern is significantly higher than in southern Vietnam. For instance, there are over 15 coal power plants in the northern as compared to 3 in the southern [100]. A recent study by Bui et al. [13] investigated HMs in $PM_{2.5}$ in Bien Hoa, an industrial city in the southern Vietnam. The authors indicated significant increases in HMs concentrations in areas adjacent

Table 2 HMs concentration (ng m^{-3}) in SEA and different sites worldwide

Site	Site desciption	Country	PM size	As	Cu	Cr	Mn	Ni	Pb	Cd	V	References
Ho Chi Minh	Urban	Vietnam	PM_{10}	2.8	16.1	14.4	45.5	6.2	44.1	–	12.7	Truong [88]
Ho Chi Minh	Urban	Vietnam	PM_2^*	0.7	4.5	0.3	5.3	0.9	19.6	–	2.9	Hien [100]
Hanoi	Urban	Vietnam	PM_2^*	3.5	5.9	0.5	13.3	0.6	56.0	1.6	1.5	Hien [100]
Hanoi	Urban	Vietnam	$PM_{2.5}$	3.9	56.8	3.3	32.0	1.3	65.7	1.8	1.2	Chifflet [70]
Bien Hoa	Industrial	Vietnam	$PM_{2.5}$	5.5	380	190	35.0	35.0	60.0	15	–	Bui [20]
Jakarta	Urban	Indonesia	Coarse	–	22.1	14.7	29.8	4.6	27.2	–	5.0	Santoso [71]
Lembang	Background	Indonesia	Coarse	0.5	2.5	0.9	3.1	0.7	1.9	–	0.87	Lestiani [40]
Bandung	Urban	Indonesia	$PM_{2.5}$	–	–	6.9	2.1	–	59.2	–	1.0	Santoso [50]
Setu	Urban	Indonesia	$PM_{2.5}$	15	6.0	2.5	6.5	3.0	197	–	5.5	Santoso [49]
Serpong	Industrial	Indonesia	$PM_{2.5}$	24	14.5	7.0	11.0	–	191	–	7.5	Santoso [49]
Makassar	Industrial	Indonesia	TSP	47	41.8	47.2	–	1.8	376	–	–	Raus [75]
Kalimantan	Industrial	Indonesia	$PM_{2.5}$	–	872	217	112	279	260	7.0	129	Betha [12]
Kuala Lumpur	Urban	Malaysia	PM_{10}	5.7	23.0	6.7	6.7	3.6	24.0	–	5.2	Rahman [65]
Jalan Semarak	Urban	Malaysia	$PM_{2.5-10}$	0.8	33.7	13.5	33.4	–	3.5	–	1.1	Sara [83]
Peninsula	Urban	Malaysia	$PM_{2.5}$	3.8	–	–	–	2.5	7.7	–	19	Alias [5]
Klang Valley	Urban	Malaysia	$PM_{2.5}$	5.7	28.3	108	4.0	17.2	21.8	0.5	5.1	Khan [44]
Klang Valley	Urban	Malaysia	Coarse	7.5	11.0	11.0	8.8	4.0	22.0	–	5.5	Rahman [73]
Manila	Urban	Philippines	$PM_{2.5}$	–	–	3.3	–	2.8	10.5	–	–	Pabroa [62]
Burgos	Urban	Philippines	$PM_{2.5}$	–	–	45.0	–	35.0	80.0	53	–	Bagtasa [7]
University	NUS	Singapore	$PM_{2.5}$	9.0	12.8	9.6	6.5	49.7	2.9	0.3	–	Betha [64]
Bangkok	Urban	Thailand	$PM_{2.5}$	1.2	4.4	1.8	7.3	1.6	9.7	–	2.6	Kayee [34]
Chiang Rai	Urban	Thailand	$PM_{2.5}$	0.5	2.6	3	6.6	1.4	6.4	–	0.5	Kayee [34]

(continued)

Table 2 (continued)

Site	Site desciption	Country	PM size	As	Cu	Cr	Mn	Ni	Pb	Cd	V	References
Khon Kaen	University	Thailand	$PM_{2.5}$	–	33.8	–	285	–	75.4	11	–	Sakunkoo [61]
Khon Kaen	Residential	Thailand	$PM_{2.5}$	–	57.3	–	428	–	210	19	–	
Khon Kaen	Industrial	Thailand	$PM_{2.5}$	–	186	–	2296	–	273	101	–	
Khon Kaen	Agricultural	Thailand	$PM_{2.5}$	–	104	–	248	–	23.9	3.5	–	
Ulsan	Industrial	Korea	$PM_{2.5}$	–	42.1	11.6	26.5	5.2	127	5.6	–	Hieu [36]
Seoul	Urban	Korea	PM_{10}	–	28.2	73.6	37.6	46.3	200	–	3.0	Kim [37]
Changhua	Urban	Taiwan	PM_{10}	3.4	15.7	30.7	20.1	9.8	21.2	–	11	Hsu [38]
Central region	Urban	Taiwan	PM_{10}	3.2	43.6	6.1	30.6	5.2	17.7	–	7.7	Wu [109]
Hyderabad	Urban	India	PM_{10}	25	56.0	287	94.9	75.0	218	3.0	–	Gummeneni [67]
Beijing	Urban	China	$PM_{2.5}$	47	72.2	41.6	164	22.7	282	5.8	0.5	Duan [18]
Zhuzhou	Urban	China	PM_{10}	19	19	42	108	39	159	5.4	5.2	ye [111]
Urumqi	Urban	China	PM_{10}	–	101	40.0	260	50.0	340	–	30	Wang [105]
Nanjing	Urban	China	PM_{10}	0.5	48.5	3.6	19.8	3.6	21.3	0.3	–	Hu [30]
Acerra	Urban	Italia	PM_{10}	1.9	24.1	14.2	50.5	11.0	55.8	4.3	8.0	Di Vaio [16]
Elche	Urban	Spain	PM_{10}	–	16.0	8.0	14.0	9.0	–	–	7.0	Galindo [29]
Silesia	Urban	Poland	PM_{10}	8.5	–	11.5	2.87	8.35	78.8	–	–	Zajusz-Zubek [94]
Coimbra	Urban	Portugal	PM_{10}	2.5	32.2	4.9	13.5	1.1	1.4	–	3.2	Alves [8]
Bogota	Urban	Colombia	PM_{10}	0.5	51.7	2.9	–	1.7	24.8	–	1.50	Ramírez [59]
Arequipa	Urban	Peru	PM_{10}	13	17.1	–	20.2	3.0	4.6	–	21	Li [41]
Elefsis	Urban	Greek	PM_{10}	2.8	17.6	21.9	23.6	13.1	34.1	0.7	25	Koukoulakis [48]
Alberta	Urban	Canada	$PM_{2.5-10}$	0.2	3.8	1 3	8.32	1.41	0.5	0.03	1.6	Landis [39]
Portland		US	PM_{10}	–	14.4	3 0	7.3	2.1	6.1	–	1.4	Miller [47]

*For soluble HMs

to industrial areas, with levels of HMs several times higher than in Hanoi and HCMC urban areas. These studies are mainly conducted in large cities and no studies have been investigated at the rural/background sites.

In Indonesia, the largest country in SEA, studies have been conducted on Java island (an Indonesian island, bordered by the Indian Ocean to the south and the Java Sea to the north). The studies have spread across this area from the Northern [82] to the southern [50]. In addition, scientific investigations have been conducted in the central region of Indonesia and on Kalimantan Island. These studies spanned from the northernmost areas, as documented by Santoso et al. [82], to the southern most regions, as explored by Lestiani et al. [50]. Besides that, some studies are found in Central Indonesia [80] and Kalimantan Island (Betha et al. 2012). The HMs concentrations were obtained in background areas (i.e. Lembang, southern Java island), in which the concentrations of As, Ni, V, Cr, Pb, Cu, and Mn were 0.5 ng m^{-3}, 0.7 ng m^{-3}, 0.87 ng m^{-3}, 0.9 ng m^{-3}, 1.9 ng m^{-3}, 2.5 ng m^{-3}, and 3.1 ng m^{-3}. This results in Lembang were considerably lower than those in other locations in Indonesia [50]. A study by [82] showed relatively homogenous HMs concentrations in $PM_{2.5}$ in the northwest region of Java island, specifically the Serpong and Setu locations. Except for As and Pb, other HMs showed less than 15% difference between Serpong and Setu (Table 2). In contrast to the northwest region of Java island, central Indonesia exhibited large discrepancies in HMs concentrations [12]. Notably, except for Ni, the concentrations of As, Cu, Cr, and Pb are 3.1, 6.9, 18.8, and 1.9 times higher than the concentration at the Setu urban location and 2–200 times higher than the concentration at the background site [50]. Except for Pb and V, the concentration of some HMs else (Cu, Cr, Mn, and Ni) in Setu is also lower 2.6, 1.9, 1.5, and 1.2 times than the concentration in the study in Jakarta [82]. Broadly, research endeavors in Indonesia have been diverse, spanning various geographical regions. The findings distinctly illustrate variations in the spatial distribution of HMs concentrations.

In Philippines, scientific investigations have been undertaken in two distinct locations, namely Manila, situated on the western coast of Luzon Island, and Burgos, located in the northern region of the country. In Burgos, extensive industrial operations are prevalent, leading to elevated concentrations of HMs [10]. Bagtasa et al. [10] reported the concentrations of Ni, Cr, and Pb are 35 ng m^{-3}, 45 ng m^{-3}, and 80 ng m^{-3} respectively. These are significantly greater about 12.5, 13.6, and 7.6 times than those observed in Manila [71]. The reason for the much higher HMs concentrations in Burgos compared to Manila is attributed to sources from industrial emissions, solid waste burning, and secondary sulfate [10].

Only one publication reported atmospheric HMs in $PM_{2.5-10}$ in Singapore which was conducted at the National University of Singapore [11]. In this study, the analysis encompassed a range of HMs including Cd, Pb, Mn, As, Cr, Cu, and Ni obtained in two distinct conditions (i.e. haze vs non-haze). The concentrations of these metals were 0.3 ng m^{-3}, 2.9 ng m^{-3}, 6.5 ng m^{-3}, 9.0 ng m^{-3}, 9.6 ng m^{-3}, 12.8 ng m^{-3}, 49.7 ng m^{-3}, respectively, during non-haze condition. Conversely, during the haze period, the concentrations of all metals, with the exception of Mn, escalated to 3.2, 2.3, 3.0, 2.0, 3.4, and 1.3 times their normal values. Furthermore, an evaluation of the potential impact on human health was conducted, focusing on the excessive lifetime cancer risk. The results revealed that the observed risks exceeded the acceptable limits set by the International Agency for Research on Cancer (IARC), as outlined in the study by Betha et al. [11].

Due to the limited studies on atmospheric Hg in SEA, knowledge regarding how atmospheric Hg is distributed across this region remains incomplete. Broadly speaking, urban regions characterized by numerous potential Hg sources such as cement production, biomass burning, and burning of fossil fuels exhibit elevated levels of atmospheric Hg compared to regions with less emissions (i.e. off the East Sea; [90]). The northern region of peninsular SEA, encompassing northern Vietnam, northern Thailand, and Myanmar, is thought to serve as substantial anthropogenic Hg emissions, particularly due to springtime biomass burning as well as multiple industrial activities [103]. The influence of the dispersion of Hg pollutants resulting from anthropogenic sources within this vicinity, covering a wide expanse influenced by prevailing winds, was illustrated in studies by Zhang et al. [114] and Nguyen et al. [64]. Consequently, the atmospheric Hg concentration in this specific area might surpass levels found in other parts of SEA. This underscores the necessity for additional research endeavors to validate this supposition.

6 Temporal Distribution of HMs in the Atmosphere Across SEA Region

Air pollution in general and HMs in particular in different regions often exhibit large seasonal variations, mainly related to changes in air mass origin and transport paths related to monsoon activities and meteorology factors (e.g., wind, air temperature, relative humidity, rainfall, etc.). In the SEA region, atmospheric HMs exhibit fluctuations influenced by the cyclic patterns of dry and rainy seasons. Furthermore, it can be observed that in SEA, the region can be geographically divided into two areas. Specifically, in Peninsular SEA (e.g., Thailand, Vietnam, the Malaysian peninsula, Myanmar, Cambodia, Laos), the dry season occurs from December to April, followed by the rainy season from May to September [13, 100, 101]. In contrast, in Maritime SEA (e.g., Eastern Malaysia, Indonesia, Philippines, Singapore, Brunei, East Timor), the dry season extends from May to September, and the rainy season typically lasts from late October to March of the following year [9, 43, 44, 75]. Existing research in this area predominantly emphasizes seasonal patterns, with a limited exploration of diurnal variation.

In Maritime SEA, several studies have reported the seasonal patterns of HMs concentrations in countries such as Philippines, Eastern Malaysia, and Indonesia. The major reason that trigger this pattern are the long-range air pollution transport from northern East Asia due to monsoonal winds. A study conducted by Bagtasa et al. [10] in Burgos, northwest Philippines, revealed significant fluctuations in atmospheric HMs concentrations, with the highest levels observed in spring. In a more detail, Pb and Cr exhibited substantial increases during spring and gradually declined in the fall. On the other hand, Ni and Cd reached peak concentrations in spring and reached their lowest levels during the summer months [10]. The authors also suggested that these toxic components might have originated from the northwest during the North-east monsoon period. In Eastern Malaysia, wind flow distinguishes the seasons (as the dry season is from middle May until middle September) namely the South-West (SW) monsoon, the North-East (NE) monsoon (early November until middle March) is expected to occur while the NE monsoon is known as the wet season [43, 44]. A study by Khan et al. [43] showed that HMs concentrations (including As, V, Mn, Ni, and Cd) in $PM_{2.5}$ particles, during the dry season were 1,5 to 10 times higher than during the rainy season. Similarly, research by Khan et al. [44] showed that Ni and Pb concentrations during the dry season were 1.5 to 1.7 times higher than during the rainy season, while other HMs like Mn, As, Cd, and V exhibited slight increases of 1.1 to 1.6 times in the rainy season compared to the dry season. Furthermore, a study conducted in Indonesia by Rauf et al. [75] in Maros Regency, showed higher (7–57 folds) HMs concentrations (e.g., As, Cr, Cu) in the dry season as compared to the rainy season. Particularly noteworthy was the significant increase in Pb concentration during the dry season, which was 108 times higher than in the rainy season. The presence of Pb was primarily attributed to coal combustion and transportation activities. Additionally, factors such as burning of fuel, higher dry deposition, and increased vehicle activites contributed to elevated HMs concentrations during the dry season.

In Peninsular SEA, studies have investigated the seasonal variations in HMs concentrations that have been reported in Vietnam. In general, most HMs exhibited similar seasonal patterns, in which higher HMs concentrations were found in the dry seasons as compared to the wet season [13, 100, 101]. The major reasons that trigger this pattern are weather conditions and air mass transport influenced by monsoon activity. In a recent study conducted by Hien et al. [100], researchers investigated the seasonal variations in metal concentrations in HCMC and Hanoi. In a recent study conducted by Hien et al. [100], researchers investigated the seasonal variations in metal concentrations in HCMC and Hanoi. The study revealed that the concentrations of Cr, Cd, Pb, and As were lower in the dry season compared to the wet season. Specifically, in HCMC, the concentrations of these metals in the dry season were significantly lower than those in the wet season. These disparities can be attributed to the influence of seasonal monsoons (SW, S, E, NE) in the dry season, coupled with higher boundary heights, resulting in lower metal concentrations during this month compared to the wet season. Additionally, Hanoi experienced complexities in atmospheric transport patterns, including contributions from air masses in northeast Vietnam, southeast China, and Hainan, leading to increased HMs levels in the region. Another study conducted by Bui et al. [13] in Bien Hoa City, Dong Nai province, highlighted higher HMs concentrations during the dry season compared to the rainy season. Specifically, at Gate 2 Bien Hoa (SB), the concentration of Cu and Co in the dry season is higher than in the rainy season. This variation can be attributed to temperature and wind conditions that influence the height of the boundary layer, causing pollutants to either disperse or accumulate. Additionally, high air humidity and rainfall during the rainy season contribute to wet deposition.

In the dry season, the region is affected by air masses from the Southwest and Southeast, coupled with wind patterns in densely populated and highly industrialized major cities. Consequently, HMs concentrations may rise due to increased emissions from these urban areas. Furthermore, Truong et al. [101] analyzed 11 HMs to investigate their concentrations in HCMC. Regarding seasonal changes, most HMs exhibited a dry > rain pattern, attributed to higher rainfall during the rainy season. In contrast, shell elements such as Al, Sr, and Mn displayed a dry > precipitation pattern, which can be explained by differences in air mass transport influenced by monsoon activity.

However, research efforts have primarily focused on seasonal variations in HMs concentrations in the air, with very few studies delving into daily fluctuations. A recent study by Truong et al. [101] involving a 2-h sample analysis revealed a clear diurnal pattern of atmospheric HMs in an urban area of HCMC, Vietnam. Peaks at rush hours (i.e. 10 a.m.) were observed for most HMs (e.g., Cr, Ni, Cu and As), with an increase of 18–52% compared to the daily averages. This pattern can be attributed to peak-hour emission followed by subsequent dilution by increasing turbulence at noon. On the other hand, Pb and Sb exhibited reversed diurnal patterns, with higher nighttime than daytime [101]. This observed pattern may be ascribed to the modulation of planetary boundary height and/or heightened emission rates during nighttime. In general, a study by Truong et al. [101] highlights the importance of considering not only seasonal variations but also the intricate day-to-day dynamics when assessing HMs pollution in SEA.

For atmospheric Hg temporal variations, observation studies in an urban megacity in southern Vietnam (HCMC) indicated that changes in air mass origins and transport paths, and factors associated with meteorological (e.g., rainfall, wind) induced seasonal patterns of both TGM and PBM [63, 65, 67]. The primary reason for a 46% increase in TGM levels in HCMC during the Northeast monsoon (October–November), in contrast to the Southwest monsoon (June–September), appears to be linked to Hg export from China [63]. Besides, PBM concentrations exhibited a February > July pattern owing to greater wet and dry depositions in the rainy season and a part of the contribution of transboundary transport air pollution from East Asia in the dry season [63]. On the other hand, diurnal variation of TGM and PBM in HCMC probably resulted from local anthropogenic activities such as residential burning and industrial processes and/or the modulation of the planetary boundary height [63, 65].

Funding The authors thank the Ho Chi Minh Department of Science and Technology for financial support in this research under Grant No. 39/2021/HĐ-QKHCN.

References

1. ASEAN (2022) ASEAN and the Growth of Regional Cooperation in Southeast Asia. https://asean.org/

2. Agarwal A, Mangal A, Satsangi A, Lakhani A, Kumari KM (2017) Characterization, sources and health risk analysis of $PM_{2.5}$ bound metals during foggy and non-foggy days in sub-urban atmosphere of Agra. Atmosp Res 197:121–131
3. Ahmed YH, Chin XG, Zhao XM (2017) Microwave assisted digestion followed by ICP-MS for determination of trace metals in atmospheric and lake ecosystem. J Environ Sci 55:1–10. https://doi.org/10.1016/j.jes.2016.06.014
4. Ahmed M, Guo X, Zhao XM (2016) Determination and analysis of trace metals and surfactant in air particulate matter during biomass burning haze episode in Malaysia. Atmosp Environ 141:219–229. https://doi.org/10.1016/j.atmosenv.2016.06.066
5. Ali MM, Ali ML, Proshad R, Islam S, Rahman Z, Tusher TR (2020) Heavy metal concentrations in commercially valuable fishes with health hazard inference from Karnaphuli river, Bangladesh. Human and Ecol Risk Assessm: an Int J 26:2646–2662. https://doi.org/10.1080/10807039.2019.1676635
6. Ali M, Ali D, Hossain D, Al-Imran S, Khan Md, Osman BM (2021) Environmental Pollution with Heavy Metals: a Public Health Concern
7. Alias NF, Khan MF, Sairi NA, Zain SM, Suradi H, Rahim HA, Latif MT (2020) Characteristics, emission sources, and risk factors of heavy metals in $PM_{2.5}$ from southern Malaysia. ACS Earth and Space Chem 4(8):1309–1323. https://doi.org/10.1021/acsearthspacechem.0c00103
8. Alves C, Rienda IC, Vicente A, Vicente E, Gonçalves C, Candeias C, Rocha F, Lucarelli F, Pazzi G, Kovâts N, Hubai K, Pio C, Tchepel O (2021) Morphological properties, chemical composition, cancer risks and toxicological potential of airborne particles from traffic and urban background sites. Atmos Res 264:105837. https://doi.org/10.1016/j.atmosres.2021.105837
9. Amil N, Latif MT, Khan MF, Mohamad M (2016) Seasonal variability of PM2.5 composition and sources in the Klang Valley urban-industrial environment. Atmospheric Chem Phys 16(8):5357–5381. https://doi.org/10.5194/acp-16-5357-2016
10. Bagtasa G, Cayetano MG, Yuan CS (2018) Seasonal variation and chemical characterization of PM2.5 in Northwestern Philippines. Atmos Chem Phys 18:4965–4980. https://doi.org/10.5194/acp-18-4965-2018
11. Betha R, Behera SN, Balasubramanian R (2014) Southeast Asian smoke haze: fractionation of particulate-bound elements and associated health risk. Environ Sci Technol 48(8):4327–4335. https://doi.org/10.1021/es405533d
12. Betha R, Pradani M, Lestari P, Joshi UM, Reid JS, Balasubramanian R (2013) Chemical speciation of trace metals emitted from Indonesian peat fires for health risk assessment. Atmosp Res 122:571–578. https://doi.org/10.1016/j.atmosres.2012.05.024
13. Bui DL, Le Hoang A, Ngo QK, Nghiem XT (2023) Chemical characterization, source apportionment, and health risk assessment nexus of $PM_{2.5}$-bound major heavy metals in Bien Hoa city, southern Vietnam. Atmosp Environ X 17:100209. https://doi.org/10.1016/j.aeaoa.2023.100209
14. Calvo AI, Pont V, Olmo FJ, Castro A, Alados-Arboledas L, Vicente AM, Fernández-Raga M, Fraile R (2012) Air masses and weather types: a useful tool for characterizing precipitation chemistry and wet deposition. Aerosol and Air Quality Res 12:856–878. https://doi.org/10.4209/aaqr.2012.03.0068
15. Carvalho Rui M, dos Santos Jéssica A, Silva Jessee AS, do Prado Thiago G, da Fonseca Adriel Ferreira, Chaves Eduardo S, Frescura Vera LA (2015) Determination of metals in Brazilian soils by inductively coupled plasma mass spectrometry. Environ Monit Assessm 187(8):535.https://doi.org/10.1007/s10661-015-4769-y
16. Chansuebsri S, Kraisitnitikul P, Wiriya W, Chantara S (2022) Fresh and aged PM2. 5 and their ion composition in rural and urban atmospheres of Northern Thailand in relation to source identification. Chemosphere 286:131803. https://doi.org/10.1016/j.chemosphere.2021.131803
17. Chifflet S, Guyomarc'h L, Dominutti P, Heimbürger-Boavida LE, Angeletti B, Louvat P, Mari X (2023) Seasonal variations of metals and metalloids in atmospheric particulate matter ($PM_{2.5}$) in the urban megacity Hanoi. Atmosp Pollut Res 101961. https://doi.org/10.1016/j.apr.2023.101961

18. Chuersuwan N, Nimrat S, Lekphet S, Kerdkumrai T (2008) Levels and major sources of $PM_{2.5}$ and PM_{10} in Bangkok Metropolitan Region. Environ Int 34(5):671–677. https://doi.org/10.1016/j.envint.2007.12.018
19. Cohen DD, Crawford J, Stelcer E, Bac VT (2009) Characterisation and source apportionment of fine particulate sources at Hanoi from 2001 to 2008. Atmosp Environ 44(3):320–328. https://doi.org/10.1016/j.atmosenv.2009.10.037
20. Das R, Wang X, Itoh M, Shiodera S, Kuwata M (2019) Estimation of Metal emissions from tropical Peatland burning in Indonesia by controlled laboratory experiments. J Geophys 124:6583–6599. https://doi.org/10.1029/2019JD030364
21. Dat ND, Nguyen VT, Vo TDH (2021) Contamination, source attribution, and potential health risks of heavy metals in street dust of a metropolitan area in Southern Vietnam. Environ Sci Pollut Res 28:50405–50419. https://doi.org/10.1007/s11356-021-14246-1
22. Ding Y (2019) Heavy metal pollution and transboundary issues in ASEAN countries. Water Policy 21(5):1096–1106. https://doi.org/10.2166/wp.2019.003
23. Driscoll CT, Mason RP, Chan HM, Jacob DJ, Pirrone N (2013) Mercury as a global pollutant: sources, pathways, and effects. Environ Sci Technol 47(10):4967–4983. https://doi.org/10.1021/es305071v
24. Duan J, Tan J, Hao J, Chai F (2014) Size distribution, characteristics and sources of heavy metals in haze episod in Beijing. J Environ Sci 26(1):189–196. https://doi.org/10.1016/S1001-0742(13)60397-6
25. Duong TT, Lee BK (2011) Determining contamination level of heavy metals in road dust from busy traffic areas with different characteristics. J Environ Manage 92(3):554–562
26. Estoque RC, Murayama Y (2013) City Profile: Baguio Cities 30:240–251
27. Evers DC, Han YJ, Driscoll CT, Kamman NC, Goodale WM, Lambert KF, Holsen TM, Chen CY, Clair TA, Butler TJ (2007) Biological mercury hotspots in the Northeastern United States and Southeastern Canada. Bioscience 57:1–15. https://doi.org/10.1641/B570107
28. Fujii Y, Mahmud M, Tohno S, Okuda T, Mizohata A (2016) A case study of $PM_{2.5}$ characterization in Bangi, Selangor, Malaysia during the Southwest monsoon season. Aerosol and Air Quality Research 16(11):2685–2691
29. Galindo N, Yubero E, Clemente A, Nicolás JF, Navarro-Selma B, Crespo J (2019) Insights into the origin and evolution of carbonaceous aerosols in a mediterranean urban environment. Chemosphere 235 636-642. https://doi.org/10.1016/j.chemosphere.2019.06.202113
30. Galvez MCD, Vallar E, Castilla RM, Mandia P, Branzuela R, Rempillo O, Orbecido AH, Beltran A, Ledesma N, Deocaris C, Morris V, Belo LP (2022) Principal component analysis of heavy metals in atmospheric aerosols from Meycauayan, Bulacan, Philippines. Preprints 2022020120. https://doi.org/10.20944/preprints202202.0120.v1
31. Ghani FA, Redzuan NIN, Nasir NFM, Salamat M (2017) Review on ASEAN transboundary haze pollution agreement 2002: problems and solutions. J Humanit Lang Culture and Business 1(1):153–161
32. Gummeneni S, Yusup YB, Chavali M, Samadi SZ (2011) Source apportionment of particulate matter in the ambient air of Hyderabad city India. Atmosp Res 101(3):752–764. https://doi.org/10.1016/j.atmosres.2011.05.002
33. Gustin MS, Hou D, Tack FMG. The term "heavy metal(s)": History, current debate, and future use. Sci Total Environ 2021 Oct 789:147951. https://doi.org/10.1016/j.scitotenv.2021.147951
34. Hai CD, Oanh NTK (2013) Effects of local, regional meteorology and emission sources on mass and compositions of particulate matter in Hanoi. Atmosp Environ 78:105–112. https://doi.org/10.1016/j.atmosenv.2012.05.006
35. Harari R, Harari F, Gerhardsson L, Lundh T, Skerfving S, Strömberg U, Broberg K (2012) Exposure and toxic effects of elemental mercury in gold-mining activities in Ecuador. Toxicol Lett 213(1):75–82. https://doi.org/10.1016/j.toxlet.2011.09.006
36. Hieu NT, Lee BK (2010) Characteristics of particulate matter and metals in the ambient air from a residential area in the largest industrial city in Korea. Atmos Res 98(2-4):526–537. https://doi.org/10.1016/j.atmosres.2010.08.019

37. Hopke PK, Cohen DD, Begum BA, Biswas SK, Ni BF, Pandit GG, Santoso M, Chung YS, Davy P, Markwitz A, Waheed S, Siddique N, Santos FL, Pabroa PCB, Seneviratne MCS, Wimolwattanapun W, Bunprapob S, Vương TB, Hien PD, Markowicz A (2008) Urbanair quality in the Asian region. Sci Total Environ 404:103–112. https://doi.org/10.1016/j.scitotenv.2008.05.039
38. Hsu CY, Chiang HC, Lin SL, Chen MJ, Lin TY, Chen YC (2016) Elemental characterization and source apportionment of PM10 and PM2. 5 in the western coastal area of central Taiwan. Sci Total Environ 541:1139–1150. https://doi.org/10.1016/j.scitotenv.2015.09.122
39. Hu X, Zhuhong D, Yun Z, Sun Yuanyuan W, Jichun CY, Hongzhen L (2013) Size distribution and source apportionment of airborne metallic elements in Nanjing, China. Aerosol and Air Quality Res 13(6):1796–1806. https://doi.org/10.4209/aaqr.2012.11.0332
40. Janta R, Chantara S (2017) Tree bark as bioindicator of metal accumulation from road traffic and air quality map: a case study of Chiang Mai, Thailand. Atmosp Pollut Res 8(5):956–967. https://doi.org/10.1016/j.apr.2017.03.010
41. Ji X, Abakumov E, Chigray S, Saparova S, Polyakov V, Wang W et al (2021) Response of carbon and microbial properties to risk elements pollution in arctic soils. J Hazard Mater 408:124430. https://doi.org/10.1016/j.jhazmat.2020.124430
42. Kayee J, Sompongchaiyakul P, Sanwlani N, Bureekul S, Wang X, Das R (2020) Metal concentrations and source apportionment of $PM_{2.5}$ in Chiang Rai and Bangkok, Thailand during a biomass burning season. ACS Earth and Space Chem 4(7):1213–1226. https://doi.org/10.1021/acsearthspacechem.0c00140
43. Khan MF et al (2016) Comprehensive assessment of $PM_{2.5}$ physicochemical properties during the Southeast Asia dry season (southwest monsoon). J Geophys Res Atmos 121:14589–14611. https://doi.org/10.1002/2016JD025894
44. Khan MF, Latif MT, Saw WH, Amil N, Nadzir MSM, Sahani M, Tahir NM, Chung JX (2016) Fine particulate matter in the tropical environment: monsoonal effects, source apportionment, and health risk assessment. Atmosp Chem Phys 16:597–617. https://doi.org/10.5194/acp-16-597-2016
45. Kim KH, Mishra VK, Kang CH, Choi KC, Kim YJ, Kim DS, Youn YH, Lee JH (2005) The metallic composition of aerosols at three monitoring sites in Korea during winter 2002. Environ Monit Assessm 121:381–399. https://doi.org/10.1007/s10661-005-9136-y
46. Kim Oanh NT, Upadhayay N, Zhuang YH et al (2006) Particulate air pollution in six cities: spatial and temporal distributions, and associated sources. Atmos Environ 40:3367–3380. https://doi.org/10.1016/j.atmosenv.2006.01.050
47. Klaus Kästle (2013) Map of Southeast Asia Region. Nations Online Project. One World Nations Online. https://www.nationsonline.org/oneworld/map_of_southeast_asia.htm
48. Koukoulakis KG, Chrysohou E, Kanellopoulos PG, Karavoltsos S, Katsouras G, Dassenakis M, Nikolelis D, Bakeas E (2019) Trace elements bound to airborne PM_{10} in a heavily industrialized site nearby Athens: seasonal patterns, emission sources, health implications. Atmosp Pollut Res 10(4):1347–1356. https://doi.org/10.1016/j.apr.2019.03.007
49. Landis MS, Patrick Pancras J, Graney JR, White EM, Edgerton ES, Legge A, Percy KE (2017) Source apportionment of ambient fine and coarse particulate matter at the Fort McKay community site in the Athabasca Oil Sands Region, Alberta, Canada. Sci Total Environ 584–585:105–117. https://doi.org/10.1016/j.scitotenv.2017.01.110
50. Lestiani DD, Santoso M, Trompetter WJ et al (2013) Determination of chemical elements in atmosphereatmospheree particulate matter collected at Lembang, Indonesia by particle induced X-ray emission. J Radioanal Nucl Chem 297:177–182. https://doi.org/10.1007/s10967-012-2348-z
51. Li J, Michalski G, Olson EJ, Welp LR, Larrea Valdivia AE, Larico JR, Zapata FA, Paredes LM (2021) Geochemical characterization and heavy metal sources in PM_{10} in Arequipa, Peru. Atmosphere 12:641. https://doi.org/10.3390/atmos12050641
52. Lian M, Wang J, Sun L, Xu Z, Tang J, Yan J (2019) Profiles and potential health risks of heavy metals in soil and crops from the watershed of Xi River in Northeast China. Ecotoxicol Environ Safety 169:442–448. https://doi.org/10.1016/j.ecoenv.2018.11.046

53. Liu P, Hu W, Tian K, Huang B, Zhao Y, Wang X (2020) Accumulation and ecological risk of heavy metals in soils along the coastal areas of the Bohai Sea and the Yellow Sea: a comparative study of China and South Korea. Environ Int 137:105519. https://doi.org/10.1016/j.envint.2020.105519
54. Liu A, Ma Y, Gunawardena JM, Egodawatta P, Ayoko GA, Goonetilleke A (2018) Heavy metals transport pathways: The importance of atmospheric pollution contributing to stormwater pollution. Ecotoxicol Environ Safety 164:696–703
55. Ma Y, Egodawatta P, McGree J, Liu A, Goonetilleke A (2016) Human health risk assessment of heavy metals in urban stormwater. Sci Total Environ 557:764–772. https://doi.org/10.1016/j.scitotenv.2016.03.067
56. Makkonen U, Hellén H, Anttila P, Ferm M (2010) Size distribution and chemical composition of airborne particles in south-eastern Finland during different seasons and wildfire episodes in 2006. Sci Total Environ 408(3):644–651. https://doi.org/10.1016/j.scitotenv.2009.10.050
57. Mamun AA, Cheng I, Zhang L, Dabek-Zlotorzynska E, Charland JP (2020) Overview of size distribution, concentration, and dry deposition of airborne particulate elements measured worldwide. Environ Rev 28(1):77–88. https://doi.org/10.1139/er-2019-0035
58. Mason Robert P (2013) Trace metals in aquatic systems (Mason/Trace Metals in Aquatic Systems)||Metal(loid)s in the atmosphere and their inputs to surface waters 167–218. https://doi.org/10.1002/9781118274576.ch5
59. Mbengue S, Alleman LY, Flament P (2014) Size-distributed metallic elements in submicronic and ultrafine atmospheric particles from urban and industrial areas in northern France. Atmospheric Res 135:35–47. https://doi.org/10.1016/j.atmosres.2013.08.010
60. Miller AJ, Raduma DM, George LA, Fry JL (2019) Source apportionment of trace elements and black carbon in an urban industrial area (Portland, Oregon). Atmosp Pollut Res 10(3):784–794. https://doi.org/10.1016/j.apr.2018.12.006
61. Mitra S, Chakraborty AJ, Tareq AM, Emran TB, Nainu F, Khusro A, Idris AM, Khandaker MU, Osman H, Alhumaydhi FA, Simal-Gandara J (2022) Impact of heavy metals on the environment and human health: Novel therapeutic insights to counter the toxicity. J King Saud Univer–Sci 34(3):101865. https://doi.org/10.1016/j.jksus.2022.101865
62. Mulware SJ (2013) Trace elements and carcinogenicity: a subject in review. 3 Biotech 3:85–96. https://doi.org/10.1007/s13205-012-0072-6
63. Nguyen LSP, Hien TT, Truong MT, Nguyen DTC, Sheu GR (2022) Atmospheric particulate-bound mercury (PBM_{10}) in a Southeast Asia megacity: sources and health risk assessment. Chemosphere 307:135707. https://doi.org/10.1016/j.chemosphere.2022.135707
64. Nguyen LSP, Huang HY, Lei TL, Bui TT, Wang SH, Chi KH, Sheu GR, Te LC, Ou-Yang CF, Lin NH (2020) Characterizing a landmark biomass-burning event and its implication for aging processes during long-range transport. Atmos Environ 241:117766. https://doi.org/10.1016/j.atmosenv.2020.117766
65. Nguyen LSP, Pham TDH, Truong MT, Tran AN (2023) Characteristics of total gaseous mercury at a tropical megacity in Vietnam and influence of tropical cyclones. Atmos Pollut Res 14(8):101813. https://doi.org/10.1016/j.apr.2023.101813
66. Nguyen LSP, Sheu GR (2019) Four-year measurements of wet mercury deposition at a tropical mountain site in Central Taiwan. Aerosol Air Qual Res 19:2043–2055. https://doi.org/10.4209/aaqr.2019.05.0250
67. Nguyen LSP, Hien TT, (2024). Long-range atmospheric mercury transport from Across East Asia to a Suburban Coastal Area in Southern Vietnam. Bull Environ Contam Toxicol 112:14. https://doi.org/10.1007/s00128-023-03842-1
68. Nour HE, El-Sorogy AS, Abd El-Wahab M, Mohamaden M, Al-Kahtany K (2019) Contamination and ecological risk assessment of heavy metals pollution from the Shalateen coastal sediments, Red Sea, Egypt. Marine Pollut Bullet 144:167–172. https://doi.org/10.1016/j.marpolbul.2019.04.056
69. Obrist D, Kirk JL, Zhang L, Sunderland EM, Jiskra M, Selin NE (2018) A review of global environmental mercury processes in response to human and natural perturbations: changes of emissions, climate, and land use. Ambio 47(2):116–140. https://doi.org/10.1007/s13280-017-1004-9

70. Paatero P, Tapper U (1994) Positive matrix factorization: a non-negative factor model with optimal utilization of error estimates of data values. Environmetrics 5:111–126. https://doi.org/10.1002/env.3170050203
71. Pabroa PCB, Racho JMD, Jagonoy AM, Valdez JDG, VII ATB, Yee JR, Cohen DD (2022) Characterization, source apportionment and associated health risk assessment of respirable atmosphere particulates in Metro Manila, Philippines. Atmosp Pollut Res 13(4):101379. https://doi.org/10.1016/j.apr.2022.101379
72. Rahman SA, Hamzah MS, Elias MS, Salim NAA, Hashim A, Shukor S, Siong WB, Wood AK (2015) A long term study on characterization and source apportionment of particulate pollution in Klang Valley, Kuala Lumpur. Aerosol Atmosphere Qual Res 15:2291–2304. https://doi.org/10.4209/aaqr.2015.03.0188
73. Rahman SA, Hamzah MS, Wood AK, Elias MS, Salim NAA, Sanuri E (2011) Sources apportionment of fine and coarse aerosol in Klang Valley, Kuala Lumpur using positive matrix factorization. Atmosp Pollut Res 2(2):197–206. https://doi.org/10.5094/APR.2011.025
74. Ramírez O, De La Campa AS, Amato F, Catacolí RA, Rojas NY, de la Rosa J (2018) Chemical composition and source apportionment of PM_{10} at an urban background site in a high–altitude Latin American megacity (Bogota, Colombia). Environ Pollut 233:142–155. https://doi.org/10.1016/j.envpol.2017.10.045
75. Rauf AU, Mallongi A, Lee K, Daud A, Hatta M, Al Madhoun W, Astuti RDP (2021) Potentially toxic element levels in atmospheric particulates and health risk estimation around industrial areas of Maros Indonesia. Toxics 9:328. https://doi.org/10.3390/toxics9120328
76. Sahu C, Basti S, Sahu SK (2021) Particulate collection potential of trees as a means to improve the air quality in urban areas in India. Environ Processes 8:377–395. https://doi.org/10.1007/s40710-021-00494-3
77. Sakunkoo P, Thonglua T, Sangkham S, Jirapornkul C, Limmongkon Y, Daduang S, Pimonsree S (2022) Human health risk assessment of $PM_{2.5}$-bound heavy metal of anthropogenic sources in the Khon Kaen Province of Northeast Thailand. Heliyon 8(6):e09572. https://doi.org/10.1016/j.heliyon.2022.e09572
78. Samontha A, Waiyawat W, Shiowatana J, McLaren RG (2007) Atmospheric deposition of metals associated with air particulate matter: fractionation of particulate-bound metals using continuous-flow sequential extraction. Sci Asia 33:421–428. https://doi.org/10.2306/scienceasia1513-1874.2007.33.421
79. Sandhya M et al (2019) Human health risk assessment of heavy metals in urban stormwater. Environ Biotechnol: For Sustain Future 103–125.https://doi.org/10.1016/j.scitotenv.2016.03.067
80. Santoso M, Hopke PK, Hidayat A (2008) Sources identification of the atmospheric aerosol at urban and suburban sites in Indonesia by positive matrix factorization. Sci Total Environ 397(1–3):229–237. https://doi.org/10.1016/j.scitotenv.2008.01.057
81. Santoso M, Lestiani DD, Markwitz A (2013) Characterization of airborne particulate matter collected at Jakarta roadside of an arterial road. J Radioanal Nucl Chem 297:165–169. https://doi.org/10.1007/s10967-012-2350-5
82. Santoso M, Lestiani DD, Mukhtar R, Hamonangan E, Syafrul H, Markwitz A, Hopke PK (2011) Preliminary study of the sources of ambient air pollution in Serpong, Indonesia. Atmosp Pollut Res 2(2):190–196. https://doi.org/10.5094/APR.2011.024
83. Sara YY, Rashid M, Chuah TG, Suhaimi M, Mohamed NN (2013) Characteristics of airborne PM2. 5 and Pm2. 5–10 in the urban environment of Kuala Lumpur. Adv Mater Res 620:502–510
84. Sara YY, Rashid M, Chuah TG, Suhaimi M, Mohamed NN (2013) Characteristics pf AtmosphereAtmospheree $PM_{2.5}$ And $PM_{2.5\text{-}10}$ in the urban environment of Kuala Lumpur. Adv Mater Res 620:502–510. https://doi.org/10.4028/www.scientific.net/AMR.620.502
85. Schroeder WH, Munthe J (1998) Atmospheric mercury—an overview. Atmos Environ 32:809–822. https://doi.org/10.1016/S1352-2310(97)00293-8
86. See SW, Balasubramanian R, Rianawati E, Karthikeyan S, Streets DG (2007) Characterization and source apportionment of particulate matter $\leq$ 2.5 μm in Sumatra, Indonesia, during

a recent peat fire episode. Environ Sci Technol 41:3488–3494. https://doi.org/10.1021/es061943k
87. See SW, Balasubramanian R, Wang W (2006) A study of the physical, chemical, and optical properties of ambient aerosol particles in Southeast Asia during hazy and nonhazy days. J Geophys Res 111:D10S08. https://doi.org/10.1029/2005JD006180
88. Seinfeld J, Pandis S (1998) In: Atmospheric chemistry and physics. Wiley & Sons Inc, New York
89. Sentian J, Herman F, Yih CY, Wui JCH (2019) Long-term air pollution trend analysis in Malaysia. Int J Environ Impacts 2(4):309–324. https://doi.org/10.2495/EI-V2-N4-309-324
90. Sheu GR, Lin NH, Lee CT, Wang JL, Chuang MT, Wang SH, Chi KH, Ou-Yang CF (2013) Distribution of atmospheric mercury in northern Southeast Asia and South China Sea during Dongsha experiment. Atmosp Environ 78:174–183. https://doi.org/10.1016/j.atmosenv.2012.07.002
91. Sheu GR, Nguyen LSP, Truong MT, Lin DW (2019) Characteristics of atmospheric mercury at a suburban site in northern Taiwan and influence of trans-boundary haze events. Atmosp Environ 214:116827. https://doi.org/10.1016/j.atmosenv.2019.116827
92. Soe PS, Kyaw WT, Arizono K, Ishibashi Y, Agusa T (2022) Mercury pollution from artisanal and small-scale gold mining in Myanmar and other Southeast Asian Countries. Int J Environ Res Public Health 19:6290. https://doi.org/10.3390/ijerph19106290
93. Srithawirat T, Latif MT (2015) Concentration of selected heavy metals in the surface dust of residential buildings in Phitsanulok, Thailand. Environ Earth Sci 74:2701–2706. https://doi.org/10.1007/s12665-015-4291-0
94. Srithawirat T, Latif MT, Sulaiman FR (2016) Indoor PM_{10} and its heavy metal composition at a roadside residential environment, Phitsanulok, Thailand. Atmósfera 29(4):311–322. https://doi.org/10.20937/ATM.2016.29.04.03
95. Stein AF, Draxler RR, Rolph GD, Stunder BJB, Cohen MD, Ngan F (2016) NOAA's HYSPLIT atmospheric transport and dispersion modeling system. Bull Am Meteor Soc 96:2059–2077. https://doi.org/10.1175/BAMS-D-14-00110.1
96. Stohl A (1998) Computation, accuracy and applications of trajectories—a review and bibliography. Atmosp Environ 32(6):947–966. https://doi.org/10.1016/j.atmosres.2011.05.002
97. Sulong NA, Latif MT, Khan MF, Amil N, Ashfold MJ, Wahab MIA, Sahani M (2017) Source apportionment and health risk assessment among specific age groups during haze and non-haze episodes in Kuala Lumpur, Malaysia. Sci Total Environ 601–602:556–570. https://doi.org/10.1016/j.scitotenv.2017.05.153
98. Tan SY, Praveena SM, Abidin EZ, Cheema MS (2018) Heavy metal quantification of classroom dust in school environment and its impacts on children health from Rawang (Malaysia). Environ Sci Pollut Res 25:34623–34635
99. Titseesang T, Wood T, Panich N (2008) Leaves of orange jasmine (Murraya paniculata) as indicators of airborne heavy metal in Bangkok, Thailand. Annals of the New York Acad Sci 1140(1):282–289
100. To H, Nguyen C, Huy D, Le Hoang A, David O, Grant F, Graham M, Alex B (2022) Soluble trace metals associated with atmospheric fine particulate matter in the two most populous cities in Vietnam. Atmosp Environ: X 15:100178. https://doi.org/10.1016/j.aeaoa.2022.100178
101. Truong MT, Nguyen LSP, Hien TT, Pham TDH, Do TTL (2022) Source apportionment and risk estimation of heavy metals in PM_{10} at a Southern Vietnam megacity. Aerosol Atmosphere Qual Res 22:220094. https://doi.org/10.4209/aaqr.220094
102. UN Environment (2019) Global Mercury Assessment (2018) UN environment programme. Chemicals and Health Branch Geneva, Switzerland
103. UNEP (2013) UNEP: global mercury assessment 2013: sources, emissions, releases and environmental transport. UNEP Chemicals Branch, Geneva, Switzerland
104. Di Vaio P, Magli E, Caliendo G, Corvino A, Fiorino F, Frecentese F, Saccone I, Santagada V, Severino B, Onorati G (2018) Heavy metals size distribution in PM_{10} and environmental-sanitary risk analysis in Acerra (Italy). Atmosphere 9:58. https://doi.org/10.3390/atmos9020058

105. Wang Q, Bi XH, Wu JH (2013) Heavy metals in urban ambient PM_{10} and soil background in eight cities around China. Environ Monit Assess 185:1473–1482. https://doi.org/10.1007/s10661-012-2646-5
106. Weatherbee DE (2014) International relations in Southeast Asia: the struggle for autonomy. Rowman & Littlefield
107. Wong CSC, Li XD, Zhang G, Qi SH, Peng XZ (2003) Atmospheric deposition of heavy metals in the Pearl River Delta. China 37(6):767–776. https://doi.org/10.1016/s1352-2310(02)00929-9
108. World Bank (2022) World Bank national accounts data and OECD National Accounts data files. https://data.worldbank.org/indicator/NY.GDP.MKTP.CD
109. Wu PC, Huang KF (2021) Tracing local sources and long-range transport of PM10 in central Taiwan by using chemical characteristics and Pb isotope ratios. Sci Rep 11(1):7593. https://doi.org/10.1038/s41598-021-87051-y
110. Yap CK, Chew WY, Tan SG (2012) Heavy metal concentrations in ceiling fan and roadside car park dust collected from residential colleges in Universiti Putra Malaysia, Serdang, Selangor. Pertanika J Trop Agric Sci 35(1):75–83
111. Ye L, Huang M, Zhong B, Wang X, Tu Q, Sun H, Wang C, Wu L, Chang M (2017) Wet and dry deposition fluxes of heavy metals in Pearl River Delta Region (China): Characteristics, ecological risk assessment and source apportionment. J Environ Sci. S1001074217325524. https://doi.org/10.1016/j.jes.2017.11.019
112. Zajusz-Zubek E, Mainka A, Korban Z, Pastuszka JS (2015) Evaluation of highly mobile fraction of trace elements in PM_{10} collected in Upper Silesia (Poland): preliminary results. Atmos Pollut Res 6:961–968. https://doi.org/10.1016/j.apr.2015.05.001
113. Zeng J, Han G, Yang K (2020) Assessment and sources of heavy metals in suspended particulate matter in a tropical catchment, northeast Thailand. J Cleaner Prod 265:121898. https://doi.org/10.1016/j.jclepro.2020.121898
114. Zhang H, Fu X, Lin CJ, Shang L, Zhang Y, Feng X, Lin C (2016) Monsoon-facilitated characteristics and transport of atmospheric mercury at a high-altitude background site in southwestern China. Atmos Chem Phys 16(20):13131–13148. https://doi.org/10.5194/acp-16-13131-2016
115. Zhu G, Noman MA, Narale DD, Feng W, Pujari L, Sun J (2020) Evaluation of ecosystem health and potential human health hazards in the Hangzhou Bay and Qiantang Estuary region through multiple assessment approaches. Environ Pollut 114791.https://doi.org/10.1016/j.envpol.2020.114791

Heavy Metals in the Ecosystem; Sources and Their Effects

Abderrahim Maftouh, Omkaltoume El Fatni, Asma Ben Moussa, Faissal Boukir, and Waliha Noor us Subha

Abstract Heavy metals are defined by their density, Metals with densities more than 5 g/cm^3 are classified as heavy metals. According to this study, the heavy metals that pose the most risk to humans are lead (Pb), cadmium (Cd), mercury (Hg), and arsenic (As). Heavy metals may pollute water sources, harming aquatic life and rendering the water hazardous for human consumption. Anthropogenic activities have contributed more to the soil's pollution with heavy metals and other contaminants than natural sources do. Chemical fertilizers, insecticides and herbicides with a chemical basis are increasingly being used in everyday agriculture activities all over the world. They are dispersing and depositing in water, and the environment and water are degrading, affecting aquatic life and causing sediments and deposits in bodies of water. If even tiny amounts of heavy metals like arsenic, mercury, lead, chromium, or cadmium are present, they are extremely dangerous. Atomic absorption spectrometry (AAS), neutron activation analysis (NAA), X-ray fluorescence, inductively coupled plasma optical emission spectrometry (ICPOES), atomic emission/fluorescence spectrometry, inductively coupled plasma mass spectrometry (ICP-MS), and anodic stripping voltammetry are just a few of the instruments available heavy metal concentrations can be determined using this method. Remote sensing spectra are useful to map the presence of toxic heavy metals (HMs) in soil because they give information on organic matter, Fe-oxides, and clays in soils and plant cover. Kemper and Sommer used airborne hyperspectral data to map the remaining hazardous heavy metals (HMs)

A. Maftouh (✉) · O. El Fatni
LPHE-Modeling and Simulations, Faculty of Sciences, Mohammed V University in Rabat, B.P. 1014 Rabat, Morocco
e-mail: abderrahim_maftouh@um5.ac.ma

A. Ben Moussa
National Office of Electricity and Drinking Water (ONEE), Rabat, Morocco

F. Boukir
WATEC Technologies, Mohammedia, Morocco

W. Noor us Subha
National University of Sciences and Technology (NUST), Islamabad 44000, Pakistan

N. Kumar (ed.), *Heavy Metal Remediation*, Earth and Environmental Sciences Library,
https://doi.org/10.1007/978-3-031-53688-5_2

in Spanish soils following a quarrying disaster. Numerous corrective action procedures are being used. Phytoremediation is one type of bioremediation that seems to be more environmentally friendly.

Keywords Heavy metals · Water pollution · Anthropogenic activities · Environmental contamination · Spectroscopy analysis · Phytoremediation

1 Introduction

Heavy metals are defined by their density, Metals with densities more than 5 g/cm^3 are classified as heavy metals. According to this study, the heavy metals that pose the most risk to humans are lead (Pb), cadmium (Cd), mercury (Hg), and arsenic (As). Heavy metals are introduced into the ecosystem through a variety of anthropogenic (caused by humans) and natural activities. When Ecosystems may be disrupted by the accumulation of heavy metals that are introduced into the environment particularly those in terrestrial and aquatic habitats. Here are some typical routes through which Heavy metals infiltrate in the environment:

1.1 Natural Sources of Heavy Metals

Two natural or geophysical processes that lead to the environmental accumulation of heavy metals include the weathering of rocks that contain metals and volcanic eruptions. The number of heavy metals in the environment that are attributed to humans has increased because of the worldwide trends of urbanization and developing cities on Earth [19]. Heavy Metals may be released from their indigenous spheres and end up in other environmental elements because of weathering processes that occur naturally. Hydroxides, sulphates, oxides, sulphates, phosphates, silicates, and organic compounds are all forms of heavy metals. The eight heavy metals that are most often found in the environment are lead (Pb), nickel (Ni), chromium (Cr), cadmium (Cd), arsenic (As), mercury (Hg), zinc (Zn), and copper (Cu). The heavy metals are present in traces, yet they nonetheless pose a major health risk to animals including humans.

1.2 Anthropogenic (Human Caused) Sources of Heavy Metals

The contaminants are also discharged into other areas of the environment via mining, metallurgical processes, runoffs, the agricultural sector, and treatment of wastewater. It has been demonstrated that anthropogenic heavy metal processes exceed natural fluxes for some elements. Most of the naturally occurring metals in wind-blown dusts come from industrial regions. The combustion of fossil fuels releases nickel, vanadium, mercury, selenium, and tin, as well as other important anthropogenic methods such as smelting, and this releases arsenic, copper, and zinc, pesticides, which emit arsenic, and car exhaust, which emits lead. Human activities have been shown to considerably contribute to environmental deterioration due to the constant manufacturing of commodities to meet the needs of a large population. Natural and anthropogenic sources of heavy metal contamination in the ecosystem is represented in Fig. 1.

Here are a few typical methods through which heavy metals enter the ecosystem:

a. **Industrial Emissions:** Through their effluents, factories and other industrial facilities frequently discharge heavy metals into bodies of water. The direct release of toxic metals into rivers, lakes, and seas can occur because of industries like mining, metal smelting, and industrial activities.
b. **Agricultural Activities:** Heavy metals may accumulate in soil as an outcome of the usage of fertilizers, insecticides, and herbicides that contain these substances. These metals can contaminate soil and water over time by dissolving into groundwater or being washed into surrounding water bodies during rain.

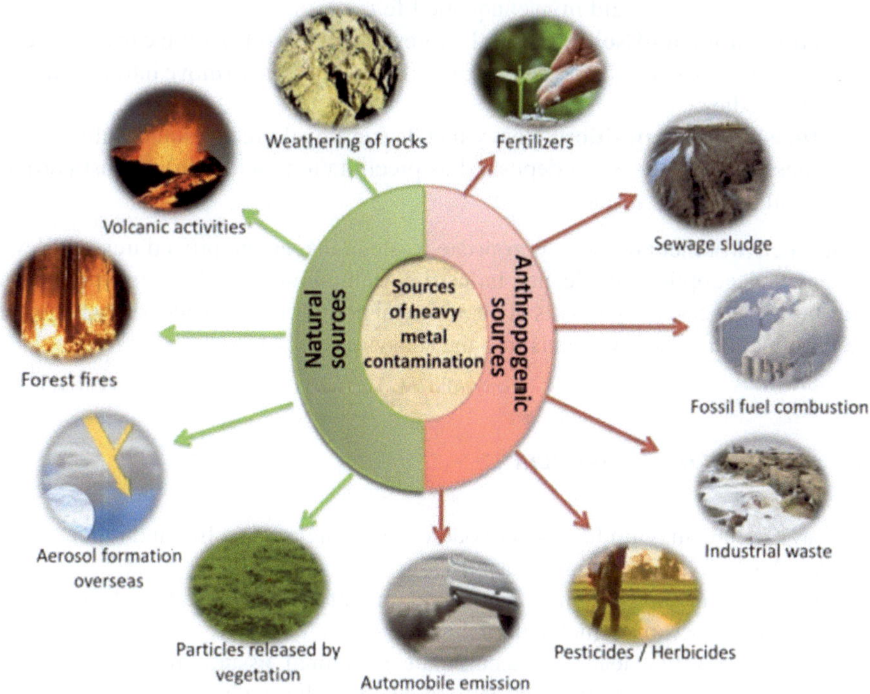

Fig. 1 Natural and anthropogenic sources of heavy metal contamination in the ecosystem

c. **Ore Extraction and Mining**: Ores containing heavy metals are exposed during mining activities. Heavy metals may be discharged during the excavation and processing of these ores, polluting the environment and the nearby soil and water.
d. **Inadequate Waste Disposal:** The leaching process of metals into soil and groundwater occurs due to improper disposal of electronic waste materials, batteries, and other items containing heavy metals. Such waste can pollute the environment when it is disposed of in landfills.
e. **Air Pollution:** Heavy metals are released into the atmosphere through burning fossil fuels, industrial operations, and emissions from automobiles. Contamination can result from these contaminants when they deposit into soil and water.
f. **Natural Mechanisms**: Heavy metals may be discharged into the ecosystem as a result of natural events including volcanic eruptions, forest fires, and rock weathering. Even while these processes are natural, they nonetheless have the potential to increase the number of heavy metals in ecosystems.
g. **Wastewater Drainage:** Heavy metals can be found in domestic sewage as well as untreated wastewater from residences and sewage treatment facilities. These metals can enter rivers, lakes, and seas if they are not properly handled.
h. **Ocean and Ships:** Copper is a common heavy metal found in anti-fouling coatings used on ships. These paints have the potential to release heavy metals into the ocean, which could impact aquatic life.
i. **Deforestation and Soil Erosion:** Deforestation and soil erosion can erode heavy metals that are naturally present in the soil, causing them to move into surrounding water bodies.
j. **Atmospheric Deposition:** Heavy metals can travel great distances through the atmosphere before being deposited as precipitation (acid rain) or dust particles on soil and water surfaces.

Strict environmental laws, appropriate waste management procedures, the development and adoption of cleaner industrial technologies, and raising public and industry awareness of the effects of heavy metal pollution are all necessary to prevent the introduction of heavy metals into the ecosystem.

1.3 Impacts on Ecosystem

- Heavy metals may pollute water sources, harming aquatic life and rendering the water hazardous for human consumption.
- Contaminated soil inhibits plant development and can lead to heavy metal absorption by crops, which enters the food chain.
- Heavy metals like lead can accumulate in animal tissues, resulting in larger quantities in species higher up the food chain, including humans.

- Heavy metals such as lead, mercury, cadmium, and arsenic are poisonous to both people and animals, causing devastation on the neurological system, kidneys, and other organs.
- Pollution from heavy metals can cause biodiversity loss because Certain organisms are more susceptible to the hazardous effects of metals than others.

The hazardous characteristics of heavy metals have been acknowledged for a while. Nonetheless, current experimental studies indicate that certain, such as nickel, copper, and zinc, are necessary for humans and are often found in nature [9]. Manganese is found all over the world and accounts for around 0.1% of the earth's crust. Heavy metals, on the other hand, come in a variety of forms of negative environmental consequences, for instance, the conversion of mercury compounds in the presence of water into methylmercury results in extremely hazardous sediments [20].

Chromium is frequently utilized in manufacturing and has the potential to cause cancer. However, a few heavy metals have been associated in the regulation of specific physiologic physiological activities [7]. Naturally occurring important heavy metals are absorbed by the body through water, food, and air. govern a variety of biological activity [31]. Most poisonous heavy metals, such as lead., thallium, cadmium., and antimony, are important and widely used in industrial operations pollutants of the environment. Although less prevalent in nature, thallium has a greater detrimental effect than other heavy metals and causes baldness in humans [17].

The advantages of heavy metals are frequently overshadowed by the hazards involved; for example, excessive antimony and chromium exposure increases carcinogenesis, while lead toxicity causes mental retardation in youngsters [32]. While Itai Itai disease is brought on by cadmium exposure, Minamata illness is caused by mercury toxicity. In addition, several human organs such as the kidneys, brain, liver, skin, and heart can become toxically affected by heavy metals. People must migrate away from industrial locations where heavy metal emissions are high to prevent hazardous impacts [5]. Heavy metal bioavailability in food chains is represented in Fig. 2.

2 Literature Review

2.1 Process of Heavy Metal Contamination in the Soil

Anthropogenic activities have contributed more to the soil's pollution with heavy metals and other contaminants than natural sources do. Chemical fertilizers, insecticides and herbicides with a chemical basis are increasingly being used in everyday agriculture activities all over the world. Because they are used over extended periods of time, these compounds greatly increase the number of heavy metals that end up in soils and plants. Apart from the essential nutrients required for plant development, most fertilizers also include several heavy metals. For example, a major issue is the presence of Cd and other heavy metal pollutants in lime and superphosphate fertilizer [19].

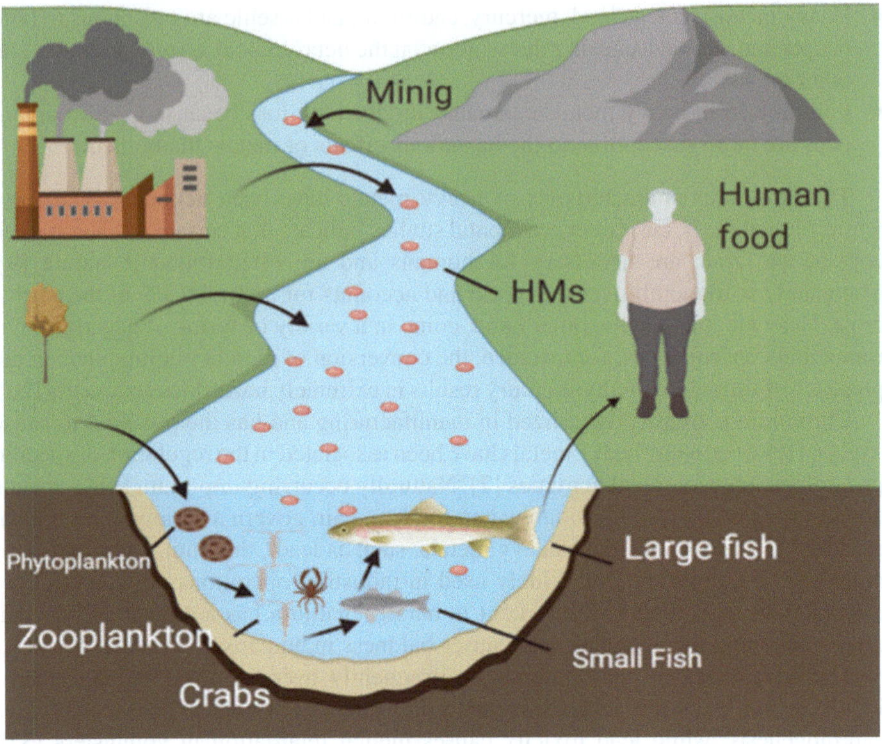

Fig. 2 Heavy metal bioavailability in food chains

Phosphorus and nitrogen fertilizers contain many hazardous heavy metals as impurities, including Cadmium (Cd), Chromium (Cr), Copper (Cu), Lead (Pb) and Zinc (Zn) with concentrations 13 mg/kg; 0.9 mg/kg, 60 mg/kg; 3.4 mg/kg, 26 mg/kg; 2 mg/kg, 13 mg/kg; 1.9 mg/kg and 236 mg/kg; 5 mg/kg respectively. As a result of plants consuming these heavy metals and moving up the food chain, a major issue occurs. In combination with other unfavorable elements like acidic atmospheric precipitation and insufficient liming, excessive metal concentration in soil contributes to soil acidification ultimately making heavy metals more accessible. Together, these exacerbate the situation for both growing plants and the creatures that consume them demonstrated Cadmium (Cd), Lead (Pb), and Arsenic (As) impacts of fertilizer usage. The levels of these heavy metals in a field of wheat were assessed both before and after gathering the crops and applying fertilizer. The analysis demonstrated that composite fertilizer had the highest concentration of Cadmium (Cd), but zinc sulfate fertilizer had the highest concentrations of Lead (Pb) and Arsenic (As), which were over the allowable limits. The use of fertilizer has led to noticeably greater levels of Cadmium (Cd), Lead (Pb), and Arsenic (As) in soils used to grow wheat. Lead (Pb) and Arsenic (As) were the two heavy metals that grew in concentration the greatest. There are several more possible sources of soil pollution besides excessive fertilizer usage and chemical pesticides/herbicides [29].

Pb and As were the two heavy metals that grew in concentration the greatest. There are several more possible sources of soil pollution besides excessive fertilizer usage and chemical pesticides/herbicides. For instance, substantial heavy metal pollution results through the direct release of industrial and other urban pollutants into water bodies or from their dispersion across the land. Various quantities of heavy metals are present in a significant percentage of municipal garbage. Additionally, sewage sludge has a high concentration of harmful metals.

2.1.1 Chemistry of Soils and the Hazards of Heavy Metals

Since heavy metals were a component of the underlying material from which the soil originated, they are naturally present in soil. Soil contains a wide variety of heavy metals, including phosphates, oxides, sulfides, and sulfates. Plants take these heavy metals as a normal part of their physiology, and they make their ascending the food chain. Once they enter the food chain, they seriously endanger human health. Upon entering the acidic environment of the human digestive tract, these metals undergo oxidation and interact with biological molecules. Free radicals are produced when biomolecules oxidize, resulting in oxidative stress and subsequent cellular damage [14].

2.1.2 Effects of Heavy Metal Exposure to Soil Environment

Human health is directly impacted by toxic metal contamination in soil and water sources. There are functional functions for some heavy metals in human physiology, including Cu, Fe, Zn, Ni, Mg, Mn, Mo, Cr, and Se, even in minute amounts. Higher concentrations of these metals may be harmful to one's health. As was previously said, a variety of human and natural activities send enormous amounts of heavy metals into the air, water, and soil, where they join the ecosystem and become a part of the environment. Human exposure to heavy metals is rising because various heavy metal sources have ingrained themselves into our daily lives. For instance, hazardous forms of arsenic, such as arsenite and arsenate, which can contaminate soil and ground water, are included in agricultural goods including herbicides, insecticides, and fungicides. Preservatives and coatings for wood may contain considerable amounts of arsenic [1].

Another highly common element in ecology is lead. The main ways to get exposed to lead are through the burning of oil, drinking water, cigarettes, and industrial wastes (particularly from the manufacturing of iron and steel), which includes car exhaust. Lead (Pb) that is discharged into the atmosphere eventually finds its way into soil and water bodies. As a result, drinking water and eating plant items high in Pb can expose humans to Pb. Due to its presence in gaseous forms, air can potentially expose humans to mercury. Dental amalgams and thermometers both contain large amounts of mercury. It also finds extensive use in the construction, metalworking, and electrical equipment sectors. As a result, meals and drinking water may include the metal form and the gaseous emission is readily inhaled. Mn, which is incorporated as MMT into fuel (methylcyclopentadienyl manganese tricarbonyl), is another metal that may be swallowed in significant quantities. Massive amounts of Mn are released in gasoline fumes [2].

Alkaline batteries, paints, pigments, and polymers all contain Cd in significant quantities. These industries make it simple for Cd to penetrate the soil and water. Humans can be directly exposed to cadmium (Cd) through the consumption of food that has been cultivated in soil containing this metal. Additionally, the paper, textile, stainless steel, and wood preservative industries use significant amounts of chromium (Cr), contributing to environmental Cr pollution, especially in its hexavalent form. Steel manufacturing processes also generate nickel (Ni). Additional environmental sources of nickel include nickel presenting, smoking cigarettes, making batteries, and producing wire and other electrical components [12].

2.2 *Process of Heavy Metal Contamination in the River Ecosystem*

The cause of the heavy metals in the organisms and water are manmade and natural processes. They are dispersing and depositing in water, and both the environment and water are degrading, affecting aquatic life and causing sedimentation and depositing in bodies of water. If even tiny amounts of heavy metals like arsenic, mercury, lead, chromium, or cadmium are present, they are extremely dangerous. The movement of heavy metals takes place along rivers. Heavy metals are now one of the main pollutants in river ecology. The nature of the substrate sediment, the metals are absorbed by water, suspended particles, and water chemistry behaves in natural waters. Because of the complexation processes, sorption, dissolution, and precipitation throughout their speciation, heavy metals go through several alternations [25].

As a result, heavy metals are employed as sensitive markers to monitor the water's alternation. The association of heavy metals with geochemical phases in sediments has an impact on the overall behavior of heavy metals in water environments. The results of geochemical distribution have also been utilized to help forecast probable pollutant mobility and bioavailability. Heavy metal contamination has become a big issue because of bioaccumulation, abundant sources, nonbiodegradability, and severe toxicity. Heavy metals in the aquatic environment can be found in both natural and manmade sources. Natural causes include forest fires, volcano eruptions, and other natural phenomena, as well as anthropogenic activity like power plants, industrial effluents, mineral extraction, and e-waste. Many heavy metals have been deposited within the sediments by coprecipitation., adsorption and hydrolysis. Heavy metal residues in sediments may alter water quality, accumulating in aquatic animals, and have negative consequences on the environment and public health.

2.2.1 Heavy Metals in Aquatic Waterbody

Different trace elements are essential to life processes, and plants and animals use various techniques to extract the right quantity traces of constituents from their surroundings. Similar systems are also in charge of absorbing non-essential metals. The definition of bioaccumulation is the metal absorption and subsequent retention by an organism.

Metal availability in the environment, as well as an organism's capacity for absorption and subsequent release, influence both essential and non-essential metal bioaccumulation. The surface of microbes tends to bond with heavy metals, allowing them to enter the cell. As part of the chemical processes used by microorganisms to digest food, heavy metals are chemically transformed into microbes. For instance, certain bacteria will convert mercury into methylmercury. Compared to mercury, methylmercury is more easily absorbed by microorganism bodies [27]. Heavy metals enter the fish if such microbes are consumed by a large creature like a fish. Instead of disappearing more quickly, heavy metals linger in fish for extended periods of time. The microbe that ate fish caused an increase in the levels of heavy metals. Additionally, when larger species consume smaller ones, a food chain is created and the quantity of heavy metals in living things rises. Bioaccumulation is the term for the elevated number of heavy metals present throughout many living things [23].

Biomagnification is the state in which an organism's concentration of a contaminant surpasses that of the contamination in its food and is a major route for the chemical to enter the body. Apart from this, biomagnifications are defined as the rise in the pollutant level in species collected from the same food web that are in an increasing tropical state. In addition, the process in which the degree of contamination is higher than the source is known as biomagnification. There are two main types of biomagnifying substances: synthetic lipophilic organic substances, which are referred to as persistent organic pollutants; these are substances that do not easily break down and metabolize because they have not been exposed previously and do not have specific detoxification and discharging mechanisms. There are two main types of biomagnifying substances: first, there are synthetic lipophilic organic substances, which are not easily broken down and metabolized because they have not been exposed before and therefore do not have specific detoxification and discharging mechanisms. These substances are referred to as persistent organic pollutants (POPs). The second kind of biomagnifying compounds are metals, which do not break down. Since metals are a naturally occurring component of the environment, organisms, especially those that exist in greater concentrations have developed ways to absorb and release metals. The primary problems arise when organisms are exposed to higher-than-average levels of metals, since they may not be able to detoxify or release the metals rapidly enough to prevent injury. Certain metals that are transported in natural forms, like POPs, such as methylmercury, organo-selenium, and organotin, may bioaccumulate rapidly [21].

In addition to being persistent, heavy metals may be fatal in minute amounts and induce long-term oxidative damage in both plants and animals. Thus, from the perspective of ecotoxicology, heavy metals are more significant. They are not biodegradable, which makes them persistent in the ecosystem. The balance of aquatic climates can be upset by heavy metal pollution in rivers, leading to a fast decline in the number of different aquatic species. Heavy metals are often released into river ecosystems as particulates, which eventually settle into the sediments. Surface sediments are the main source of heavy metals and other toxic substances. The aquatic organism and other aquatic macrophytes may absorb them. Huge concentrations of trace metals seep into aquatic ecosystems, eventually settling in the lowest sediments and severely affecting microorganisms. The diatoms have changed due to an increase in contamination by heavy metals in the ecosystems of rivers. The current study is focused on the pollution in fish modifying as a metabolic processes regulate heavy metal dispersion, danger, and accumulation [22]. Heavy metals initially created a special fish enzyme that has the ability to interfere with metabolism and subsequently manifest as cellular toxic effects, necrosis, and tissue death. Heavy metals initially created a unique enzyme in the fish, which can disrupt the metabolism and subsequently manifest as cellular toxicity, necrosis, and tissue death. Contaminants generated permanent alterations in biota through histological and histopathological changes in organs and tissues. Toxicity in tissue has been detected using histological approaches. Heavy metal buildup in fish is caused by a variety of ways. Heavy metal physiology and biochemistry, as well as geochemistry, are responsible for fluctuations in metal concentrations seen in distinct populations in aquatic environments. Several bioaccumulation investigations have indicated heavy metal transport into gill cells and interior epithelium. Biomagnification is the state in which a living creature has a contaminant level above the contamination level in its diet and a major channel for a chemical to enter the organism from food. Apart from these biological amplifications, which are defined as the rise in pollutant level in species tested from the same food web in increasing tropics. In relation to this, biomagnification is described as a method whereby the contamination level exceeds the source. There are two main categories of biomagnifying substances: first, synthetic lipophilic organic substances, also known as persistent organic pollutants, which do not easily disintegration and metabolize due to insufficient prior exposure and lack of specific detoxification and discharge mechanisms, and second, metals that are not deteriorated. Given that metals are a fundamental element of the environment, living things have developed ways to acquire and discharge them, especially those that naturally exist in environments with greater metal concentrations [26].

2.2.2 Effects of Heavy Metals Exposure to Aquatic Life

The primary problems arise if organisms are exposed to higher levels of metals than usual because they may not be able to detoxify or release them rapidly enough without damaging. Some metals, such POPs, organotin and methylmercury, which are transmitted in organic forms, may bioaccumulate fast. The balance of aquatic climates can be upset by heavy metal pollution in rivers, which causes a fast decline in the number of different aquatic animals. Heavy metals are often released as particulate matter into the river environment, where they eventually settle as silt. Surface sediments are a significant source of heavy metals and other toxicants. The aquatic organism and other aquatic macrophytes may take them up. The enormous volumes of trace metals that fall into aquatic environments eventually linger in the lowest sediments and have a deleterious impact on microorganisms [3]. The increased level of heavy metal contamination in the river environment is what has caused the change in diatoms. The focus of current study is on the transformation of pollutants in fish since the metabolic processes control the distribution, danger, and accumulation of heavy metals. Heavy metals initially produced a particular enzyme in fish, which can alter metabolism and subsequently become visible as cellular toxicity, necrosis, and eventually tissue death. Aquatic life is negatively impacted by heavy metals. Fish is necessary to maintain human health since it has high levels of omega-3 and proteins. However, when fish is consumed by humans, the dangerous heavy metals that the fish have ingested are transferred to the human body. Heavy metals accumulate in soft tissues and become highly toxic when they are not metabolized by the body [33]. When people are exposed to heavy metals in industrial, agricultural, and pharmaceutical production, they can enter their bodies through water, air, skin absorption, and food. Adults typically receive medications through industrial contact. The typical route of delivery for heavy metals in children is ingestion. Children exposed to polluted soil and eating objects like paint chips or dirt have greater hazardous levels because of general hand-to-mouth activity. After being ingested, the heavy metal is subsequently carried throughout the body to several pivot points, where it may be directed: (1) storage centers, (2) action centers, (3) excretion centers, (4) action sites, where a poisonous metal interacts with an endogen macromolecule or a specific cellular structure, potentially producing harmful consequences on the organism as a whole (5), these are the metabolism centers, where the metal is metabolized by detoxified enzymes.

2.3 Genotoxicity Due to Heavy Metals Exposure

DNA is a prominent target of environmental toxicity in living organisms among the numerous molecular elements of the cell.

Loss of DNA integrity can result in mutations and other irreversible consequences. Heavy metal exposure in earthworms can result in DNA damage and changes, disrupting normal cellular processes and, as a result, affecting cell viability. Heavy metals have the ability to covalently attach to DNA and create "DNA adducts," lower the methylation level of certain bases, or break DNA strands. In their investigation of metal-induced mutagenicity, Hirano and Tamae saw the synthesis of 8-oxoguanine (8-Oxo-Gua), a critical precursor to oxidative DNA damage that results in GC-to-TA point mutations, which in turn trigger carcinogenesis. It has been noted that heavy metals such as As and Cd increase the buildup of 8-Oxo-Gua in earthworms.

3 Methodology

3.1 Analyzing and Monitoring the Toxicity of Heavy Metals in the Environment

Analyzing the amount of metal in the soil, estimating how much is taken by plants, and measuring the metal accumulation in the plant are the steps involved in monitoring metal toxicity. An estimate of the degree of soil pollution may be obtained by measuring the number of heavy metals in the soil and comparing it to a standard reference. Based on this evaluation, metal levels can be predicted to be hazardous, normal, or inadequate. The availability of these elements cannot be inferred from measurements of the content of heavy metals in soil. A plant sample's heavy metal concentration should be determined to track the bioavailability.

Atomic absorption spectrometry (AAS)., neutron activation analysis (NAA), X-ray fluorescence, inductively coupled plasma. optical emission spectrometry (ICPOES), atomic emission/fluorescence spectrometry, inductively coupled plasma mass spectrometry (ICP-MS), and anodic stripping voltammetry are just a few of the instruments available for determining the concentration of heavy metals [10]. AAS is the most used quantitative approach among these methods, measuring over 70 distinct components. This method's main benefit is the extremely small sample size needed. The utmost precision can also be achieved when measuring metals that are accessible in parts per million or even parts per billion levels. It measures the amount of radiation absorbed by free atoms. The sample splits into free atoms at high temperatures, and the amount of radiation absorbed by these atoms determines how much is absorbed. Higher concentrations of these heavy metals are accumulated by plants employed in phytoremediation. The computation of two crucial factors—the translocation factor (TF) and the bioconcentration factor (BCF)—determines their potential to hyperaccumulate these metals. The capacity of plants to accumulate metals from their substrate is known as the BCF, and it may be calculated using the formula below.

$$\textbf{Bioconcentration Factor}(\textbf{BF}) = \textbf{Heavy Metal}(\textbf{Plant Part})/\textbf{Metal}(\textbf{Substrate})$$

Any portion of the plant, such as the stem, leaf, root, etc., can have its BCF determined.

The phytoremediation potential of a plant is described by the TF. It conveys the extent of metal buildup inside plant components. The formula below may be used to compute this:

$$\textbf{Translocation Factor} = \textbf{Heavy Metal (Plant shoot)}/\textbf{Metal (Plant root)}$$

3.2 Utilizing Remote Sensing to Assess the Levels of Heavy Metals in Various Soil Types

Monitoring heavy metals in soil is crucial for safeguarding both the environment and public health. Hyperspectral soil Heavy Metal determination is a low-cost yet challenging approach because of the influence of soil properties and intricate terrain. Modern remote sensing techniques have made it possible to study the contamination of soil quickly and affordably by heavy metals. We create models that help map heavy metal concentrations across large areas using space-supported hyperspectral sensors. Pollutants are spread by floods onto agricultural and residential areas, where they remain for many years before being removed by soil erosion. Heavy Metals in contaminated soils have a negative impact on food items, the environment, and human health. In light of growing urbanization and a dearth of high-quality food items on the market, remote sensing plays a critical role in the detection and management of heavy metal-polluted soils [11, 15].

By assessing it rapidly, remote sensing helps to comprehend the risk of contamination by heavy metals. Conventional techniques for identifying heavy metals (HMs) involve soil collection, chemical digestion, and interpolation to create a map of the region's dangers. These approaches are typically costly and time-consuming [16]. In heavy metal soil pollution research, advanced hyperspectral remote sensing techniques are widely employed as they provide a quicker, more affordable, and more expansive map of contamination. For mapping the distribution of heavy metals in mining sites, hyperspectral imaging has been widely used. Hyperspectral imaging remote sensing techniques may be used to evaluate soil properties and heavy metals' content. Soil contamination may be mapped using continuous, information-rich spectra obtained from hyperspectral imaging. Spectrophotometry is less expensive than conventional techniques [18].

3.3 Utilizing Remote Sensing to Assess the Levels of Heavy Metals in Sediments and Waterbodies

The rapid processes of industrialization, urbanization, and militarism are to blame for the poisonous heavy metal (HM) poisoning of soil. Serious health issues, including abnormalities of the neurological and digestive systems, arise when humans are exposed to high quantities of toxic heavy metals (HMs), particularly chromium (Cr), nickel (Ni), arsenic (As), zinc (Zn), lead (Pb), and copper (Cu) in urban and industrial regions [6]. Anemia, acne, arthritis, allergies, anorexia, sciatica, schizophrenia, chronic bronchitis, asthma, and issues with the kidneys and liver can all occur in those exposed to heavy metal pollution. In recent years, there has been a lot of interest in the mapping of harmful metals using remote sensing techniques that use satellite or aerial views. Finding the origins of harmful metals in soil is facilitated by remote sensing. When present in low to moderate quantities, toxic heavy metals

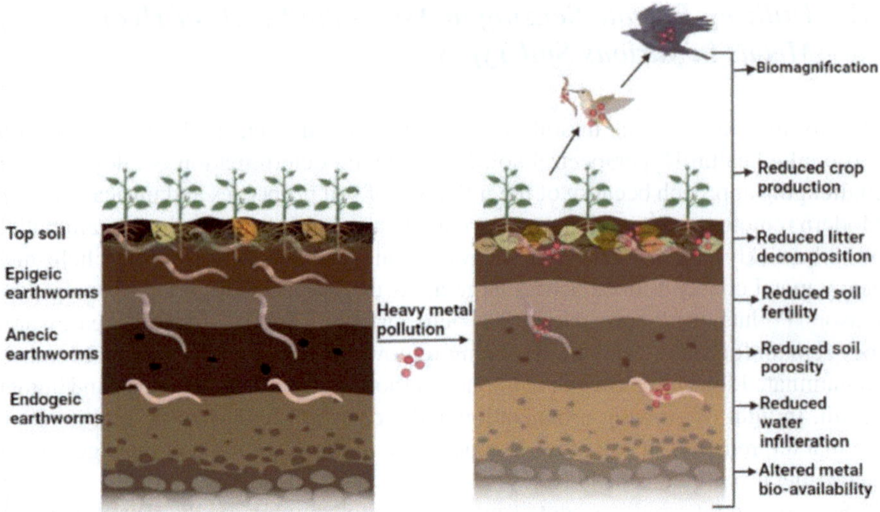

Fig. 3 Toxic impacts of heavy metals pollution on soil layers

(HMs) have a bland spectrum appearance and easily bond to minerals and organic materials. Since the chemical composition of minerals affects the locations of their spectral assignments, it is easy to forecast the presence of toxic metals in soil using the spectral information of such minerals [13]. Remote sensing spectra are useful for mapping the distribution of harmful heavy metals (HMs) in soil because they give information on organic matter, Fe-oxides, and clays in soils and plant cover. Kemper and Sommer used airborne hyperspectral data to map the remaining hazardous heavy metals (HMs) in Spanish soils following a quarrying disaster [28]. Toxic impacts of heavy metals pollution on soil layers is depicted in Fig. 3.

3.4 *Techniques for Remediating Soil to Remove Heavy Metals*

In agriculture, contaminated soil presents significant challenges. Many strategies have been used to mitigate the negative effects of heavy metal pollution. These remediation procedures can be carried out either straight on the original site or by transferring the contamination from the original site area and performing them somewhere else. The term "in-situ" refers to the former technique and "ex-situ" to the latter. While many ex-situ remediation techniques are highly effective and quick to clean up, in situ techniques are thought to be more environmentally and economically feasible. Historically, physical, chemical, and biological techniques are used in the heavy metal cleanup process [24].

3.4.1 Physical Remediation

Physical techniques, such as surface capping and thermal treatments, are typically used to halt contamination or repair soil degradation. The process of surface capping entails replacing polluted soil with uncontaminated soil. Either uncontaminated dirt is spread over the top layer of contaminated soil, or the two soils are mixed, to complete the process. While the second approach might just lower the soil's heavy metal level, the former could be helpful in preventing heavy metals from getting into the food chain. Pollution isolation and dispersion avoidance are other common practices. While this approach does not directly remove heavy metals from soil, it can lessen the degree of pollution in the soil or water. Thermal treatments are used to remove volatile elements like mercury. Different heat treatments (up to 600% °C) are applied to the subsurface soil to remove contaminants with high vapor pressure. Because thermal treatments can selectively repair contaminants and soil physio-chemical characteristics are claimed to be changed at such high temperatures, this approach is less desirable than other more appropriate methods [24].

3.4.2 Chemical Remediation

Chemical remediation procedures, in contrast to physical approaches, entail the extraction of toxins using chemical reagents. Numerous techniques, including vitrification, solidification, and/or soil flushing and washing, can be used to accomplish this. Applying water containing washing agents to the soil during the flushing and washing process helps make the pollutants more soluble. Following their removal from the soil, these pollutants build up in the groundwater. It becomes less dangerous if groundwater is removed and treated further. The main factors influencing the selection of washing chemicals are hydrogeologic characteristics including soil type and moisture content.

Numerous employees have suggested using a variety of cleaning chemicals, including water, chelating agents, saponin, and organic acids. Remedial methods such as vitrification and solidification are based on limiting the mobility of heavy metals in soil and lowering their bioavailability. Vitrification is the more sophisticated and favored procedure amongst the two. Simple solidification entails combining different materials.

Electro kinetics is another cutting-edge method for eliminating heavy metal contamination from soil. This technique uses low-intensity electric current flowing through electrodes buried in the ground. Charged pollutants are eliminated by a variety of processes, including electrolysis, electromigration, and electroosmosis. Both ex situ and in situ techniques can benefit from the use of this technique [30].

Bioremediation and phytoremediation are two biological techniques for heavy metal cleanup. These methods, which employ plants and microbes, are somewhat recent. There are different levels of innate ability in some microorganisms and plants to take up and retain these metals. For bioremediation, a variety of bacteria, fungi, and algae are used. Phytoremediation is the term for using plants for this kind of treatment [4].

There is a broad category of plants that could accumulate considerable levels of several heavy metals without suffering any negative consequences. Nevertheless, phytoremediation is ineffective in removing them from soil's deeper layers [8].

4 Conclusions and Recommendations

Heavy metals enter the food chain and are ingested by humans and other animals after polluting the atmosphere, hydrosphere, and agricultural areas. Fast industrialization and environmentally harmful farming methods are adding to the ecosystem's overall toxic burden. More research is required to fully comprehend (i) the processes governing the dispersion, bioaccumulation, and absorption of heavy metals; (ii) the variables influencing the metabolic pathways in earthworms exposed with metals; and (iii) the detoxifying responses of earthworms. Subsequent investigations must concentrate on identifying the molecular catalysts responsible for reduced fertility or unfavorable outcomes like genotoxicity and epigenetic modifications in subsequent generations. To effectively combat heavy metal pollution and shield soil organisms from the toxicity of heavy metals, innovative approaches and environmentally friendly solutions must be found. Finding a sustainable approach to maximize soil faunal diversity with biomass and their effects on soil quality is also a big problem for the future. Marine food contamination must be mapped rapidly and precisely to prevent metal pollutants from entering the food chain and to develop effective remediation procedures. For moderately polluted water, biological treatment can be an ecologically beneficial and cost-effective alternative. People are comparatively more exposed to these heavy metals in metropolitan settings. The quality of the air has deteriorated significantly due to industrial discharges and vehicle exhausts, particularly in metropolitan areas. People are compelled to breathe polluted air and unintentionally breathe in a variety of contaminants, some of which are extremely harmful and cancer-causing. Numerous corrective action procedures are being used. Phytoremediation is one type of bioremediation that seems to be more environmentally friendly. Many metals and plants have their absorption processes identified. The concept of creating genetically modified plants to increase the bioaccumulation of harmful metals is beginning to show results. Phytoremediation does, however, also have several shortcomings. For instance, compared to physical or chemical methods, it takes a long time to remove contaminants from sites, and its application is limited in that it requires background knowledge about the site and its characteristics as well as remediation objectives, efficiency, cost-effectiveness, public acceptability, and time. Research is being conducted to screen native plant species for the purpose of remediating soil contaminated with hazardous metals and to evaluate the availability of different phytoremediation parameters. Furthermore, some success has been made in the application of biotechnological methods in plants that have been modified for improved phytoremediation of heavy metals.

Genes involved in the absorption and transport of metals have been engineered to create many transgenic plants. To remove all contamination from the soil, a single remediation procedure would not be adequate. Toxic heavy metals and other dangerous pollutants can be effectively removed from soil by using a variety of remediation techniques.

References

1. Ahmed MK, Shaheen N, Islam MS, Habibullah-al-Mamun M, Islam S, Mohiduzzaman M, Bhattacharjee L (2015) Dietary intake of trace elements from highly consumed cultured fish (Labeo rohita, Pangasius pangasius and Oreochromis mossambicus) and human health risk implications in Bangladesh. Chemosphere 128:284–292
2. Ali H, Khan E, Ilahi I (2019) Environmental chemistry and ecotoxicology of hazardous heavy metals: environmental persistence, toxicity, and bioaccumulation. J Chem
3. Baki MA, Hossain MM, Akter J, Quraishi SB, Shojib MFH, Ullah AA, Khan MF (2018) Concentration of heavy metals in seafood (fishes, shrimp, lobster and crabs) and human health assessment in Saint Martin Island, Bangladesh. Ecotoxicol Environ Saf 159:153–163
4. Balamurugan D, Udayasooriyan C, Kamaladevi B (2014) Chromium (VI) reduction by Pseudomonas putida and Bacillus subtilis isolated from contaminated soils. Int J Environ Sci 5(3):522
5. Bhattacharya S (2018) Medicinal plants and natural products can play a significant role in mitigation of mercury toxicity. Interdiscip Toxicol 11(4):247–254
6. Chibuike GU, Obiora SC (2014) Heavy metal polluted soils: effect on plants and bioremediation methods. Appl Environ Soil Sci
7. Coetzee JJ, Bansal N, Chirwa EM (2020) Chromium in environment, its toxic effect from chromite-mining and ferrochrome industries, and its possible bioremediation. Exposure and Health 12:51–62
8. Dhankher OP, Pilon-Smits EA, Meagher RB, Doty S (2012) Biotechnological approaches for phytoremediation. In: Plant biotechnology and agriculture, Elsevier, pp 309–328
9. Engwa GA, Ferdinand PU, Nwalo FN, Unachukwu MN (2019) Mechanism and health effects of heavy metal toxicity in humans. Poisoning in the Modern World-New Tricks for an Old Dog 10:70–90
10. Fard RS, Matinfar HR (2016) Capability of vis-NIR spectroscopy and Landsat 8 spectral data to predict soil heavy metals in polluted agricultural land (Iran). J Geosci 9(20):745
11. Foulds SA, Brewer PA, Macklin MG, Haresign W, Betson RE, Rassner SME (2014) Flood-related contamination in catchments affected by historical metal mining: an unexpected and emerging hazard of climate change. Sci Total Environ 476:165–180
12. Godwill EA, Jane IC, Scholastica IU, Marcellus U, Eugene AL, Gloria OA (2015) Determination of some soft drink constituents and contamination by some heavy metals in Nigeria. Toxicol Rep 2:384–390
13. Grunwald S, Vasques GM, Rivero RG (2015) Fusion of soil and remote sensing data to model soil properties. Adv Agron 131:1–109
14. Guzzi G, Ronchi A, Pigatto P (2021) Toxic effects of mercury in humans and mammals. Chemosphere 263:127990
15. Henke JM, Petropoulos GP (2013) A GIS-based exploration of the relationships between human health, social deprivation and ecosystem services: the case of Wales, UK. Appl Geogr 45:77–88
16. Hiller E, Lachká L, Jurkovič Ľ, Ďurža O, Fajčíková K, Vozár J (2016) Occurrence and distribution of selected potentially toxic elements in soils of playing sites: a case study from Bratislava, the capital of Slovakia. Environ Earth Sci 75:1–13

17. Karbowska B (2016) Presence of thallium in the environment: sources of contaminations, distribution and monitoring methods. Environ Monit Assess 188:1–19
18. Karim Z, Qureshi BA, Mumtaz M, Qureshi S (2014) Heavy metal content in urban soils as an indicator of anthropogenic and natural influences on landscape of Karachi—a multivariate spatio-temporal analysis. Ecol Ind 42:20–31
19. Khan I, Ghani A, Rehman A, Awan S, Noreen A, Khalid I (2016) Comparative analysis of heavy metal profile of *Brassica campestris* (L.) and *Raphanus sativus* (L.) irrigated with municipal waste water of sargodha city. J Clin Toxicol 6(307):2161–0495.1000307
20. Kim J-J, Kim Y-S, Kumar V (2019) Heavy metal toxicity: an update of chelating therapeutic strategies. J Trace Elem Med Biol 54:226–231
21. Kumar R et al. (2022) Eisenia fetida as protein source for growth enhancement of Heteropneustes fossilis. Egypt J Aquat Biol Fisheries 26(2):577–588
22. Kumar R, Sharma P, Gupta R, Kumar S, Sharma MMM, Singh S, Pradhan G (2020) Earthworms for eco-friendly resource efficient agriculture. In: Resources use efficiency in agriculture, pp 47–84
23. Liu M, Xu Y, Nawab J, Rahman Z, Khan S, Idress M, Ali A, Ahmad R, Khan SA, Khan A (2020) Contamination features, geo-accumulation, enrichments and human health risks of toxic heavy metal (loids) from fish consumption collected along Swat river Pakistan. Environ Technol Innov 17:100554
24. Maftouh A, El Fatni O, El Hajjaji S, Jawish MW, Sillanpää M (2023) Comparative review of different adsorption techniques used in heavy metals removal in water. Biointerface Res Appl Chem 13:397
25. Mohiuddin K, Zakir H, Otomo K, Sharmin S, Shikazono N (2010) Geochemical distribution of trace metal pollutants in water and sediments of downstream of an urban river. Int J Environ Sci Technol 7:17–28
26. Moiseenko T, Gashkina N (2020) Distribution and bioaccumulation of heavy metals (Hg, Cd and Pb) in fish: influence of the aquatic environment and climate. Environ Res Lett 15(11):115013
27. Morkunas I, Woźniak A, Mai VC, Rucińska-Sobkowiak R, Jeandet P (2018) The role of heavy metals in plant response to biotic stress. Molecules 23(9):2320
28. Naderi A, Delavar MA, Kaboudin B, Askari MS (2017) Assessment of spatial distribution of soil heavy metals using ANN-GA, MSLR and satellite imagery. Environ Monit Assess 189:1–16
29. Palmer-Wilson K, Donald J, Robertson B, Lyseng B, Keller V, Fowler M, Wade C, Scholtysik S, Wild P, Rowe A (2019) Impact of land requirements on electricity system decarbonisation pathways. Energy Policy 129:193–205
30. Parmar S, Singh V (2015) Phytoremediation approaches for heavy metal pollution: a review. J Plant Sci Res 2(2):135
31. Roohani N, Hurrell R, Kelishadi R, Schulin R (2013) Zinc and its importance for human health: an integrative review. J Res Med Sci 18(2):144
32. Sun H, Brocato J, Costa M (2015) Oral chromium exposure and toxicity. Current Environ Health Reports 2:295–303
33. Vu CT, Lin C, Yeh G, Villanueva MC (2017) Bioaccumulation and potential sources of heavy metal contamination in fish species in Taiwan: assessment and possible human health implications. Environ Sci Pollut Res 24:19422–19434

Management of New Methods Used in the Reclamation of Soils Contaminated with Heavy Metals for Sustainable Agriculture: The Case of Biochar

Ahmet Çelik, Korkmaz Bellitürk, and Mehmet Fırat Baran

Abstract The rise in population and industrialization is increasing the strain on natural resources. This results in a greater need for industrial and agricultural production, which subsequently puts pressure on soil and water resources. The production of solid and liquid waste from industrial activity can contaminate soil and water sources. Such pollution is harmful to the ecosystem, including flora and fauna and human health. Heavy metal pollution poses a particularly severe environmental concern. With this study, how to manage new methods used in the reclamation of soils contaminated with heavy metals for sustainable agriculture and the importance of this process are revealed. Heavy metals, like copper, cadmium, nickel, lead, chromium, arsenic, and mercury, frequently contaminate natural assets. Heavy metal pollution can diminish crop yields and imperil the quality of soil and the environment. Biochar is soil amendments that offer practical solutions to the negative effects of heavy metals on soil, as well as to environmental health and fertility. Biochar has received significant attention in recent years due to its economical production cost and unique physicochemical properties. It is an inert substance generated from the pyrolysis of biomass, leaving behind a solid organic residue loaded with carbon. Biochar can be synthesised using various biomass sources, such as wood, crop remains, and animal waste. The typical method of obtaining biochar is pyrolysis, where organic materials undergo thermal decomposition in oxygen-depleted conditions. Organic materials are heated to temperatures exceeding 400 °C using electric heating or high-temperature media in an inert atmosphere. Notable benefits of biochar for soil and the environment include reducing heavy metal pollution, increasing soil fertility, soil quality and water retention, enhancing soil microbial activity, and decreasing

A. Çelik (✉)
Department of Soil Science and Plant Nutrition, Faculty of Agriculture, Adıyaman University, Adıyaman, Türkiye
e-mail: ahmetcelik@adiyaman.edu.tr

K. Bellitürk
Department of Soil Science and Plant Nutrition, Faculty of Agriculture, Tekirdağ Namık Kemal University, Tekirdag, Türkiye

M. F. Baran
Department of Biosystem Engineering, Faculty of Agriculture, Siirt University, Siirt, Türkiye

N. Kumar (ed.), *Heavy Metal Remediation*, Earth and Environmental Sciences Library,
https://doi.org/10.1007/978-3-031-53688-5_3

greenhouse gas emissions. Accordingly, biochar presents an extraordinary potential as a soil amendment for addressing a range of environmental issues. It is a feasible and cost-efficient technique for heightening the health and fertility of soil while concurrently restricting heavy metal contamination.

Keywords Heavy metal pollution · Biochar · Soil quality · Sustainable · Natural resources

1 Introduction

Water and soil are the two most vital elements for natural life. As the pressure on nature intensifies daily, the world population and industrialisation tend to rise in direct correlation. The demand for increased agricultural and industrial production persists with the expanding population. As industrial production rises, solid and liquid waste constantly jeopardises water and soil assets [1]. Since these contaminants seriously harm the environment, wildlife, humans and other components, it is a grave concern. According to [2] report, wastewater represents an increasing danger to soil quality, ecology and health. The adverse effects of hazardous waste on water and soil assets are reported to pose a risk to sustainable development, particularly in the agricultural sector and for food security. Heavy metal pollution (HM) is considered a critical global environmental problem [3, 4], with heavy metals (HMs) such as Copper (Cu), Cadmium (Cd), Nickel (Ni), Lead (Pb), Chromium (Cr), Arsenic (As), and Mercury (Hg) being the most common pollutants of natural assets. Heavy metal pollution (HM) is considered a critical global environmental problem [3, 4], with heavy metals (HMs) such as Copper (Cu), Cadmium (Cd), Nickel (Ni), Lead (Pb), Chromium (Cr), Arsenic (As), and Mercury (Hg) being the most common pollutants of natural assets. Alongside these, nutrient anions also contribute. Ammonium (NH_4^+), nitrate (NO_3^-), phosphate (PO_4^{3-}), and organic pollutants like antibiotics and dyes [5] are known to contaminate agricultural land. According to a study by Chinese researchers, heavy metals are estimated to pollute over 20 million hectares of agricultural land, which constitutes 20% of the country's total land mass [6].

Heavy metal (HM) toxicity is a significant abiotic stressor, leading to a substantial reduction in crop yields and posing a significant threat to both soil and environmental quality. As such, sustainable measures must be taken to reduce the adverse effects of HMs on soil, plants, people, and the environment. Biochar (BC) has been identified as a viable method of soil amendment, yielding beneficial outcomes in reducing the hazardous effects of HMs while simultaneously enhancing soil health, fertility and environmental quality. In recent years, the scientific community has increasingly focused on biochar due to its unique physicochemical properties and cost-effectiveness, facilitating versatile practical applications. One of the most promising areas of research in which biochar has shown significant results is the development of biotechnological strategies aimed at mitigating heavy metal pollution in the environment. The positive influence of biochar on the bacteria responsible for these activities has been investigated extensively, indicating its potential benefits [7].

Biochar application decreases the uptake and translocation of HMs to plant parts by forming complexes and precipitates [8, 9]. It also enhances soil pH, fertility, and cation exchange capability (CEC), and increases HMs adsorption, ultimately constraining their mobility and availability to plants. This study reveals how to manage new techniques used to recondition heavy metal-contaminated soil to promote sustainable agriculture. The significance of this process is highlighted.

2 What is Biochar?

Despite the relatively recent coinage of the term "biochar", the practice of using biochars has persisted throughout history in various forms. Its primary function has been for soil improvement. The use of biochars for soil improvement can be traced back at least 2000 years, as evidenced by historical records [10, 11]. In the Amazon Basin, extensive evidence suggests the usage of biochar in the fertile soils called Terra Preta and Terra Mulata. These soils were fabricated by prehistoric native societies in order to improve the productivity of soil in their immediate surroundings [11]. To date, the terra preta soils in the region remain highly fertile when compared to other soils in the area, including Oxisols and Ultisols. These soils often have up to four times more organic matter in the top 30 cm layer [12].

Biochar is an inert material that has undergone extensive scientific research over the past two decades. Its usefulness in promoting soil health and sustainable ecosystems has led to its prioritisation in research. While there are varying definitions, biochar can be defined as a solid organic substance created by the pyrolysis of biomass. Biochar is an organic substance derived through pyrolysis at a wide temperature range and partial or anaerobic conditions from various forms of organic biomass, including wood, crop residues, and animal manures [13, 14]. Biochar typically encompasses macronutrients, including hydrogen (H), oxygen (O), magnesium (Mg), nitrogen (N), phosphorus (P), potassium (K), and carbon (C), as its primary component. The technical term abbreviations will be clarified upon first use.

Pyrolysis is the preeminent thermochemical technique for manufacturing biochar. The following sections will provide a detailed explanation. This method can be employed on a small scale using a cooking burner or on a large scale with a pyrolysis system. Pyrolysis is a chemical process driven by thermal energy that converts biomass into biochar. This process is conducted in a low-oxygen environment at temperatures ranging from 350 to 700 °C [15]. Technical abbreviations will be explained upon first usage. The resulting carbon-rich solid material is commonly known as biochar or charcoal. Common academic sections will be included in the structure of the text and a balanced, objective, and value-neutral writing style will be maintained throughout. Moreover, the language used will have a formal register and avoid contractions, colloquial words, informal expressions, and unnecessary jargon, whilst ensuring grammatical correctness and logical flow of information in concise sentences. Additionally, pyrolysis also produces a liquid phase, partially consisting of condensed volatiles, referred to as bio-oil or tar. Pyrolysis processes fall into three categories depending on the conditions: slow pyrolysis (temperatures below 300 °C), moderate pyrolysis (temperatures ranging between 300 and 500 °C), and fast pyrolysis (high heat transfer, accelerated in a short time, temperatures above 500 °C) [16]. Technical term abbreviations are explained when first used. The language is clear, value-neutral, and avoids bias. The structure is logical and causal connections between statements are established. The text is free from grammatical errors and follows British spelling conventions. Primary decomposition and secondary fragmentation and the production of oxygen functional groups begin at around 400–500 °C during biomass pyrolysis, as previously suggested [17–19].

The presence of cellulose, hemicellulose and lignin in diverse raw materials can influence the yield as well as the elemental composition and ash content of biochar [20–22]. As the chemical and elemental characteristics of the feedstocks can vary. For instance [20], noted a greater ash content in straw biochar in comparison to other biochar, attributed to the high Si amount in straw. Biochar produced at the same pyrolysis temperature reveals a reduced yield with increased residence time.

To achieve maximum biochar yield and production, the chosen method must be suitable for the specific type of biomass. Optimum process conditions, including heating rate, temperature, and residence time, are also crucial. These conditions can significantly impact the physical and chemical properties of biochar during production [23].

Various factors, such as biomass type, age, components (including cellulose, semicellulose, lignin and extractives), plant species, parts of plants, and the processes involved in the uptake and transport/storage of certain compounds from soil-air to plant tissues, have a direct and indirect impact on biochar's characteristics [24, 25]. Meanwhile, aspects such as raw material, temperature, particle size and heating rate mainly influence the properties of biochar. These factors are more likely to influence the yield of biochar rather than its quality.

3 Preparation and Utilization of Biochar

Pyrolysis is a commonly employed method for generating biochar, entailing the thermal breakdown of organic substances devoid of oxygen. Generally, organic materials are heated to temperatures higher than 400 °C using electric heating or high temperature media in an inert atmosphere. There are numerous factors that impact the physicochemical characteristics of biochar, including feedstock, reaction temperature, heating rate, residence time, and ambient conditions. It is possible to pyrolyze any organic material to produce biochar [26, 27]. Owing to the extensive presence of biomass, abundant sources of waste are utilized as raw materials, including corn and wheat straw, rice husks, peanut and almond shells, tobacco stems and waste, sawdust from trees, cooking wastes, various animal manures, and sludge [22, 24, 28] (Fig. 1).

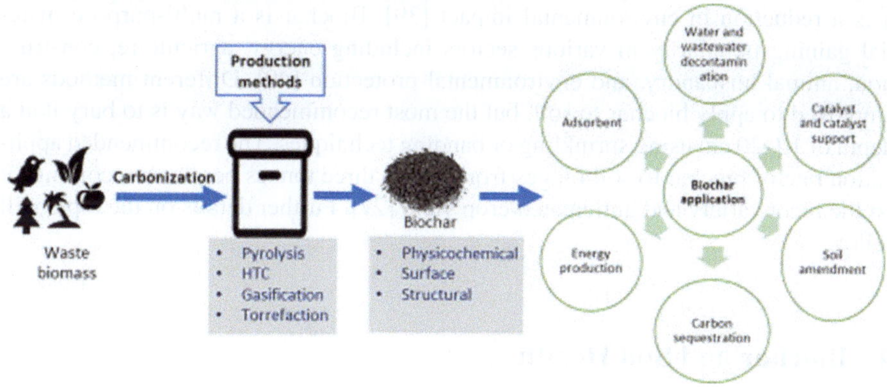

Fig. 1 Biochar production process [29]

The temperature range for producing biochar through high-temperature anoxia, hydrothermal synthesis, and flash carbonization are 400–900, 180–250, and 300–600 °C, correspondingly [22, 30]. Biochar production and the number of acidic functional groups (–COOH, –OH) dropped as the pyrolysis temperature increased, while the quantity of alkaline functional groups, ash content, and pH increased. Altogether, a rise in pyrolysis temperature caused biochar production and the count of acidic functional groups (–COOH, –OH) to decrease, alongside an increase in alkaline functional groups, ash content, and pH. Furthermore, it is worth mentioning that the effect of pyrolysis temperature on surface area and pore volume holds great importance [31].

The impact of pyrolysis temperature on surface area and pore volume is crucial. Additionally, the biochar yield declines with increased residence time at a constant pyrolysis temperature. Biochar is manufactured using pyrolysis [32], hydrothermal carbonisation (HTC), and microwave carbonisation techniques [33]. The preparation methods have an impact on the physical and chemical properties of biochar, such as its yield, ash content, specific surface area, pore structure, number and type of functional groups, and cation exchange capacity. HTC, compared to pyrolysis, does not require a drying step and offers higher biochar yields [34]. Microwave carbonisation boasts controllable process, rapid heating and energy efficiency advantages [35]. In addition to the aforementioned methods of biomass conversion, such as pyrolysis, hydrothermal carbonization, and microwave carbonization [36], there are additional methods such as flash carbonisation and roasting. Flash carbonisation involves the ignition of flash fire at high pressure (1–2 Mpa) on the bed filled with biomass to convert it into gas and solid phase products [30].

The utilization of agricultural waste to produce biochar has been suggested as a cost-efficient and eco-friendly approach for treatment and resource management [37, 38]. An additional benefit of reusing the remaining biomass instead of discarding it is a reduction in environmental impact [39]. Biochar is a multi-purpose material gaining popularity in various sectors including energy, agriculture, construction, animal husbandry, and environmental protection [40]. Different methods are employed to apply biochar to soil, but the most recommended way is to bury it at a depth of 10–20 cm using sprinkling or banding techniques. The recommended application rate of biochar to soil ranges from one to three tonnes per decare, contingent on the plant variety and anticipated crop yield [27]. Further details on the topic will follow.

4 Biochar and Soil Health

Soil quality plays a vital role in determining agricultural productivity and sustainable food production. While food security and nutrition are global concerns, they pose a particularly significant challenge in developing countries due to the population growth and depletion of natural resources such as soil, water, and other assets [41, 42].

As a results of intensive and unconscious practices on the land, the agricultural process, nutrient cycling, and nutrient release and uptake damage natural soil systems. This leads to a decrease in organic matter levels and soil fragility from excessive soil usage [10]. As a result of lowered plant nutrients available for growth and development, the plant's productivity and growth rate are reduced. Recent scientific research suggests that the use of biochar has resulted in advantageous effects on soil quality, carbon sequestration, and mitigation of greenhouse gas emissions [43]. Moreover, these studies have shown that the use of biochar leads to an increase in soil fertility, quality, and crop yield. Applying its properties to soil, organic waste benefits the soil pH, water retention capacity, microorganisms' activity, cation exchange capacity, and nutrient retention [10, 44]. Furthermore, it maintains the stability of heavy metals [43, 45] and reduces the adverse impacts of temperature, drought, and salinity stress on plants. On average, numerous scientific studies have shown that the use of bichar can lead to an increase in crop yields of approximately 20% [41] (Fig. 2).

One reason for the increase in crop yield when biochar is applied is the enhanced utilisation of nitrogen from fertilisers [46, 47]. This occurs due to a decrease in nitrogen loss from the soil through the increased cation exchange capacity (CEC) of the soil from biochar application [48]. Alternatively, the inhibitory effect of biochar on the conversion of nitrate with fertiliser is known to enhance nitrogen utilisation [10, 47]. Once again, the use of biochar facilitates biological nitrogen fixation in leguminous plants [49], while simultaneously contributing towards soil quality parameters through carbon sequestration.

Biochar has a strong capacity for adsorbing ammonia, phosphate, and potassium by means of its functional group. This feature has the ability to simultaneously reduce nitrogen emissions and leaching losses whilst promoting a habitat for diverse microorganisms. As a vital component in soil, biochar plays a significant role in fertiliser release, enhancing nutrient utilisation efficacy and minimising losses. Biochar is a highly regarded soil amendment due to its potential to sequester

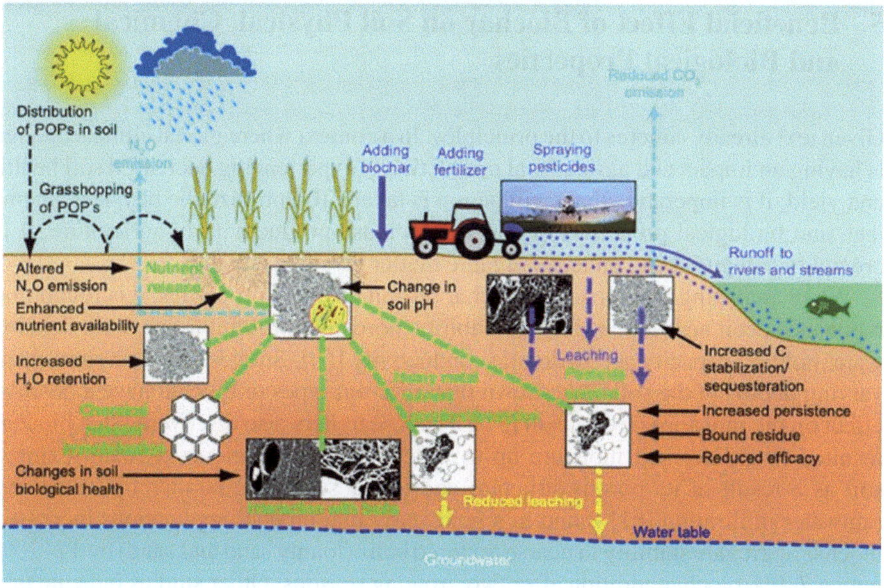

Fig. 2 Effect of biochar use in agriculture on soil health [50]

carbon, serve as a nutrient source, improve water retention, provide aeration, and enhance microbial activity. This, in turn, leads to an overall improvement in soil health. [51–53]. In comparison to using biochar alone or in combination with other substances, incorporating biochar with manure or organic fertiliser/compost has been demonstrated to enhance nutrient use efficiency, enrich soil properties, and ultimately improve crop yields by increasing nutrient availability to plants [52, 54–56].

Biochar applications enable greater nutrient retention, reducing the need for conventional fertilisers to achieve a desired crop yield [57] study into the potential of biochar to reduce greenhouse gas emissions suggests a potential reduction of 10–30% in nitrogen fertilizers. In relation to the use of nitrogen fertilisers, [58] estimate a CO_2 emission rate of approximately 13.5 tonnes per 1 tonne of nitrogen fertiliser produced and utilised. This demonstrates the significance of biochar for eco-friendly and farming purposes [10].

5 Beneficial Effect of Biochar on Soil Physical, Chemical and Biological Properties

Given text already adheres to the principles. In a context where global climate change is having an impact and agricultural production is experiencing decreased soil health and yield, it is imperative that swift action is taken. To optimize the physical, chemical, and biological properties of soil and increase productivity in agriculture, it is crucial to safeguard and enhance organic matter levels [31, 59].

When analysing the impact of biochar on soil physical properties, it is found that it enhances soil aeration, aggregate stability, porosity, infiltration rate, bulk density, water holding capacity and hydraulic conductivity [60]. Studies on water contents at varying depths in the soil have shown that different types of biochar have a positive effect on the moisture level of soil [31]. The researchers also discovered that biochar resulted in greater soil moisture up to a depth of 35 cm in sandy loam textured soil as a result of its porous structure and larger surface area, thereby enhancing rainwater infiltration [61] found in a field study that biochar applications increased biochar aggregate stability in coarse textured soils (loamy sand and sand) by 17–20% with a 7–9% biochar addition after two growing seasons. Other studies indicate that while total porosity and available water capacity increased, base saturation decreased. However, soil monotypic structure should not be disregarded in such research. Additionally, [62] found that applying biochar improves macroporosity, water availability for plants, relative field capacity, and dry bulk density of soil. As biochar has a lower volume weight compared to mineral soils, its use can increase soil volume per unit weight, leading to a reduction in soil volume weight. Furthermore, a reduction in the bulk density of soil treated with biochar may indicate improved soil structure, flocculation, and aeration [63].

Biochar has a notable impact on soil chemical characteristics, including pH, EC, lime, organic matter, and the absorption of different nutrients. Some variations, including the type and age of the biomass, its constituent components (cellulose, semi-cellulose, lignin, and extractives), the species and parts of plants, and the processes by which plants absorb particular compounds from soil and air and transport them to plant tissues [24], all impact the chemical structure and composition, surface area, and mineral structure of biochar due to its elemental properties [64].

The rise in soil pH following biochar application is largely attributed to the alkaline pH of biochar. In a comparable study, Çelik et al. [24] observed that almond biochar is suitable for application in alkaline soils with high pH, whilst tobacco biochar is effective in agricultural soils with low pH. When added to soil, alkaline biochars ($pH > 7$) are believed to elevate soil pH [65]. Moreover, the addition of certain biochar species can raise the electrical conductivity (EC) of low EC soils. Importantly, increasing EC in acidic, red soils does not harm crops according to some studies [31].

The nitrogen level in soils treated with biochar was discovered to be greater than that of untreated soils. This is largely attributed to the enhanced capacity of soil to retain moisture and promote root growth resulting in increased nitrogen fixation within the soil [66]. Besides releasing nitrogen from different biochars with significant nitrogen content, biochars can successfully adsorb ammonia (NH_3) [67]. Furthermore, they act as a binder for soil ammonia and have the potential to decrease the volatilisation of ammonia from the soil surface. Biochars can retain more nutrients, leading to a higher nitrate retention and superior nutrient supply to plants while also reducing nutrient loss through soil leaching [68]. The rise in available phosphorus via soil application of biochar is likely owing to the latter's high potential for adsorption and desorption. Similarly, the increase in potassium in ploughed or cultivated soils is explainable by the K-rich ash's presence in biochar. Furthermore, the boost in the cation exchange capacity of soil treated with biochar is primarily due to the presence of additional hydroxyl and carboxyl groups, as well as an increased surface area and variable charges within the biochar [69].

Organic carbon is a very important parameter of soil quality. Several studies show that biochar promotes carbon sequestration in the soil. This is attributed to the high carbon content and resilient structure of biochar, which significantly contributes to soil health. The addition of varying amounts of biochar to the soil was found to increase the OC content between 33 to 35% [70].

In addition to biochar carbon, microbial carbon, decomposition and leaching in the rhizosphere have contributed to the augment in soil carbon stock. The inclusion of biochar has diverse advantageous impacts on soil biological characteristics, including amplified soil microbial activity, improved soil nutrient uptake by plants [71], and augmented nutrient availability in the soil [72]. Additionally, the physical structure of biochar offers a secure habitat for microbial colonies [13, 73]. Moreover, enhanced soil moisture, improved aeration, and enhanced soil aggregation promote microbial proliferation in soil treated with biochar [27].

6 Beneficial Effect of Biochar on Heavy Metal Toxicity

Biochar (BC) is a vital soil amendment material that helps mitigate the adverse effects of heavy metals (HM) (Hg, Pb, Cd, As, Ti and Li) and reduces the pressure on soil–water resources. Heavy metals have long-lasting effects and cannot biodegrade in polluted soils. The remediation of heavy metals from contaminated soils is both time consuming and expensive. Biochar can stabilise heavy metals, enhance the quality characteristics of contaminated soil and significantly decrease the absorption of various metals in plants [74]. Its adsorption properties are beneficial in limiting the mobility and subsequent availability of heavy metals for plants. The high surface area and porous structure of biochar allow it to readily adsorb heavy metals on its surface. In this instance, biochar can decrease the bioavailability of heavy metals, thus inhibiting their absorption by plants. This is particularly effective in soils that are contaminated and acidic. The type of biochar and the presence of exchangeable cations, including Ca^{2+}, K^+, Mg^{2+}, and Na^+, can influence the cation exchange with heavy metals. Biochar can enhance both the sorption and stabilisation processes [75]. Under field conditions, research has demonstrated that the sorption process resulting from biochar in soils contaminated with heavy metals is primarily dependent on the soil type and the cations in both biochar and soil. Thus, the remediation of metals in contaminated soils may vary.

Biochar can facilitate the precipitation of heavy metals into insoluble forms, which could potentially enhance their immobilisation and minimise their availability in the soil. According to [76], the mineral components of biochar, including phosphates and carbonates, are essential in stabilising metals in the soil as salts are able to co-precipitate alongside metals, thereby reducing their bioavailability. Biochar application reduces the uptake and translocation of heavy metals to plant parts by forming complexes and precipitation [22].

Heavy metals become less toxic when biochar forms complexes with them. For instance, biochar complexation can make arsenic less bioavailable for uptake by plants. Ionic forms of metals that are in a dissolved state in the soil solution become available for uptake by plants. During the process, both acidic functional groups along with basic functional groups are crucial in binding HMs to the biochar surface and the internal pores.

The pH levels of biochars derived from different biomasses exhibit variation. The response of numerous biochars to heavy metals changes in accordance with alterations in soil reaction. Biochar has an alkaline nature and mitigates heavy metal mobilisation in polluted soils through mulching effects [77]. Nevertheless, the removal capacity of a given type of biochar differs for distinct categories of heavy metals. Low pH soil can heighten the mobility of certain heavy metals, particularly Cd and Pb [78]. Furthermore, biochar plays a role in the oxidation or reduction of heavy metals, causing a change in their form and associated toxicity. A notable example is that biochar is capable of reducing chromium (VI) to the less toxic chromium (III).

Biochar's impact on heavy metal toxicity is augmented by its ability to enhance soil quality by decreasing the uptake of HMs by plants. Biochar can increase soil pH, reducing the bioavailability of some heavy metals as described earlier. In certain cases, it may reduce bioavailability by up to 90%. Additionally, biochar can enhance soil fertility, boosting plant growth resilience against diseases and pests and reducing heavy metal intake per unit biomass [79]. Thus, biochar application presents a plausible remedy for heavy metal-contaminated soil remediation.

7 Examples of Biochar Application to the Sustainable Management of Heavy Metal-Contaminated Soil and Water

Biochar is a promising technology for sustainably managing soil and water that is affected by heavy metals. As mentioned earlier, biochar is an agricultural and environmentally friendly technique that can immobilise heavy metals in the soil, remediate polluted water, reduce absorption of heavy metals by plants and improve soil fertility. Heavy metal immobilisation in the soil happens when we introduce biochar to the soil by employing varied mechanisms such as precipitation, adsorption and complexation. Biochar has demonstrated effectiveness in immobilising heavy metals, including lead, cadmium, copper, zinc and arsenic. Furthermore, biochar has been shown to have the capacity to remove heavy metals from polluted water, owing to its sizable surface area and the existence of functional groups that can attach themselves to these metals. Several examples have demonstrated biochar's effectiveness in sustainably managing heavy metal-contaminated soil and water. Nevertheless, further extensive research is indispensable.

Biochar offers a variety of advantages both environmentally and agriculturally. These include sequestering carbon, rehabilitating degraded lands, reducing greenhouse gas emissions, adsorbing pollutants, and eradicating stream and groundwater pollution [80–82].

The subsequent sections consider the advantages of applying biochar for sustainable agricultural and environmental practices. Biochar has the potential to restore polluted soil and water in situ, meaning that it can be directly administered to the contaminated region. This environmentally friendly and cost-effective approach has been demonstrated to work at industrial sites, agricultural land and mine tailings. Biochar is believed to have the potential to mitigate global warming by inducing positive changes in soil and stabilising water pollution through the removal of organic pollutants [83, 84]. Consistent citation and footnote style have been maintained in accordance with formal writing conventions. In addition, research indicates that biochar has the ability to disinfect Escherichia coli, convert 95% of 2-chlorobiphenyl through advanced oxidation processes, and thus aid water purification efforts [84]. Technical abbreviations such as E. coli will be explained upon first use. The role of biochar in promoting sustainable water management is crucial for eliminating water pollution and supporting the agricultural water cycle. Biochar's high surface area allows it to effectively sorb heavy metals, making it a key feature in this regard. Various studies indicate that biochar can effectively remove contaminants such as lead, cadmium, copper and zinc from polluted water [85] achieved positive outcomes in their research on eliminating heavy metals from water-based solutions with magnetic biochar developed from marine macroalgae waste biomass. They suggested that this method could be viable for various applications.

The sequestration of carbon in agricultural land via biochar is a vital strategy to combat climate change as soil organic carbon (SOC) reserves in most agricultural areas have been depleted. Including biochar in land cultivation holds potential for carbon sequestration through the presence of decomposition-resistant carbon, ultimately reducing greenhouse gas emissions. This method has been researched extensively by [86, 87, 35]. Biochar is a stable substance that has the potential to store soil carbon for a few centuries after being incorporated into the soil. Additionally, it offers other advantages such as improved water retention and nutrient availability [88, 89]. Sequestering carbon in the soil serves as a means of long-term storage for CO_2 or other carbon variants, thereby reducing or slowing down global warming. The principal outcome of this technique is evident through the augmented carbon stock within the soil via the employment of biochar [86, 90]. Lehmann's [91] study demonstrates that the application of biochar to soil improves its quality and increases its ability to retain 20% more CO_2 from the atmosphere. The longevity of biochar in the soil, estimated to exceed 1000 years, highlights its value in sustainable agricultural practices [92–94]

Incorporating biochar into agricultural soils presents several benefits, including a significant reduction in greenhouse gas (GHG) emissions, as well as the absorption of pollutants. Furthermore, it can enhance soil fertility and increase crop productivity [95–97]. The primary long-lived greenhouse gases that cause global warming are CO_2, CH_4, and N_2O present in the atmosphere. Thus, it is imperative to decrease the emissions of these gases in the environment immediately [90].

The properties of biochars discussed above significantly impact the soil ecosystem and human health indirectly. Furthermore, the incorporation of biochars enhances water retention in the soil, promoting a better water balance, thereby improving nutrient availability. Biochar with a high adsorbing capacity is capable of capturing toxic compounds such as pesticides and herbicides when introduced to soil, thereby preventing their entry into the food chain [98]. Additionally, planting biochar with an alkaline pH has been shown to enhance the phytoavailability of essential nutrients including sodium (Na), potassium (K), calcium (Ca), magnesium (Mg) and molybdenum (Mo) in plants [63]. In a previous study, [99] found that biochar can raise soil pH, thereby reducing aluminium (Al) toxicity and improving CEC in ultisols. As a result, P and base cations in soil can become more bioavailable.

Biochars made from diverse raw materials exhibit discrepancies in their ability to adsorb soil minerals and soil organic matter, greatly impacting the soil biota's accessible pore space [100, 101]. Biochars from woody feedstocks are typically coarse and highly recalcitrant, with C contents that can reach up to 80% [102]. Consequently, the mineralisation of C depends on the source of the biochar feedstock.

Biochar enhances crop yields not only under typical conditions but also during adverse conditions like drought and salinity [103]. The application of biochar at significant levels has the potential to alleviate the detrimental impact of salt stress on the growth of plants [104] found that the application of biochar led to an increase in crop productivity in salt-affected soils. It was observed that biochar addition reduced salt stress by adsorbing Na^+ and increasing K^+ content in the xylem, resulting in an increased tuber yield in potatoes. Biochar has the potential to reduce the negative impact on mineral uptake caused by salinity. Hence, it could be considered a new approach to alleviate salinisation effects on the soil polluted by agricultural activities [105, 106].

8 Beneficial Effect of Biochar on Economic and Environmental in Sustainable Agriculture

Biochar has the potential to provide multiple economic and environmental advantages to promote sustainable agriculture holistically. Negative consequences resulting from technological advancements in this field are now evident. These undesirable effects comprise climate change, soil and water pollution, waste production, depletion of fossil fuels, and food shortages, among others. Recently, sustainable development goals have been established to address diverse social, economic, and environmental challenges. Biochar deployment presents a highly promising approach for tackling multiple issues simultaneously. Furthermore, applying biochar is one of the sustainable development objectives. Sustainable Development Goals (SDGs) constitute a set of 17 goals established by the United Nations General Assembly in 2015 to address environmental, economic, and socio-political challenges. The goals aim to promote sustainable development across the globe [107]. As more research on biochar is conducted, it is poised to deliver greater benefits with superior outcomes.

In the face of rising environmental pressures in agriculture, the application of biochar in sustainable farming presents considerable environmental benefits. These vital ecosystem services guarantee a smooth and sustainable continuation of life. They comprise of reduced soil erosion, improved water quality, lowered greenhouse gas emissions, increased biodiversity, waste reduction, enhanced soil health, and climate change mitigation. However, the widespread implementation of biochar faces significant obstacles due to limitations in resource availability and technological development, leading to higher initial production costs [102]. The economic viability of any new innovation plays a crucial role in its eventual large-scale adoption. The integration of biochar into reformed agricultural practices through investments of varying intervals could facilitate improved welfare for producers, both in the short and long term. It is believed that employing biochar can offer numerous benefits to farmers, such as higher product yield and quality, lower fertiliser and pesticide expenses, enhanced soil fertility and reduced soil depletion, decreased greenhouse gas emissions, and heightened land value. Nevertheless, potential drawbacks associated with the use of biochar must also be considered. In developing countries, farmers who solely rely on agriculture for their livelihood and have limited education are unlikely to adopt eco-friendly technology without clear economic incentives due to input costs and preconceptions. The commercial application costs of using biochar in the field have not been definitively established. However, the cost of biochar is closely linked to the expenses of the raw materials, their collection and transportation, the processing technique employed, and the worth of the by-products, as shown in the calculations. The cost of producing biochar using agricultural and green waste, poultry litter and sewage sludge varies [10].

Thermochemical methods utilised for the production of biochar, particularly in rural areas, foster regional growth, generate revenue for farmers through small and medium-sized enterprises' creation of sufficient energy, and offer solutions for agricultural waste management [96].

9 Conclusion

Biochar is a potentially valuable material for managing soil and water that is contaminated with heavy metals. It has the ability to immobilise heavy metals in soil, improve plant growth in contaminated soils, decrease heavy metal uptake by plants, address heavy metal contaminated water, and remediate contaminated soil and water on site. Furthermore, biochar being a low-cost material can be produced from various biomass raw materials.

Biochar is a durable material that can persist in soil for centuries, making it a valuable resource for improving soil health, increasing yields, reducing greenhouse gas emissions and mitigating the effects of climate change.

Adopting biochar is a key strategy for advancing the circular economy. Biochar plays a significant role in the reclamation of troublesome water and soil resources, and contributes to increased yields and quality standards. Its capacity for carbon sequestration is an asset in the reduction of greenhouse gas emissions and supports environmentally and agriculturally-friendly processes. While there have been extensive studies on biochar, there are still aspects yet to be discovered.

For the sustainable life of all living things, we have no choice but to live nature-friendly and in harmony with nature.

References

1. Nkoh JN, Ajibade FO, Atakpa EO, Abdulaha-Al Baquy M, Mia S, Odii EC, Xu R (2022) Reduction of heavy metal uptake from polluted soils and associated health risks through biochar amendment: a critical synthesis. J Hazard Mater Adv 6:100086
2. UNEP (2023) Wastewater—Turning Problem to Solution. UNEP—UN Environment Programme [WWW Document]. Retrieved from https://www.unep.org/resources/report/wastewater-turning-problem-solution. Accessed on 06 Oct 2023
3. Saxena G, Purchase D, Mulla SI, Saratale GD, Bharagava RN (2020) Phytoremediation of heavy metal-contaminated sites: eco-environmental concerns, field studies, sustainability issues, and future prospects. Rev Environ Contam Toxicol 249:71–131
4. Xie Y, Fan J, Zhu W, Amombo E, Lou Y, Chen L, Fu J (2016) Effect of heavy metals pollution on soil microbial diversity and bermudagrass genetic variation. Front Plant Sci 7:755
5. Henao SG, Ghneim-Herrera T (2021) Heavy metals in soils and the remediation potential of bacteria associated with the plant microbiome. Front Environ Sci 9:1–17
6. Sun Y, Li H, Guo G, Semple KT, Jones KC (2019) Soil contamination in China: current priorities, defining background levels and standards for heavy metals. J Environ Manag 251:109512
7. Amabilis-Sosa LE, Valenzuela EI, Quezada-Renteria JA, Pat-Espadas AM (2022) Biochar-assisted bioengineered strategies for metal removal: mechanisms, key considerations, and perspectives for the treatment of solid and liquid matrixes. Sustainability 14(24):17049
8. Haiying T, Shubin W, Ying L, Hassan MU, Ying S, Huang G, Mostafa YS (2022) Biochar: a promising soil amendment to mitigate heavy metals toxicity in plants. Notulae Botanicae Horti Agrobotanici Cluj-Napoca 50(3):12778–12778
9. Khan Z, Fan X, Khan MN, Khan MA, Zhang K, Fu Y, Shen H (2022) The toxicity of heavy metals and plant signaling facilitated by biochar application: implications for stress mitigation and crop production. Chemosphere 308:136466

10. Filiberto DM, Gaunt JL (2013) Practicality of biochar additions to enhance soil and crop productivity. Agriculture 3(4):715–725
11. O'Neill B, Grossman J, Tsai MT, Gomes JE, Lehmann J, Peterson J, Neves E, Thies JE (2009) Bacterial community composition in Brazilian Anthrosols and adjacent soils characterized using culturing and molecular identification. Microb Ecol 58:23–35
12. Glaser B, Haumaier L, Guggenberger G, Zech W (2001) The "Terra Preta" phenomenon: a model for sustainable agriculture in the humid tropics. Naturwissenschaften 88:37–41
13. Alkharabsheh HM, Seleiman MF, Battaglia ML, Shami A, Jalal RS, Alhammad BA, Al-Saif AM (2021) Biochar and its broad impacts in soil quality and fertility, nutrient leaching and crop productivity: a review. Agronomy 11(5):993
14. Weber K, Quicker P (2018) Properties of biochar. Fuel 217:240–261
15. Mazarji M, Bayero MT, Minkina T, Sushkova S, Mandzhieva S, Tereshchenko A, Keswani C (2021) Realizing united nations sustainable development goals for greener remediation of heavy metals-contaminated soils by biochar: Emerging trends and future directions. Sustainability 13(24):13825
16. Liu M, Almatrafi E, Zhang Y, Xu P, Song B, Zhou C, Zeng G, Zhu Y (2022) A critical review of biochar-based materials for the remediation of heavy metal contaminated environment: applications and practical evaluations. Sci Total Environ 806:150531
17. Huang H, Reddy NG, Huang X, Chen P, Wang P, Zhang Y, Huang Y, Lin P, Garg A (2021) Effects of pyrolysis temperature, feedstock type and compaction on water retention of biochar amended soil. Sci Rep 11:7419
18. Khawkomol S, Neamchan R, Thongsamer T, Vinitnantharat S, Panpradit B, Sohsalam P, Werner D, Mrozik W (2021) Potential of biochar derived from agricultural residues for sustainable management. Sustainability 13:8147
19. Li S, Chan CY, Sharbatmaleki M, Trejo H, Delagah S (2020) Engineered biochar production and its potential benefits in a closed-loop water-reuse agriculture system. Water 12:2847
20. Enders A, Hanley K, Whitman T, Joseph S, Lehmann J (2012) Characterization of biochars to evaluate recalcitrance and agronomic performance. Bioresour Technol 114:644–653
21. Williams PT, Besler S (1993) The pyrolysis of rice husks in a thermogravimetric analyser and static batch reactor. Fuel 72:151–159
22. Yang X, Zhang S, Ju M, Liu L (2019) Preparation and modification of biochar materials and their application in soil remediation. Appl Sci 9(7):1365
23. Yaashikaa PR, Kumar PS, Varjani S, Saravanan A (2020) A critical review on the biochar production techniques, characterization, stability and applications for circular bioeconomy. Biotechnology Rep 28:e00570
24. Celik A, İnan M, Sakin E (2019) Elementary analyses and comparison of SEM characteristics of biochars obtained from tobacco and almond residues applied to soil. Harran J Agric Food Sci 23(4):500–510
25. Veiga T, Lima JT, Dessimoni A, Pego M, Soares JR, Trugilho PF (2017) Different plant biomass characterizations for biochar production. Cerne 23(4):529–536
26. Laird DA, Brown RC, Amonette JE, Lehmann J (2009) Review of the pyrolysis platform for coproducing bio-oil and biochar. Biofuels Bioprod Biorefining 3:547–562
27. Ortaş I (2022) Development of corn plant under long-term biochar and mycorrhiza applications and its effects on carbon emission to the atmosphere and carbon sequestration budget Ç.Ü. Scientific Research Projects (BAP) Coordination Unit, Adana
28. Tan Z, Lin CS, Ji X, Rainey TJ (2017) Returning biochar to fields: a review. Appl Soil Ecol 116:1–11
29. Amalina F, Abd Razak AS, Krishnan S, Zularisam AW, Nasrullah M (2022) A comprehensive assessment of the method for producing biochar, its characterization, stability, and potential applications in regenerative economic sustainability–a review. Clean Mater 3:100045
30. Cha JS, Park SH, Jung SC, Ryu C, Jeon J-K, Shin M-C, Park Y-K (2016) Production and utilization of biochar: a review. J Ind Eng Chem 40:1–15
31. Pandian K, Subramaniayan P, Gnasekaran P, Chitraputhirapillai S (2016) Effect of biochar amendment on soil physical, chemical and biological properties and groundnut yield in rainfed Alfisol of semi-arid tropics. Archiv Agron Soil Sci 62(9):1293–1310

32. Zwieten LV, Kimber S, Morris S, Chan KY, Downie A, Rust J, Joseph S, Cowie A (2010) Effects of biochar from slow pyrolysis of papermill waste on agronomic performance and soil fertility. Plant Soil 327:235–246
33. Liu Z, Quek A, Hoekman SK, Balasubramanian R (2013) Production of solid biochar fuel from waste biomass by hydrothermal carbonization. Fuel 103:943–949
34. Sabio E, Álvarez-Murillo A, Román S, Ledesma B (2016) Conversion of tomato-peel waste into solid fuel by hydrothermal carbonization: influence of the processing variables. Waste Manag 47:122–132
35. Liu J, Wu J, Li Y, Su Y, Ge T, Jones DL (2014) Effects of biochar amendment on the net greenhouse gas emission and greenhouse gas intensity in a Chinese double rice cropping system. Eur J Soil Biol 65:30–39
36. Chen D, Zheng Z, Fu K, Zeng Z, Wang J, Lu M (2015) Torrefaction of biomass stalk and its effect on the yield and quality of pyrolysis products. Fuel 159:27–32
37. Jatav HS, Rajput VD, Minkina T, Singh SK, Chejara S, Gorovtsov A, Kalinitchenko VP (2021) Sustainable approach and safe use of biochar and its possible consequences. Sustainability 13(18):10362
38. Varalta F, Sorvari J (2020) In Hettiarachchi H et al organic waste composting through nexus thinking: practices, policies, and trends, Springer International Publishing, pp 213–232
39. Pinotti L, Luciano A, Ottoboni M, Manoni M, Ferrari L, Marchis D, Tretola M (2021) Recycling food leftovers in feed as opportunity to increase the sustainability of livestock production. J Clean Prod 294:126290
40. Burachevskaya M, Minkina T, Bauer T, Lobzenko I, Fedorenko A, Mazarji M, Rajput VD (2023) Fabrication of biochar derived from different types of feedstocks as an efficient adsorbent for soil heavy metal removal. Sci Rep 13(1):2020
41. Agegnehu G, Srivastava AK, Bird MI (2017) The role of biochar and biochar-compost in improving soil quality and crop performance: a review. Appl Soil Ecol 119:156–170
42. Lal R (2015) Restoring soil quality to mitigate soil degradation. Sustainability 7(5):5875–5895
43. Murtaza G, Ahmed Z, Usman M, Tariq W, Ullah Z, Shareef M et al (2021) Biochar induced modifications in soil properties and its impacts on crop growth and production. J Plant Nutr 44:1–15. https://doi.org/10.1080/01904167.2021.1871746
44. Zheng J, Han J, Liu Z, Xia W, Zhang X, Li L et al (2017) Biochar compound fertilizer increases nitrogen productivity and economic benefits but decreases carbon emission of maize production. Agric Ecosyst Environ 241:70–78. https://doi.org/10.1016/j.agee.2017.02.034
45. Gasco G, Cely P, Paz-Ferreiro J, Plaza C, Mendez A (2016) Relation between biochar properties and effects on seed germination and plant development. Biol Agric Hortic 32:237–247. https://doi.org/10.1080/01448765.2016.1166348
46. Steiner C, Teixeira WG, Lehmann J, Nehls T, Vasconcelos de Macedo JL, Blum WEH, Zech W (2007) Long term effects of manure, charcoal and mineral fertilization on crop production and fertility on a highly weathered Central Amazonian upland soil. Plant Soil 291:275–290
47. Widowati UW, Soehono LA, Shi DZ, Guritno B (2011) Effect of biochar on the release and loss of nitrogen from urea fertilization. J Agric Food Tech 1:127–132
48. Masulili A, Utomo WH, Syechfani MS (2010) Rice husk biochar for rice based cropping system in acid soil. J Agric Sci 3:25–33
49. Osman AI, Fawzy S, Farghali M, El-Azazy M, Elgarahy AM, Fahim RA et al (2022) Biochar for agronomy, animal farming, anaerobic digestion, composting, water treatment, soil remediation, construction, energy storage, and carbon sequestration: a review. Environ Chem Lett 20:2385–2485. https://doi.org/10.1007/s10311-022-01424-x
50. Allohverdi T, Mohanty AK, Roy P, Misra M (2021) A review on current status of biochar uses in agriculture. Molecules 26(18):5584
51. Purakayastha TJ, Bera T, Bhaduri D, Sarkar B, Mandal S, Wade P et al (2019) A review on biochar modulated soil condition improvements and nutrient dynamics concerning crop yields: pathways to climate change mitigation and global food security. Chemosphere 227:345–365
52. Rashmi I, Kumawat A, Munawery A, Karthika KS, Sharma GK, Kala S, Pal R (2022) Soil amendments: an ecofriendly approach for soil health improvement and sustainable oilseed production. In: Oilseed crops-uses, biology and production. IntechOpen

53. Sharma GK, Jena RK, Hota S, Kumar A, Ray P, Fagodiya RK et al (2020) Recent development in bioremediation of soil pollutants through biochar for environmental sustainability. In: Singh J, Singh C (eds) Biochar applications in agriculture and environment management. Springer, Cham, pp 122–140
54. Graber ER, Meller Harel Y, Kolton M, Cytryn E, Silber A, Rav David D, Elad Y (2010) Biochar impact on development and productivity of pepper and tomato grown in fertigated soilless media. Plant Soil 337:481–496
55. Guo XX, Liu HT, Zhang J (2020) The role of biochar in organic waste composting and soil improvement: a review. Waste Manage 102:884–899
56. Silber A, Levkovich I, Graber ER (2010) PH-dependent mineral release and surface properties of corn straw biochar: agronomic implications. Environ Sci Technol 44:9318–9323
57. Gaunt JL, Cowie A (2009) Biochar, greenhouse gas accounting and emissions trading. In: Lehmann J, Joseph S (eds) Biochar for environmental management: science and technology, Earthscan, London, UK, pp 317–340
58. Zhang WF, Dou ZX, He P, Ju XT, Powlson D, Chadwick D, Norse D, Lu YL, Zhang Y, Wu L et al (2013) New technologies reduce greenhouse gas emissions from nitrogenous fertilizer in China. Proc Natl Acad Sci USA 110:8375–8380
59. Srinivasarao C, Venkateswarlu B, Lal R, Singh AK, Kundu S (2013) Sustainable management of soils of dryland ecosystems for enhancing agronomic productivity and sequestering carbon. Adv Agron 121:254–329
60. Foster EJ, Hansen N, Wallenstein M, Cotrufo MF (2016) Biochar and manure amendments impact soil nutrients and microbial enzymatic activities in a semi-arid irrigated maize cropping system. Agric Ecosyst Environ 233:404–414. https://doi.org/10.1016/j.agee.2016.09.029
61. Obia A, Mulder J, Martinsen V, Cornelissen G, Børresen T (2016) In situ effects of biochar on aggregation, water retention and porosity in light-textured tropical soils. Soil Tillage Res 155:35–44
62. Castellini M, Giglio L, Niedda M, Palumbo AD, Ventrella D (2015) Impact of biochar addition on the physical and hydraulic properties of a clay soil. Soil Tillage Res 154:1–13
63. Atkinson CJ, Fitzgerald JD, Hipps NA (2010) Potential mechanisms for achieving agricultural benefits from biochar application to temperate soils: a review. Plant Soil 337:1–18
64. Lehmann J, Rondon M (2006) Bio-char soil management on highly weathered soils in the humid tropics. In: Uphoff N, Ball AS, Herren H, Husson O, Laing M, Palm C, Pretty J, Sanchez P, Sanginga N, Thies J (eds) Biological approaches to sustainable soil systems. Taylor & Francis, Boca Raton, pp 517–530
65. Roberts DA, Paul NA, Dworjanyn SA, Bird MI, de Nys R (2015) Biochar from commercially cultivated seaweed for soil amelioration. Sci Rep 5:9665–9666. https://doi.org/10.1038/srep09665
66. Egamberdieva D, Li L, Ma H, Wirth S, Bellingrath-Kimura SD (2019) Soil amendment with different maize biochars improves chickpea growth under different moisture levels by improving symbiotic performance with Mesorhizobium ciceri and soil biochemical properties to varying degrees. Front Microbiol 10:2423
67. Iyobe T, Asada T, Kawata K, Oikawa K (2004) Comparison of removal efficiencies for ammonia and amine gases between woody charcoal and activated carbon. J Health Sci 50:148–153
68. Chan KY, Van Zwieten L, Meszaros I, Downie A, Joseph S (2007) Agronomic values of green waste biochar as a soil amendment. Aust J Soil Res 45:629–634
69. Van Zwieten L, Kimber S, Morris S, Ky C, Downie A, Rust J, Joseph S, Cowie A (2010) Effects of biochar from slow pyrolysis of paper mill waste on agronomic performance and soil fertility. Plant Soil 327:235–246
70. Shenbagavalli S, Mahimairaja S (2012) Characterization and effect of biochar on nitrogen and carbon dynamics in soil. Int J Adv Bio Res 2:249–255
71. Zornoza R, Moreno-Barriga F, Acosta JA, Munoz MA, Faz A (2016) Stability, nutrient availability and hydrophobicity of biochars derived from manure, crop residues, and municipal solid waste for their use as soil amendments. Chemosphere 144:122–130. https://doi.org/10.1016/j.chemosphere.2015.08.046

72. Yin X, Penuelas J, Sardans J, Xu X, Chen Y, Fang Y et al (2021) Effects of nitrogen-enriched biochar on rice growth and yield, iron dynamics, and soil carbon storage and emissions: A tool to improve sustainable rice cultivation. Environ Pollut 287:117565. https://doi.org/10.1016/j.envpol.2021.117565
73. Guo M (2016) Application of biochar for soil physical improvement. Agricult Environ Appl Biochar: Adv Barr 63:101–122
74. Krzyszczak A, Dybowski MP, Czech B (2021) Formation of polycyclic aromatic hydrocarbons and their derivatives in biochars: the effect of feedstock and pyrolysis conditions. J Anal Appl Pyrol 160:105339
75. Chang YC, Xiao XF, Huang HJ, Xiao YD, Fang HS, He JB, Zhou CH (2021) Transformation characteristics of polycyclic aromatic hydrocarbons during hydrothermal liquefaction of sewage sludge. J Supercrit Fluids 170:105158
76. Mahmud U, Salam MB, Khan AS (2021) Ecological risk of heavy metal in agricultural soil and transfer to rice grains. Discov Mater 1:10
77. Eckert E, Kovalevska O (2021) Sustainability in the European Union: analyzing the discourse of the European green deal. J Risk Financ Manag 14:80
78. Awad M, El-Sayed MM, Li X, Liu Z, Mustafa SK, Ditta A, Hessini K (2021) Diminishing heavy metal hazards of contaminated soil via biochar supplementation. Sustainability 13(22):12742
79. Murtaza G, Ahmed Z, Eldin SM, Ali I, Usman M, Iqbal R, Tariq A (2023) Biochar as a green sorbent for remediation of polluted soils and associated toxicity risks: a critical review. Separations 10(3):197
80. Lehmann J, Gaunt J, Rondon M (2006) Bio-char sequestration in terrestrial ecosystems—a review. Mitig Adapt Strateg Glob Change 11:403–427
81. Lehmann J, Joseph S (2009) Biochar for environmental management: an introduction. In: Lehmann J, Joseph S (eds) Biochar for environmental management: science and technology, 1st edn. Earthscan, London, pp 1–12
82. Mohan D, Sarswat A, Ok YS, Pittman CU (2014) Organic and inorganic contaminants removal from water with biochar, a renewable, low cost and sustainable adsorbent—a critical review. Bioresour Tech 160:191–202
83. Barrow CJ (2012) Biochar: potential for countering land degradation and for improving agriculture. Appl Geogr 34:21–28
84. Inyang M, Dickenson E (2015) The potential role of biochar in the removal of organic and microbial contaminants from potable and reuse water: a review. Chemosphere 134:232–240
85. Son EB, Poo KM, Chang JS, Chae KJ (2018) Heavy metal removal from aqueous solutions using engineered magnetic biochars derived from waste marine macro-algal biomass. Sci Total Environ 615:161–168
86. McHenry MP (2010) Carbon-based stock feed additives: a research methodology that explores ecologically delivered C biosequestration, alongside live weights, feed use efficiency, soil nutrient retention, and perennial fodder plantations. J Sci Food Agric 90:183–187
87. Spokas KA, Koskinen WC, Baker JM, Reicosky DC (2009) Impacts of woodchip biochar additions on greenhouse gas production and sorption/degradation of two herbicides in a Minnesota soil. Chemosphere 77:574–581
88. Lal R (2008) Black and buried carbons' impacts on soil quality and ecosystem services. Soil Tillage Res 99:1–3
89. Sohi SP, Krull E, Lopez-Capel E, Bol R (2010) A review of biochar and its use and function in soil. Adv Agron 105:47–82
90. Zhang M, Ok YS (2014) Biochar soil amendment for sustainable agriculture with carbon and contaminant sequestration. Carbon Manage 5:255–257
91. Lehmann J (2007) A handful of carbon. Nature 447:143–144
92. Lehmann J, Skjemstad J, Sohi S, Carter J, Barson M, Falloon P, Coleman K, Woodbury P, Krull E (2008) Australian climate-carbon cycle feedback reduced by soil black carbon. Nat Geosci 1:832–835

93. Liang B, Lehmann J, Solomon D, Sohi S, Thies JE, Skjemstad JO, Luizão FL, Engelhard MH, Neves EG, Wirick S (2008) Stability of biomass-derived black carbon in soils. Geochim Cosmochim Acta 72:6078–6096
94. Nguyen BT, Lehmann J (2009) Black carbon decomposition under varying water regimes. Org Geochem 40:846–853
95. Cayuela ML, Sánchez-Monedero MA, Roig A, Hanley K, Enders A, Lehmann J (2013) Biochar and denitrification in soils: when, how much and why does biochar reduce N_2O emissions? Sci Rep 3:17–32
96. Hussain M, Farooq M, Nawaz A, Al-Sadi AM, Solaiman ZM, Alghamdi SS, Siddique KH (2017) Biochar for crop production: potential benefits and risks. J Soils Sedimen 17:685–716
97. Stavi I, Lal R (2013) Agroforestry and biochar to offset climate change: a review. Agron Sustain Dev 33:81–96
98. Cao X, Harris W (2010) Properties of dairy-manure-derived biochar pertinent to its potential use in remediation. Bioresour Technol 101:5222–5228
99. Peng X, Ye L, Wang C, Zhou H, Sun B (2011) Temperature and duration-dependent rice straw-derived biochar: characteristics and its effects on soil properties of an Ultisol in Southern China. Soil Tillage Res 112:159–166
100. Kasozi GN, Zimmerman AR, Nkedi-Kizza P, Gao B (2010) Catechol and humic acid sorption onto a range of laboratory-produced black carbons (biochars). Environ Sci Technol 44:6189–6195
101. Lehmann J, Rillig MC, Thies J, Masiello CA, Hockaday WC, Crowley D (2011) Biochar effects on soil biota—a review. Soil Biol Biochem 43:1812–1836
102. Duku MH, Gu S, Hagan EB (2011) Biochar production potential in Ghana—a review. Renew Sust Energ Rev 15:3539–3551
103. Haider G, Koyro HW, Azam F, Steffens D, Müller C, Kammann C (2014) Biochar but not humic acid product amendment affected maize yields via improving plant-soil moisture relations. Plant Soil 395:141–157
104. Akhtar SS, Andersen MN, Liu F (2015) Biochar mitigates salinity stress in potato. J Agron Crop Sci. https://doi.org/10.1111/jac.12132
105. Kim HS, Kim KR, Yang JE, Ok YS, Owens G, Nehls T, Wessolek G, Kim KH (2016) Effect of biochar on reclaimed tidal land soil properties and maize (Zea mays L.) response. Chemosphere 142:153–159
106. Thomas SC, Frye S, Gale N, Garmon M, Launchbury R, Machado N, Melamed S, Murray J, Petroff A, Winsborough C (2013) Biochar mitigates negative effects of salt additions on two herbaceous plant species. J Environ Manage 129:62–68
107. Kumar A, Bhattacharya T (2021) Biochar: a sustainable solution. Environ Dev Sustain 23:6642–6680

Management of Soil and Water Contaminated with Heavy Metals with Sustainable Green Technologies: The Case of Vermicomposting

Ahmet Çelik, Korkmaz Bellitürk, and Mehmet Fırat Baran

Abstract Countries are searching for alternative energy sources because of the global impact of greenhouse gases and the possibility of an energy crisis. In this regard, innovative technologies like composting are becoming more and more popular around the globe as a means of getting rid of municipal solid waste in agricultural areas. Composting is a biological process that creates an earth-like structure called compost by means of microorganisms converting organic materials, such as animal excreta, sludge from biological treatment plants, leaves, paper, and food waste, among others. One benefit of composting is that it helps to keep the environment clean. Organic material is stabilized through composting. It is well known that earthworms improve the fertility of the soil and the quality of the produce that is grown there. By passing soil through their bodies, they are able to attain this increase. The worm's body goes through a number of processes with the soil that it sends into the digestive tract along with the nutrients it absorbs. The worms provide the soil with essential nutrients by releasing their excrement, which is a highly valuable fertilizer into the surrounding environment. Practically all of the minerals required for plant growth are present in this vermicompost. Vermicomposting and composting differ primarily in that vermicomposting does not undergo the thermophilic phase of composting. Additionally, as the worms' digestive tracts process the waste to be processed, the heavy metals in it are stabilized and separated from the fertilizer during the vermicomposting process. For all of these reasons, research on vermicompost (also known as vermicomposting) and its use in rural areas has expanded both domestically and internationally in recent years. The issue of declining organic matter in agricultural soils is addressed in developed nations through the use of

A. Çelik (✉)
Department of Soil Science and Plant Nutrition, Faculty of Agriculture, Adıyaman University, Adıyaman, Türkiye
e-mail: ahmetcelik@adiyaman.edu.tr

K. Bellitürk
Department of Soil Science and Plant Nutrition, Faculty of Agriculture, Tekirdağ Namık Kemal University, Tekirdağ, Türkiye

M. F. Baran
Department of Biosystem Engineering, Faculty of Agriculture, Siirt University, Siirt, Türkiye

N. Kumar (ed.), *Heavy Metal Remediation*, Earth and Environmental Sciences Library,
https://doi.org/10.1007/978-3-031-53688-5_4

conventional organic fertilizers; however, vermicompost fertilizers are also an effective solution. In actuality, everything here is about preserving a healthy ecosystem for next generations by preventing chemical pollution of agricultural products and soils.

Keywords Composting · Vermicomposting · Soil · Ecosystem

1 Introduction

Human and animal habitats generate a large amount of organic waste, the decomposition products of which affect the quantity and quality of soil, air and water. Since most organic waste contains a number of pathogenic microorganisms, the likelihood of contamination is high in any environment. If they are released into the environment without prior disinfection, this causes health and environmental risks. Animal and urban waste; It differentiates the unique distinctive features of the soil, such as pH, bulk density, conductivity, water retention capacity and increase in organic carbon content [85].

Ensuring the management of biosolid waste has become important not only to ensure a hygienic and healthy environment, but also to further increase productivity, which is the first to improve soil quantity. Composting is a very common and accepted process for recycling organic waste. It is clearly known that the direct application of raw wastes or poorly stabilized materials to the land causes toxicity and pathogenicity in the soil [22].

According to Westerman and Bicudo [170] organic wastes are used to bring the physical and also chemical quantities of the soil to higher levels.

Nair et al. [119] reported that worm manure is also a strong tool for the fertilizer option that does not contain pathogens, thanks to the reduction in the bulk amount of waste.

Shamini et al. [154] stated that more than 70% of solid waste is generated in Malaysia, 95% of which is destroyed on land, is not managed as it should be, and causes various problems. Since it is not possible to prevent some of the waste from turning into methane without oxygen; causes global warming. Relatively; vermicompost is used effectively in organic waste management without causing environmental pollution.

According to Sumi et al. [160], one of the major environmental issues is solid waste management. Both terrestrial and aquatic ecosystems are contaminated by animal feces and other organic wastes. Such wastes are known to be processed by biogas plants, microbial composting, and vermicomposting, among other methods. It undergoes processes to become environmentally significant substances. This has led to the discovery that vermicompost helps plants germinate, bloom, and produce fruit earlier. Practically speaking, the micro flora (colonies of bacteria and fungi) and the nitrogen, phosphorus, and potassium contents of the plants that use vermicompost technology are found to be higher than those of the witness plant (which does not use vermicompost).

Dhimal et al. [40] stated that most of the zoo waste consists of animal manure, garbage and litter; he said that it was converted into compost using worms and the composting time interval was less than conventional compost.

Jain [77] said that organic solids management is an important trouble in developing countries. Bad management of large amounts of waste is called environmental pollution.

We can state that land is a valuable resource that people rely on to make a living as well as an asset that they can pass on to future generations because it is known that agricultural activities are conducted on the land. Because of this, an individual's value system for their surroundings in relation to the land influences both the macro-level lifespan of future generations and their own well-being and health in relation to the environment. Because people rely on the land to both support themselves and make a living. The development of science and technology, urbanization and the increase in human population have constantly increased the rate of use of mineral springs, and indirectly, the classification of mines as agricultural or non-agricultural has caused soil infertility. It is also known to lead to the degradation of the land and ecology. When a land is destroyed, it means that there is a hidden power or that the land cannot sustainably produce goods, services, or food for an extended period of time. Its ecosystem is also at risk of dying out. As a living thing, soil is made up of an intricate mixture of inorganic and organic materials that were created artificially by combining various elements in predictable ways with various results and reactions. This being said, it serves as a natural environment for plant growth, a support system, and a significant source of stored or accumulated water because it provides minerals, nutrients, and water to plants. It has the mineral and nutrient components required for plant growth. It is a fact, albeit an indirect one, that the land will be devastated if it is unable to produce crops and other goods and services [121].

According to Mbagwu [108], excessive grazing, deforestation, improper farming methods, excessive consumption of wood that is burned for heat and causes desertification, and other human activities like mining that cause the land's productive capacity to decline or lag behind are the main causes of land degradation, stated to no longer be effective. The drilling and refining of oil, the disposal of waste and sludge, the diversion of rivers, and other factors are the causes of the previously mentioned soil degradation.

In this section, the importance of vermicompost, brief information about vermicomposting and various examples of areas where vermicompost is used are focused on.

2 What is Compost and Vermicompost?

Composting is an activity that takes place in a heat-relieving oxygen environment that involves transforming organic wastes into stable materials at various rates [63]. Composting is the phase that allows the decomposition of organic matter by microorganisms under controlled conditions, exposing the organic matter to a thermophilic phase (45–65 °C) that allows the waste to be sterilised by inactivating various pathogenic microorganisms [100, 110]. During composting, a decrease in the amount of waste has been observed, which is due to the production of heat, carbon dioxide and water by microorganisms [4].

Vermitechnology, in contrast to composting, is a mesophilic (< 35 °C) stage of worm life where high temperatures are typically not necessary for the worm to survive. Vermicomposting, to put it briefly, is the essential process of organic matter being broken down by worms working with microbial communities to break down organic materials and leftovers. There is evidence that the worm composting process efficiently gets rid of human pathogens and improves the process of disinfecting the processed material, even though the precise mechanisms are unknown [48]. Schematic representation of the factors that play an important role in the compost and vermicompost process is represented in Fig. 1

Vermicompost contains a higher concentration of nutrients than compost and is also less susceptible to soil salinization. In addition, vermicomposts have unusual biological properties and have been shown to harbour significantly larger and more diverse microbial populations than conventional composts [54].

In vermicomposting, the complete destruction of pathogens occurs through intestinal motility as well as the powerful activity of intestinal fluids and worms such as selective grazing [42] and the release of coelomic fluids from their bodies with antibacterial properties which emerges through its activities [90, 146, 149]. Brown and Mitchell [26] also reported scientifically supported evidence that stimulation of rare bacteria by worm activities can lead to recruitment of pathogens through competition or antagonism.

While vermicomposting and composting are potentially helpful techniques for reducing or eliminating pathogens, there may be instances where there is no connection between pathogen population decline and potential mechanisms involving temperature, worm activity, and composting duration. It prompted an investigation into the ability of two distinct composting processes to sanitize. Despite the fact that composting and the production of vermicompost have been the subject of numerous academic studies, little is known about the key mechanisms that neutralize pathogens during the process of making vermicompost as well as the process variables required to effectively reduce them [157]. Some recommended values for important factors in the composting and vermicomposting process is tablulated in Table 1.

Many microorganisms, enzymes, and hormones found in earthworm guts aid in the quick (about 4–8 week) conversion of the semi-digested material into vermicompost [66, 116]. The vermicomposting process mentioned above appears to be far more beneficial than the conventional composting method, which solely relies on

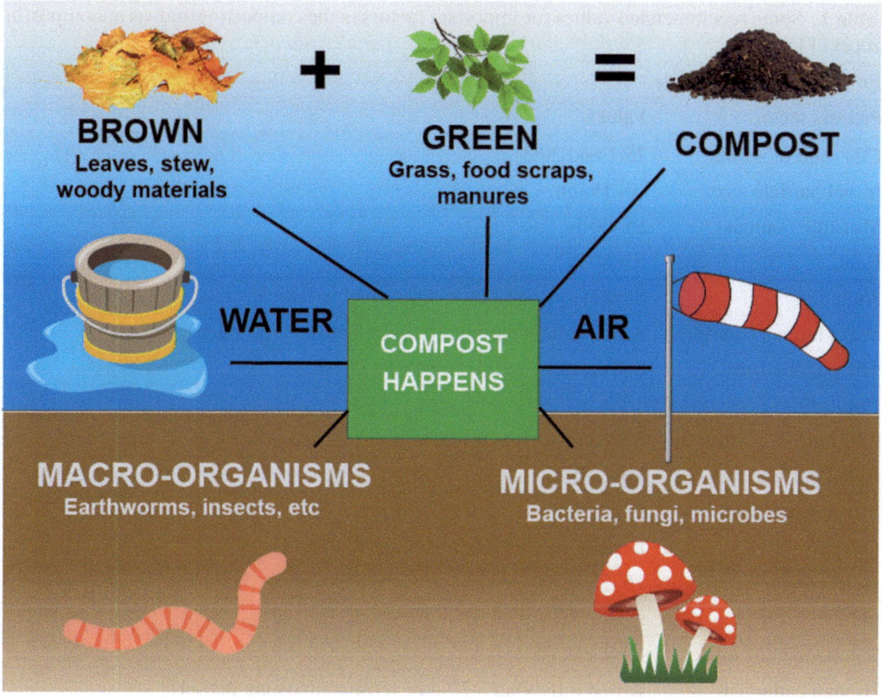

Fig. 1 Schematic representation of the factors that play an important role in the compost and vermicompost process

the superiority of microbes and consequently takes a long time (roughly 20 weeks) to produce compost [18, 143].

Vermicomposting has altered the waste's initial microbial composition in a number of ways. Nitrobacter, Azotobacter, Rhizobium, phosphate solubilizers, and actinomycetes are among the more than 10–10/g of bacteria present in the vermicompost [159].

Because vermicompost contains macro and micro nutrient elements, vitamins, enzymes, and hormones, it can have a substantial impact on the growth and profitability of plants [50, 80, 81, 106, 148]. In addition to having nutrients that plants can use, such as nitrate, available P, soluble K, Ca, and Mg, vermicompost also has macrospecific qualities that provide a large number of microsurfaces for microbial activity and robust nutrient attachment [54, 123]. Its surface area is sizable [156]. For instance, when fertilizer and vermicompost were applied together, the rice (*Oryza sativa*) plant absorbed the most N, P, K and Mg [75].

The suppression of disease may be due in part to microbial opposition, which increases microbial population and range in conjunction with organic differences. Non-thermophilic vermicomposts are rich in microbial spectrum and activity and contain a wide variety of antagonistic bacteria, whereas conventional thermophilic composts only support a limited number of microbes. As a result, they function

Table 1 Some recommended values for important factors in the composting and vermicomposting process [45]

Properties of composting	
Process factor	Values
C/N ratio of waste	25:1 and 30:1
Initial particle size	10–15 mm
Moisture content	55–60%
Heat	55–60 °C
Agitation	Simple systems don't stir during periodic cycles
Some human pathogens	Killed following the process's thermophilic stage
The passing time	Self-heating and thermophilic phase
Properties of vermicomposting	
C/N ratio of waste	25:1 and 30:1
Initial particle size	10–20 mm
Moisture content	80–85%
Oxygen	Earthworms maintain aerobic conditions
Heat	15–25 °C
pH	5–9
Salt content of waste	Low < 0.5%
Human pathogens	Killed following a 70-day vermicomposting period
Time taken	Batches of 4–12 months and continuous reactor systems of 30–60 days

as efficient biocontrol agents that aid in the suppression of illnesses brought on by soil-dwelling plant pathogenic fungi [30, 150, 153].

Pathma and Sakhivel [126] describe vermicompost as a non-thermophilic bio-oxidative process that is mediated by microbes and worms. Compost has a higher degree of elevated soil fertility due to its high concentration of soil biodiversity, water retention capacity, and growth-regulating hormones.

According to Dhimal et al. [40] and Mehta and Karnwal [109], compost has a high microbial content and all the elements needed for agricultural practices, including pH, organic matter, moisture content, N, P, K and C/N. This is what he said to explain it.

3 Vermicomposting and Soil Health

Earthworms are used as adaptable natural bioreactors in vermicomposting, an efficient method of recycling nutrients that breaks down organic materials. Vermicompost's rich nutrient content and microorganism activity improve soil productivity, promote healthy plant growth, and stop the spread of pest populations and plant diseases. According to Lavelle and Martin [95], earthworms are a very productive source of microbial enzymes as well as micro- and macronutrients.

Earthworms have been shown to have up to five times the number of microorganisms in their guts [49] and up to 1000 times the number of bacteria and "actinomycetes" in the material they ingest [51]. The number of total bacteria, proteolytic bacteria, and actinomycetes has also been found to increase as they move through the earthworms' digestive tract [36, 38, 125, 128].

According to reports, the consumption of organic wastes, which are very productive in terms of nutrients and energy that offer the right conditions and energy for the development of microbes in the earthworm's digestive system, as well as acting as a reactive state of a molecule for the development of microorganisms, can lead to an increase in the microbial activity [166].

Vermicompost has been scientifically shown to improve soil pH, microbial population, and soil enzyme activities [102]. It also dramatically lowers the amount of water-soluble chemicals that have the potential to pollute the environment [113]. The incorporation of vermicompost enhances the soil's air–water ratio, expands the macropore area, which ranges from 50 to 500 μm, and benefits plant growth [105].

Furthermore, earthworm activity improves soil structure, reducing the amount of chemicals released into surface water. With age, earthworm remains become more stable, significantly lowering the risk of erosion. In this particular context, it is considered that earthworms positively impact soil function and soil health. They indicate that the soil is sound and functioning properly [61].

Earthworms have been shown to improve soil physical properties, such as soil aggregation and the movement of rainwater or irrigation water from the surface downward by the action of gravity along the soil profile, as well as promote microbial decomposition by breaking down organic waste residues and increasing soil fertility to higher levels. But the main thing limiting the earthworm population is the availability of food. An effective way to boost the number of earthworms is to use crop residues and cover crops as a source of feed, and to leave the ruling class on the soil's surface through preservation farming techniques [47].

The increasing global population suggests that we need both an establishment arrangement that produces more output and a comprehensive and inclusive land result [142].

Sustainable outcomes are at risk due to nutrient inequality and soil toxicity brought on by the overuse of synthetic fertilizers. Overall, as a result of the current food chain imbalance [140].

According to Pimental et al. [131] and Rasmussen et al. [141], there is evidence that humans and farm animals are impacted by land chemical compound residues in food crops on an equal footing.

The total amount of reliable organic waste generated worldwide by humans, plants, and bovine animals is 38 billion m^3 [39].

Rice straw, which is produced in large quantities and is insufficient as animal feed, has a high content of silica and lignocellulose; it is quite difficult to handle or completely eliminate [59].

Most farmers are not aware that they cause air pollution such as smoke when they destroy this waste in the field with fire [132].

Large volumes of paper waste are produced by businesses and educational institutions, and this waste is typically dumped in landfills. Enhancing soil health can be achieved in part by recycling appropriate industrial and agricultural waste. Since organic matter makes up a small percentage of soil in temperate regions, it plays a crucial role in preserving fertility and enhancing soil quality [74].

When viewed as an efficient vermicomposting management strategy, rice straw, cow dung, and paper waste can all be potential substrates [155].

Earthworms and other microorganisms can stabilize organic material through vermicomposting, which is a potential substitute for managing organic solid waste [164].

Vermicompost usually has average values of 1.5–2.2% N, 1.8–2.2% P and 1.0–1.5% K. The organic carbon can vary between 9.15 and 17.98% and micro nutrient elements [3].

The integrated use of plant nutrient origin is ensured by worms turning organic plant nutrients such as farm manure, crop residues, household waste, and paper waste into vermicompost. This process has garnered significant attention in the context of sustainable agriculture, or sustainable crop production [165]. Sources and Usage Areas of Vermicompost is represented in Fig. 2.

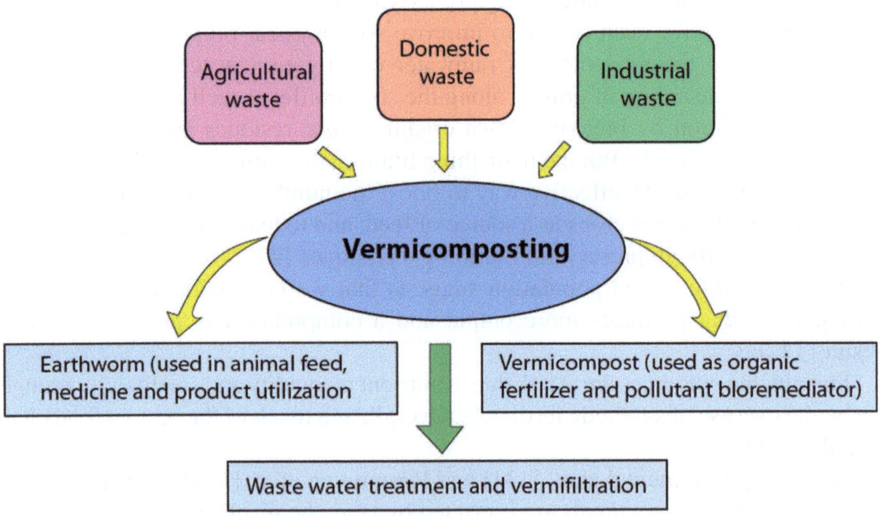

Fig. 2 Sources and usage areas of vermicompost

4 Preparation and Use of Vermicompost

Choosing the appropriate kind of worm is the first step in creating worm compost. Choosing and preparing waste that the worms can readily consume is the second step. During this procedure, proper worm food made from a blend of plant and animal waste must be prepared.

Based on their distinct ecological and trophic roles, earthworms are divided into three species: epigeic, anegic, and endogeic (Fig. 3).

Dominguez and Edwards [45] describe epigeic earthworm species as having a high rate of consumption, digestion, and assimilation of organic matter; they also have a short life cycle, high reproduction rates, and the natural ability to colonize organic waste. Because of their longevity and use tolerance, these species have good potential for vermicomposting. Only five species of earthworms are commonly used in vermicomposting: *Eisenia andrei* (Savigny), *Eisenia fetida* (Bouche), *Dendrobaena veneta* (Savigny), and to a lesser extent Perionyx excavatus (Perrier) and *Eudrilus eugeniae* (Kingberg). Very few species of earthworms possess all of these qualities.

Smaller in size, uniformly pigmented body, brief life cycle, high rate of reproduction, and regeneration characterize epigean earthworms. They feed on the surface substrates and mineralize them while living in the substrates and on the soil's superficial surface. Their gizzard, which is active, aids in the rapid conversion of organic matter into vermicompost. Furthermore, epigeic worms are resilient to spoiling, efficient biodegraders, and nutrient excretors. They can be effectively used for vermicomposting because they aid in the breakdown and early decomposition of organic waste [24, 27].

Pre-composting: It is well known that the presence of pathogens and toxic substances like ammonia or salts in the manure during vermicomposting makes pre-composting essential to preventing earthworm death [68].

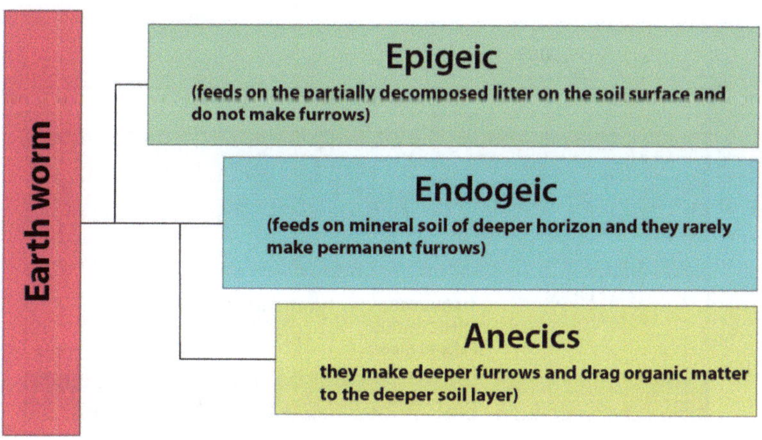

Fig. 3 Types of worms in the worm world

Vermicomposting is a biological waste management technique wherein the organic portion of waste is broken down by worms and microorganisms in regulated environments to the point where it can be handled, stored, and applied to fields of agriculture without harming the environment [1].

Table 2 provides some analysis results about vermicompost (VC) made from worm food made from different combinations of vegetal and animal waste.

Vermicompost is made using a wide range of techniques, though they differ from nation to nation. Nonetheless, continuous flow systems are the most favored option among them (Fig. 4).

Table 2 Results of vermicompost analysis showing variations in properties

Parameters	Cattle manure VC (Arancon and [58]	Food + garden waste VC [94]	Paper waste VC [13]	Cow manure VC [67]	Olive leaf waste VC [14]	Hazelnut shell + cow manure VC [16]
N, %	1.90	1.81	1.11	2.42	1.35	0.76
P, %	4.70	1.01	0.14	1.10	0.29	0.29
K, %	1.40	1.04	0.32	1.25	1.81	0.34
Ca, %	23.25	0.28	25.82	3.27	4.96	0.49
Mg, %	0.58	0.21	0.33	0.63	0.52	0.13
Fe, ppm	3454.00	1440.00	–	8500.00	14,427.00	290.21
Mn, ppm	160.00	346.00	131.50	380.39	510.67	45.30
Zn, ppm	516.00	387.20	48.80	379.95	81.84	24.37
B, ppm	58.00	–	–	75.75	63.68	–
Org. Mat. %	44.00	30.30	39.00	56.62	49.11	37.22
pH	–	–	–	6.97	8.93	7.60

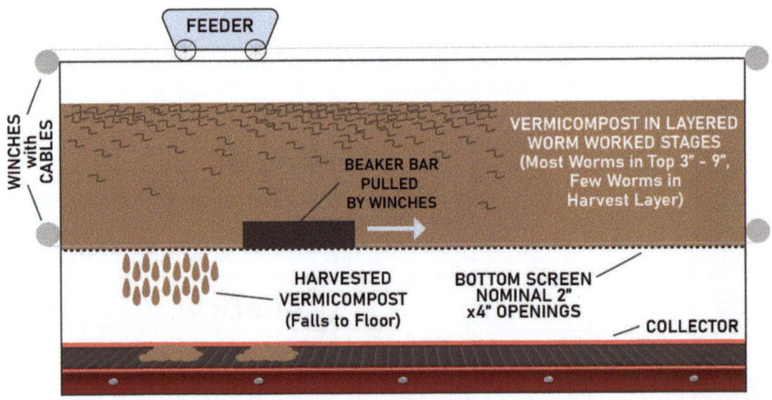

Fig. 4 Continuous flow reactor system

Fig. 5 Advantages and benefits of vermicompost

We know that worm excrement, which is high in nutrients and humus, is what makes up vermicompost. Worms can be artificially cultivated in a confined space or near trees, particularly garden trees. We can raise the quantity of worm excrement in the necessary amounts by feeding these worms biomass and closely observing the nutrients (biomass) of the worms.

Among the advantages of vermicompost are the following (Fig. 5):

1. Gibberellin, auxin, etc. It has extremely valuable hormones, enzymes, and vitamins.
2. It has an extraordinary effect on the growth of plants. It has an incredible motivation, especially in the formation of new formations such as shoots or leaves, and increases the quality of the crop in question and prolongs its shelf life.
3. It totally stops soil erosion and enhances the soil's structure, moisture content, air exchange, and water-retention ability.
4. It saves money on chemical fertilizers, which reduces food waste and allows for more work to be done with less fertilizer.
5. Vermicompost has an extremely high nutrient content because it includes all of the necessary plant nutrients.
6. Vermicompost boosts the number density and activity of worms in the soil by containing a variety of worm cocoons.
7. Vermicompost does not contain substances such as pathogenic microorganisms, toxic elements, and weed seeds.

8. Vermicompost ensures that the buffering power of the soil is brought to neutral levels.
9. Vermicompost is simple to apply, work with and maintain, and there is no chemical process that will create a bad odor during vermicomposting.
10. During crop production, worm excrement shortens the time it takes for harmful organisms and diseases to recur.

The nutrient content of vermicompost is as follows.

Calcium—0.5–1.0%

Magnesium—0.2–0.3%

N—1.5–2.5%

P—0.9–1.7%

K—1.5–2.4%

S—0.4–0.5%

Materials required for the production of vermicompost:

1. Biodegradable part of urban and rural waste.
2. Crop residues
3. Grass biomass
4. Hotel waste
5. Vegetable waste
6. Leaf litter
7. Wastes from agricultural industries.

5 Vermicomposting and Its Properties

Through the interaction of worms and related microbes, vermicomposting is a non-thermophilic biological oxidation process that turns organic matter into worm compost, a substance that resembles peat and has high porosity, oxygenation, water displacement, water holding capacity [6, 10, 54]. Vermiculture is also recognized as an affordable method of managing waste in an ecologically friendly manner [8, 12].

Earthworms are naturally occurring invertebrates in the Lumbricidae family that are found in the agricultural ecosystem. An earthworm possesses both a male and a female reproductive system. Moreover, they are hermaphrodites. Sexually mature worms grow a unique ring called a clitellum around the time of spawning. This ring contains glandular cells that secrete material to form a viscous, girdle-like structure called a cocoon. The number of fertilized eggs in each cocoon contains one to twenty lumbricid worms per individual. Worms are thought to be found in approximately 3000 different species worldwide [78].

The earthworm's intestine is a straight tube that starts at the mouth and ends with the anus, corresponding digestive glands, the muscular gizzard, the esophagus, the foregut, the midgut, and the hindgut. Mucus, which makes up the intestine, is composed of organic and mineral materials, proteins, polysaccharides and amino acids [115].

As a result, earthworms live as organic waste's natural bioreactors, dissolving it into its constituent parts and altering all of its characteristics. By building compact nests that soften and increase the porosity of the soil, earthworms accomplish the task of shaping the soil. This porosity enhances the soil's rooting, drainage, oxygenation, and water absorption. Unquestionably, soil aggregates play a crucial role in the soil-air ecosystem. They are created by earthworms and the microbes that live in their burrows and cave walls. These aggregates are granules of mineral bound together in a way that keeps the materials in the soil from eroding and from penetrating one another in both wet and dry conditions. Because they store beneficial microflora in their reservoir, earthworms accelerate soil regeneration and enable the soil to become extremely productive [117].

The guts of earthworms are home to "nitrogen-fixing" and "degrading" microbes, which they excrete in their feces along with nutrients [152]. Earthworms improve the supply of oxygen through their burrowing activity, which starts microbial activities and speeds them up. They also increase the population of soil microorganisms [19], the amount of microbes [52].

The microflora that survives in the worm gut and growth habitat works in concert with the worms to produce the vermicomposting process [70]. Through bioturbation and oxygenation, the worms quicken the composting process. However, the end product they produce is less rich in organic matter and nitrogen but has a higher yield of nitrenins that are available to the plants.

The process of vermicomposting differs greatly from that of conventional composting. This is primarily due to the worm's digestive system breaking down the organic material. The excrement that makes its way through the digestive tract of earthworms is utilized to enhance the soil's physical characteristics and fertility. Through physical and biochemical processes, earthworms actively contribute to the breakdown of organic matter in this process. The physical processes involved in the breakdown of organic substrates, such as movement and oxygenation, cause fragmentation, which increases the surface area even more. Biochemical alterations in this context include the removal of inorganic and organic material to another location, increased efficiency with nitrogen excretion, and the degradation of organic material by enzymatic digestion by microorganisms [135].

According to Pathma and Sakthivel [126], aerobic biooxidation and stabilization is a non-thermophilic method of decomposing organic waste that relies on earthworm-mediated microbial promotion, mixing, and decomposition. According to Edwards et al. [55] and Atiyeh et al. [11], vermicompost is a unique material that resembles peat and has high porosity, oxygenation, drainage, water retention capacity, and microbial activity. As a result, the plants are more resilient to illnesses and pests.

According to Masullo [107], the utilization of worms facilitates the transformation of waste materials into vermicompost following anaerobic digestion. Plant development is encouraged and irrigation frequency is decreased when vermicompost is applied to the field.

There are no differences in microbial aggregation or decomposition activity when the earthworm population is cultivated in high concentrations in vermicomposting facilities [52], and the gut and decomposition processes are the main factors involved in the quality of the microbial biomass [43].

Vermicomposts are materials high in phenolic compounds and humic acid. Due to their aphrodisiac properties, phenolic compounds can significantly reduce the risk of harmful attacks [25, 73, 89, 91, 101, 134, 161]. Indeed, polychlorinated phenols and their metabolites are present in soils that have been exposed to earthworms [93]. The earthworm *L. rubellus* contains an endogenous phenol oxidase that, when activated, generates harmful phenols like p-nitrophenol [124].

In the intestine of earthworms, humic acids have the ability to absorb monomeric phenols [168]. The plant tissue absorbs soluble phenolic compounds from the worm compost, changing them into an unappealing form that hinders pest reproduction and lowers their chances of surviving [56, 57].

Because different earthworm species supply nutrients in different ways, they live in different parts of the soil and affect the soil habitat in different ways. They belong to three distinct ecological groups based on the way they feed and build their nests. Earthworms that live in litter, or epigeic earthworms, rely on it for both survival and food. They move in a horizontal manner through the compost or leaf litter, consuming very little soil. Due to their small size, these worms are not found in soils that have little organic matter in them. An epigeic species is exemplified by Lumbricus rubellus.

Earthworms that are endogenous, or flat-living, are present in the mineral topsoil and its organic matter. They eat a lot of soil and fashion their nests into a three-dimensional labyrinth. It is possible to classify the worm species Aporrectodea and Diplocardia as having endogenous habits. Anecic (deep burrowing) worms persist in their life cycle by excavating nearly vertical burrows that extend several meters into the ground.

They can carry this debris into their nests and use it as food when they find it on the surface. One example of an anecic species is *Lumbricus terrestris* [34]. After settling into their burrows, earthworms improve the functionality of the soil. These are:

- They improve soil stability, air permeability, and moisture retention by sifting and gathering the soil;
- They promote microbial decomposition and the creation of nutrients, which are necessary for the soil's presence and are vital to other living things, by breaking down the waste remnants;
- They create materials that are highly productive in terms of nutrients such as potassium, phosphorus, and nitrogen;
- They improve root growth by creating nutrient-filled tunnels for the plant roots to follow;
- By forming tunnels and encouraging soil aggregation, they increase water infiltration;

- They turn the dirt in two directions: up and over. In addition, they lower the risk of disease by transporting organic matter deeper into the soil by stripping it all the way down to the layer where it meets the atmosphere.

It has been observed that adding organic matter to supplements can boost earthworm populations by two or three times during their lifespan. Certain liquid fertilizers have a liquid and seasoning content that is detrimental to earthworms. Their populations can essentially quickly maintain their previous levels and even start to grow in spite of this negative impact [52].

6 Beneficial Effect of Vermicompost on the Physical, Biological and Chemical Properties of Soil

Earthworms play an important role in the carbon cycle, soil formation, cellulose decomposition and humus storage. The activities and studies of earthworms have a great influence of the soil. Earthworms are voracious eaters of organic waste. They ingest only a small portion of this waste for their development and excrete most of this waste from their body in a semi-digested form [49, 76, 79].

When the organic matter passes through the earthworm's stomach, it is converted into a fine powder, followed by digestive enzymes, microorganisms and other fermenting substances that aid decomposition in the gut, and finally the molds are converted into a mature plant, "vermicompost", by the microbes associated with the earthworm gut They are released into the external environment [44].

Earthworm gut; increase in organic carbon, total organic carbon and nitrogen and moisture content, through the activity of non-motile microbes and germination of endospores, etc. It provides the most suitable habitat for a variety of digestive enzymes such as amylase, cellulase, protease, lipase, chitinase and urease, which are known from the digestive tract of earthworms. It has been reported that gut microbes are responsible for the activities of cellulase and mannose [115].

By decomposing the substrate, the earthworms increase the surface area for microbial decomposition and thus the active phase of worm composting begins. This decomposed organic material mixes with gut microbes and digestive enzymes during intestinal passage and eventually comes out as "moulds" in partially digested but not fully digested form, and these microbes then take over the further decomposition process during the developmental phase [96].

Certain microbial communities have been shown to be part of the earthworms' nutritional program, as evidenced by the destruction of certain microbes as they pass through the earthworms' digestive system [51].

It is emphasized that earthworms are important drivers of the process, as they compress and decompose the soil substrate while supplying it with oxygen, resulting in highly variable microbial activity and biodegradation potential [62, 96].

The effects of vermicomposting on soil pH, EC, C:N ratio and other nutrients were demonstrated. Worm activity reduced the pH and C:N ratio in the manure [10,

64]. Chemical analyses showed that worm manure had lower pH, EC, organic carbon [5, 112, 118] and C:N ratio [5, 139] compared to the source material.

The pH values of vermicompost are slightly lower compared to conventional compost, the nitrogen and phosphorus mineralization and the conversion of organic material into intermediate organic acids, fulvic acids and humic acids through microbial decomposition [5, 31, 96, 158] and the associated CO_2 production [60, 65].

Considering the advantages of the variety of bacteria related to earthworms, the following can be said. We can say that the ability to increase the availability of plant nutrients in earthworms probably depends on the activity of the earthworms' gut microflora. Earthworms indirectly influence the dynamics of chemical processes in the soil by subjecting litter to decomposition and influencing the activity of the soil microflora [52, 97, 129].

7 Beneficial Effect of Vermicompost on Heavy Metal Toxicity

Modernization has led to an increase in the anthropogenic content of metals, which easily penetrate various aspects of the environment. Metals are the major contributors to environmental pollution. However, the presence of these metals in various forms and hence the occurrence of ecotoxicity is a cause for concern and should be considered a priority. It is known that the term heavy metals refers to naturally occurring elements that have a concentration greater than 4 g/cm^3, have a high atomic weight and are toxic even at low concentrations [163]. In addition, heavy metals are accepted as trace elements if they occur in the environment in a lower concentration range [92, 163]. Essential heavy metals such as copper (Cu), cobalt (Co), iron (Fe), manganese (Mn) and zinc (Zn) are essential for physicochemical and biochemical functions in both plants and animals. These metals also form components of various proteins (prosthetic groups), enzymes (cofactors, activators) and are involved in various redox reactions, electron transfer and nucleic acid metabolism [111]. However, the presence of these metals in large quantities can lead to toxicity due to the narrow concentration range between the beneficial and toxic effects of the metals. In addition, metals such as (Hg) and nickel (Ni) are non-essential metals at the cellular level that have no established biological functions and have lethal effects on metal-sensitive enzymes that cause cell death [163].

As reported by Morgan and Morgan [114], the worm's digestive system and the relevant microbial groups secrete various hydrolytic enzymes during the production of worm compost, which cause the decomposition of organic material, the uptake of essential nutrients and the storage of some, but not all, metals in the tissue. The digestion of decomposable organic material releases metals bound to ions and carbonates, resulting in a higher content of mobile metal fractions. The mobile fractions mentioned above are easily absorbed through the skin into the skin tissue of the worms and stored. The bioaccumulation of metals in the tissues leads to a stress reaction with which the worms react when they come into contact with waste contaminated with heavy metals. Most worm species have adapted to this fiction by establishing a logical balance between external food supply and waste production during metabolism. For this reason, to cope with the stress caused by high metal levels, the excretion rate of metals from the body increases as the worms bypass the uptake process in the tissues [98].

The constant movement of heavy metals and their bioavailability depend on their specific chemical properties, unlike the total content of large amounts of waste [173]. There are few reports that vermicomposting can significantly reduce the total content of heavy metals while limiting their degree of dissolution and potential bioavailability in the environmental medium [82, 172]. Because of this, it has been demonstrated by science that while converting organic waste into worm compost, heavy metal diversity and bioavailability are crucial. In order to look into the speciation of metals during the vermicomposting of organic wastes, sequential extraction was used. It entails utilizing various extractants and environmental factors in a specific plan for the efficient separation of various metal fractions [32]. By maintaining pH, P, K, Na, Ca, and microorganisms, vermicompost preserved the soil's fertility and decreased its cadmium concentration [29].

Vermicompost, a mixture of wood sawdust and biochar derived from sewage sludge, was found to lower Zn and Cd levels and increase the rate of reproduction of earthworms (cocoon, progeny) [104].

In addition to helping metal ions be chewed up, earthworm microorganisms mineralize organic matter [33, 133]. Because the intestines of earthworms *L. terrestris*, *Allolobophora caliginosa*, and *Allolobophora terrestris* contain more aerobes than the soil, it is known that this process occurs [125].

By granulating clay particles and further grinding silt and sand particles, worms also boost hydraulic conductivity and natural oxygenation by expanding the surface area. Furthermore, the earthworm's body serves as a "biofilter," lowering chemical oxygen demand (COD) and biological oxygen demand (BOD) from wastewater by "absorbing" heavy metals, sediments, and organic debris through their body walls and "ingestion" and biodegradation. They guarantee the elimination of 90%, 80–90%, 90–92%, and 90–95% of the necessary COD, total dissolved solids (TDS), and total suspended solids (TSS), respectively [147].

According to a different study, some earthworms can also get rid of harmful pathogens by eating them and then excreting heavy metals and antibacterial coelomic fluid (via bioaccumulation) [149].

Vermicomposting is a technique that efficiently reduces, recycles, and stabilizes heavy metals found in waste by utilizing the ecological roles of earthworms and microbial diversity [151]. Enzymes mineralize organic wastes during vermicomposting (GAPs and CAPs) to produce easier or shorter-chain organic acids and to increase the moisture content of the wastes.

Vermicompost has a helpful buffering ability and lessens the damage caused by excessive amounts of nutrient materials, which would otherwise have a toxic effect on plants, according to a study that evaluated various organic and inorganic supplemental materials related to the growth and development of raspberries [158]. Because of this,

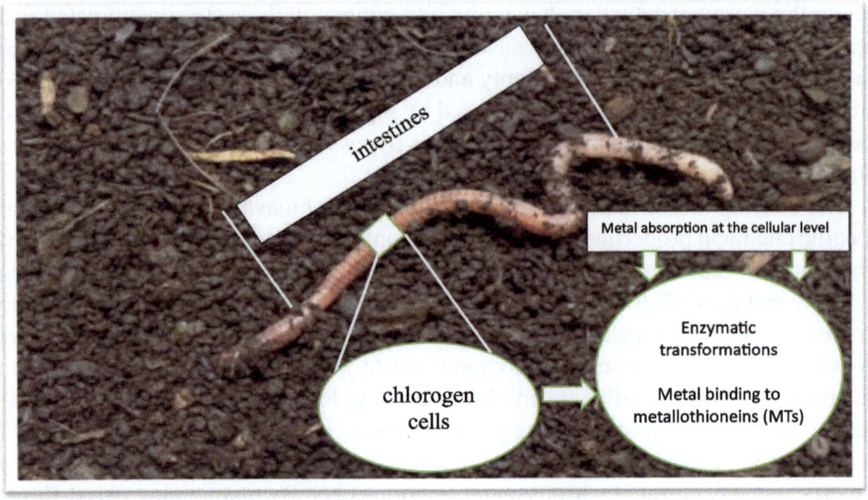

Fig. 6 Earthworms' biotransformation and heavy metal accumulation [17]

vermicompost functions as a fertilizer that penetrates the soil more slowly and as a soil conditioner [5, 9]. It is important to make sure that heavy metals, which are toxic to plants, reach the plant at a very low level during vermicomposting because they form complex aggregates with humic acids and other polymerized organic fractions [44]. Research has shown that soil amended with vermicompost yields higher-quality produce with lower levels of nitrate and heavy metals than soil treated with mineral fertilizers [87] (Fig. 6).

8 Economic and Environmental Benefits of Vermicompost in Sustainable Agriculture

According to Maji et al. [103], there was an increase in plant height, fresh weight, and dry weight in quantity when the vermicompost had a higher humic acid richness. In addition, when compared to chemical fertilizer, the maximum quantity and density of microbes (fungi and bacteria) were documented. It is well known that vermicompost is made using a variety of worm species. In sustainable agriculture, vermicompost fertilizer is an extremely potent biofertilizer that enables lowering the quantity of chemicals used in chemical agriculture. Worms are known to be involved in waste management, specifically in the recycling of organic waste, which makes this a significant factor [23].

Worms ground the organic solid into vermicompost, which had an annual economic value of 9.36 lakhs. According to Sequeira and Chandrashekar [145], Eudrilus sp. can be found in cow dung, paper products, vegetables, and garden waste (leaf and grass). Vermicompost conversion is made possible by the high yield and beneficial microbial communities found in compost, which include fungi, bacteria, actinomycetes, pseudomonads, P-solubilizers, and N2 fixers. Synthetic chemicals used carelessly have detrimental effects on organisms other than those intended and create numerous issues for the agricultural production system.

Vermicomposting is a crucial process that uses worms to convert organic waste into compost, which is highly nutrient-efficient and doesn't harm the good bacterial population. Vermicompost for sustainable agriculture has been the subject of more research lately, with many scholars conducting more of these studies. Vermicompost was created from organic wastes; in this case, the compost had high concentrations of growth hormone, beneficial microorganisms, carbon, nitrogen, phosphorus, and potassium [3].

Animal excrement from farms is now being managed using vermicompost, a sustainable technique. Composting would be a waste of energy (manpower) if worms weren't used, but when they are, the compost breaks down quickly, becomes recyclable, and contains a lot of nutrients and microbial biomass [120].

Considering the most recent developments in environmental policy, the following advice is recommended: The process of purifying organic waste by worms that move in harmony with microbial populations—known as vermicomposting—has shown to be an environmentally friendly technology. In developed nations like Canada, the UK, Spain, Italy, the US, Hong Kong, Australia, New Zealand, and Japan, vermicomposting facilities are currently engaged in both domestic and industrial marketing (Edwards et al. 2011). Furthermore, because worm compost has a high value and low operating costs, worm-based waste decomposition is becoming more and more popular in developing nations [35, 90].

Comparing worm composting to traditional thermophilic composting, worm composting is a more economical and ecologically friendly waste management technique that is beneficial to associated microbes as well as worms. The physiochemical and biological qualities of agricultural soil are enhanced by the addition of vermicomposts, which are great sources of biofertilizers. Vermi-composts are great for soil reclamation, increasing soil fertility, plant development, and sustainable agriculture. They also have excellent physico-chemical properties, the ability to buffer, and the ability to convert all nutrient materials into forms that plants can use. Additionally, vermi-composts are more productive when they contain antagonistic and plant growth-supporting bacteria. These are amazing organic substances that have the ability to heal all illnesses [126].

Tomato seedlings were used in a greenhouse experiment with: 1. Raw milk manure and tobacco residue; 2. Garden leaves; 3. Sewage sludge + rice husk; 4. Sewage sludge + garden leaves and vermicompost made from raw milk manure; the witness reported that all of the experiments demonstrated significant growth [72].

According to Lim et al. [99], different rates of *E. eugeniae* were studied in relation to waste management in relation to soybean husks and papaya wastes. It has been suggested that different 1:1 ratios work best for vermicompost.

Vermicompost, or cattle manure, was found to be effective in increasing leaf size, plant height, and yield in *Petroselinum crispum* plants [130]. *Eisenia andrei* earthworms were used by Kaouachi and his colleagues [84] to turn olive waste into vermicompost. Vermicompost that results raises nitrogen, phosphorus, potassium, calcium, magnesium, sodium, and the required EC and C/N ratio, according to observations.

A specific change in plumle length was observed in corn seedlings treated with vermicompost water as opposed to regular water, suggesting the presence of hormones that promote plant growth in vermicompost [116].

When vermicompost fertilizer is applied, tomatoes (*Lycopersicon esculentum*) [116], eggplants (*Solanum melangona*) [71], sunflowers (*Helianthus annuus*) [127], and sorghum (*Sorghum bicolor*) [37]. Abelmoschus esculentus, the okra [69], hyacinth bean (*Lablab purpureas*), grapes [21], and cherries [169] were all successfully cultivated.

Red clover and cucumber shoot tissues grown in soil mixed with vermicompost fertilizer showed an increase in mineral contents, including Ca, Mg, Cu, Mn, and Zn [144].

According to Wilson and Carlile's study, vermicompostted duck waste outperformed untreated waste in terms of tomato, lettuce, and pepper growth and development. [171].

By immersing cabbage roots in a mixture of clay and vermicompost, cabbage rot caused by *P. brassicae* was avoided [162].

Tomato and lettuce plants grew better when the parent material was vermicomposted; this was not the case with unprocessed parent material [10].

Compared to plants treated with inorganic fertilizer, potato plants treated with vermicompost showed reduced susceptibility to *P. infestans* [88].

Vermicompost has been shown to decrease the presence of spider mites, jassids (*Empoasca kerri*), aphids (*Aphis craccivora*), leaf miner (*Apoaerema modicella*), Spodoptera litura, and Helicoverpa armigera in peanuts [136–138] (*Heteropsylla cubana*) psyllids (*Leucaena leucocephala*) [20].

9 Conclusions

An important part of recycling organic waste is earthworms. We can also state that they are highly helpful in agricultural production because of high-humic organic fertilizers, which support the fertility, oxygenation, and structure of the soil. Humic acid fertilizer has bioactive ingredients that influence a variety of physiological processes, including metabolism, growth, productivity, and seed germination. These qualities are absent from regular fertilizers, even though it raises the quality levels of the properties. In addition to the benefits described above, humic acid fertilizer can effectively strengthen crops' resistance to cold and drought as well as shield them from pathogenic bacteria, insect pests, and underground plant diseases [135].

The application of vermicompost fertilizer results in the absorption of advanced nutrients by the plant as well as the transfer of plant hormones and humic materials.

In this case, soil health and crop productivity both rise as a result of the enhanced microbial activity in vermicompost. Plant pathogens should be decreased by using vermicompost and its derivatives, which are safe for the environment [167].

Vermicomposting helps to improve soil and agricultural productivity, reduces agricultural diseases, and purifies the aqueous environment by adsorbing or decomposing polluting materials [122].

In terms of agriculture, organic waste storage can be managed sustainably by creating oxygenated conditions without harming the ecosystem. Organic waste storage is hazardous to the environment worldwide. Biogas is produced from waste in order to provide energy for daily use. Solid waste management has grown to be a massive and unsolvable issue in our nation; waste can be turned into agricultural fertilizer—which is known to have many advantages by employing worms [86].

In this particular context, worm composting is thought to play a significant role in the breakdown of organic material by maintaining and even boosting soil fertility to higher levels. The assessment under consideration is a biotechnological procedure in which earthworms natural bioreactors take part. Worms promote plant growth and recycle organic waste. Worm castings are even more significant because they serve additional useful purposes [135].

Using various organic materials with high nutritional value and combining potential worm varieties with soil-friendly microbes are well-known techniques in vermiculture technology. In addition, the well-known effective dose-specific application of vermicompost in the vermiculture system, the combination of vermicompost with other inorganic fertilizers, the interactions between worms and microbes, etc. Worm manure has a bright future, according to numerous studies and studies on studies [41].

References

1. Aira M, Monroy F, Domínguez J, Mato S (2002) How earthworm density affects microbial biomass and activity in pig manure. Eur J Soil Biol 38:7–10
2. Aira M, Monroy F, Domínguez J (2007) Earthworms strongly modify microbial biomass and activity triggering enzymatic activities during vermicomposting independently of the application rates of pig slurry. Sci Total Environ 385:252–261
3. Adhikary S (2012) Vermicompost, the story of organic gold: a review. Agric Sci 3:905–917
4. Alidadi H, Parvaresh AR, Shahmansouri MR, Pourmoghadas H (2005) Combined compost and cermicomposting process in the treatment and bioconversion of sludge. Iran J Environ Health Sci Eng 2:251–254
5. Albanell E, Plaixats J, Cabrero T (1988) Chemical changes during vermicomposting (*Eisenia fetida*) of sheep manure mixed with cotton industrial wastes. Biol Fertil Soils 6:266–269
6. Arancon NQ, Edwards CA, Atiyeh R, Metzger JD (2004) Effects of vermicomposts produced from food waste on the growth and yields of greenhouse peppers. Bioresour Technol 93:139–144
7. Arancon N, ve Edwards CA (2011) The use of vermicomposts as soil amendments for production of field crops. In: Edwards CA, Arancon NQ, Sherman R (eds) Vermiculture technology. CRC Press, Taylor and Francis Groop, Chapter 10, pp 129–151

8. Asha A, Tripathi AK, Soni P (2008) Vermicomposting: a better option for organic solid waste management. J Hum Ecol 24:59–64
9. Atiyeh RM, Dominguez J, Subler S, Edwards CA (2000) Changes in biochemical properties of cow manure during processing by earthworms (*Eisenia andrei*, Bouché) and the effects on seedling growth. Pedobiologia 44:709–724
10. Atiyeh RM, Subler S, Edwards CA, Bachman G, Metzger JD, Shuste W (2000) Effects of vermicomposts and composts on plant growth in horticulture container media and soil. Pedobiologia 44:579–590
11. Atiyeh RM, Lee S, Edwards CA, Arancon NQ, Metzger JD (2002) The influence of humic acids derived from earthworm-processed organic wastes on plant growth. Biores Technol 84:7–14
12. Banu JR, Logakanthi S, Vijayalakshmi GS (2001) Biomanagement of paper mill sludge using an indigenous (*Lampito mauritii*) and two exotic (*Eudrilus eugineae* and *Eisenia fetida*) earthworms. J Environ Biol 22:181–185
13. Bellitürk K, Zahmacıoğlu A, Şerif E, ve Top M (2015a) Utilizationof paper waste by making vermicompost project. Trhracian Land. J Fertile Lands 1(3):21–24
14. Bellitürk K, Görres JH, Bağdatlı MC, Göçmez S, Turan HS, Eker M, Aslan S (2015b) Utilization of olive pruning waste as vermicompost: micro elements. Agric Vis Mag 1(1):7–12
15. Bellitürk K (2016) Vermicompost technology for solid waste management in sustainable agricultural production. Çukurova J Agric Food Sci 31(3):1–5
16. Bellitürk K, Soytürk Ö (2020) Can vermicompost obtained from *Eisenia foetida* fed by nutshell and cow manure mix be an organic fertilizer? Fresenius Environ Bull 29(12A):11273–11284
17. Bellitürk K, Goldmann Benardete B (2020) Miraculous creatures of nature (worms that have served the fertility of the soil and environmental health for centuries). Filmon Printing Solutions, Eco Reform Publications, 100p
18. Bernal MP, Faredes C, Sanchez-Monedero MA, Cegarra J (1998) Maturity and stability parameters of composts prepared with a wide range of organic wastes. Bioresour Technol 63:91–99
19. Binet F, Fayolle L, Pussard M (1998) Significance of earthworms in stimulating soil microbial activity. Biol Fertil Soils 27:79–84
20. Biradar AP, Sunita ND, Teggel RG, Devaradavadgi SB (1998) Effect of vermicompost on the incidence of subabul psyllid. Insect Environ 4:55–56
21. Buckerfield JC, Webster KA (1998) Worm-worked waste boosts grape yields: prospects for vermicompost use in vineyards. Australian New Zealand Wine Indus J 13:73–76
22. Butler TA, Sikor LPM, Steinhilber L, Douglass LW (2001) Compost age and sample storage effects on maturity indicators of biosolids compost. J Environ Qual 30:2141–2148
23. Bhat SA, Singh J, Vig AP (2017) Earthworms as organic waste managers and biofertilizer producers. Waste Biomass Valorization 1–15
24. Bhatnagar RK, Palta RK (1996) Earthworm-vermiculture and vermicomposting. Kalyani Publishers, New Delhi
25. Bhonwong A, Stout MJ, Attajarusit J, Tantasawat P (2009) Defensive role of tomato polyphenoloxidases against cotton bollworm (*Helicoverpa armigera*) and beet army worm (*Spodoptera exigua*). J Chem Ecol 35:28–38
26. Brown BA, Mitchell MJ (1981) Role of the earthworm, *Eisenia fetida*, in affecting survival of Salmonella enteriditis ser. typhimurium. Pedobiologia 22:434–438
27. Brown GG (1995) How do earthworms affect microfloral and faunal community diversity? Plant Soil 170:209–231
28. Chou C (2010) Role of allelopathy in sustainable agriculture: use of allelochemicals as naturally occurring bio-agrochemicals. Allelopath J 2010(25):3–16
29. Champar-Ngam N, Iwai CB, Ta-Oun M (2010) Vermicompost: tool for agroindustrial waste management and sustainable agriculture. Int J Environ Rural Dev 1–2:38–43

30. Chaoui H, Edwards CA, Brickner M, Lee S, Arancon N (2002) Suppression of the plant diseases, pythium (damping off), rhizoctonia (root rot) and verticillum (wilt) by vermicomposts. In: Proceedings of Brighton crop protection conference–pests and diseases, vol II(8B-3), pp 711–716
31. Chan LPS, Griffiths DA (1988) The vermicomposting of pretreated pig manure. Biol Wastes 24:57–69
32. Chen M, Li X, Yang Q, Zeng G, Zhang Y, Liao D, Guo L (2008) Total concentrations and speciation of heavy metals in municipal sludge from Changsha, Zhuzhou and Xiangtan in Middle-South Region of China. J Hazard Mater 160(2–3):324–329
33. Canellas LP, Olivares FL, Okorokova FAR (2002) Humic acids isolated from earthworm compost enhance root elongation, lateral root emergence and plasma membrane H+– ATPase activity in maize roots. Plant Physiol 130:1951–1957
34. Coleman DC, Crossley DA Jr (1996) Fundamentals of soil ecology. Academic Press, San Diego
35. Davison L, Pont D, Bolton K, Headley T (2006) Dealing with nitrogen in subtropical Australia: seven case studies in the diffusion of ecotechnological innovation. Ecol Eng 28:213–223
36. Daniel O, Anderson JM (1992) Microbial biomass and activity in contrasting soil material after passage through the gut of earthworm *Lumbricus rubellus* Hoffmeister. Soil Biol Biochem 24:465–470
37. Devi D, Agarwal SK, Dayal D (1998) Response of sunflower (*Helianthus annuus*) to organic manures and fertilizers. Indian J Agron 43:469–473
38. Devliegher W, Verstraete W (1995) *Lumbricus terrestris* in a soil core experiment: nutrient-enrichment processes (NEP) and gut-associated processes (GAP) and their effect on microbial biomass and microbial activity. Soil Biol Biochem 27:1573–1580
39. Diacono M, Persiani A, Testani E, Montemurro F, Ciaccia C (2019) Recycling agricultural wastes and by-products in organic farming: biofertilizer production, yield performance and carbon footprint analysis. Sustainability 11:3824
40. Dhimal M, Gautam I, Tuladhar R (2013) Effectiveness of vermicomposting in management of organic wastes using *Eisenia foetida* and *Perionyx favatus* in Central Zoo Jawalakhel, Nepal. J Nat Hist Mus 27:92–106
41. Dhurva P, Gauchan, Lamichhane J (2017) Vermicompost and its role in plant growth promotion Anon Chaulagain. Department of Biotechnology, School of Science. Int J Res 4(8):859–864
42. Domínguez J, Edwards CA (1997) Effects of stocking rate and moisture content on the growth and maturation of *Eisenia andrei* (*Oligochaeta*) in pig manure. Soil Biol Biochem 29:743–746
43. Domínguez J (2004) State of the art and new perspectives in vermicomposting research. In: Edwards CA (ed) Earthworm ecology, 2nd edn. CRC Press, Boca Raton, pp 401–425
44. Dominguez J, Edwards CA (2004) Vermicomposting organic wastes: a review. In: Shakir Hanna SH, Mikhail WZA (eds) Soil zoology for sustainable development in the 21st century, Cairo, pp 369–395
45. Dominguez J, Edwards CA (2011) Relationships between composting and vermicomposting. In: Edwards CA, Arancon, NQ, Sherman R (eds) Vermiculture technology. CRC Press, Taylor and Francis Groop, Chapter 2, p 20
46. Dominguez J, Edwards CA (2011) Biology and ecology of earthworm species used for vermicomposting. Edwards CA, Arancon NQ, Sherman R (eds) Vermiculture technology. CRC Press, Taylor and Francis Groop, Chapter 3, pp 27–40
47. Donahue S (2001) Agricultural management effects on earthworm populations. In: Soil quality–agronomy technical note no 11, pp 1–8
48. Eastman BR, Kane PN, Edwards CA, Trytek L, Gunadi B, Stermer AL, Mobley JR (2001) The effectiveness of vermiculture in human pathogen reduction for USEPA biosolids stabilization. Compost Sci Util 9:38–49
49. Edwards CA, Lofty R (1977) The biology of earthworms. Chapmann and Hall, London
50. Edwards CA (1988) Breakdown of animal, vegetable and industrial organic wastes by earthworms. In: Edwards CA, Neuhauser EF (eds) Earthworms in waste and environmental management SPB. The Hague, Netherlands, pp 21–31

51. Edwards CA, Fletcher KE (1988) Interaction between earthworms and microorganisms in organic matter breakdown. Agric Ecosyst Environ 20:235–249
52. Edwards CA, Bohlen PJ (1996) Biology and ecology of earthworms. Chapman and Hall, London, p 426
53. Edwards CA, Bohlen PJ (1996) Biology and ecology of earthworms, 3rd edn. Chapman & Hall, London, UK
54. Edwards CA (1998) The use of earthworms in the breakdown and management of organic wastes. In: Edwards CA (ed) Earthworm ecology. CRC Press, Boca Raton, pp 327–354
55. Edwards CA, Dominguez J, Neuhauser EF (1998) Growth and reproduction of Perionyxex-cavatus (Perr.) (*Megascolecidae*) as factors in organic waste management. Biol Fertil Soils 27:155–161
56. Edwards CA, Arancon NQ, Bennett MV, Askar A, Keeney G, Little B (2010) Suppression of green peach aphid (*Myzus persicae*) (Sulz.), citrus mealybug (Planococcus citri) (*Risso*), and two spotted spider mite (*Tetranychus urticae*) (Koch.) attacks on tomatoes and cucumbers by aqueous extracts from vermicomposts. Crop Prot 29:80–93
57. Edwards CA, Arancon NQ, Bennett MV, Askar A, Keeney G (2010) Effect of aqueous extracts from vermicomposts on attacks by cucumber beetles (*Acalymna vittatum*) (Fabr.) on cucumbers and tobacco hornworm (*Manduca sexta* L.) on tomatoes. Pedobiologia 53:141–148
58. Edwards CA (2011) Human pathogen reduction during vermicomposting. In: Edwards CA, Arancon NQ, Sherman R (eds) Vermiculture technology: earthworms, organic wastes and environmental management. CRC Press, Boca Raton, pp 249–261
59. El-Dewany C, Zaghloul A, Awad F (2018) Utilization of rice straw as a low-cost natural by-product in agriculture. Int J Environ Pollut Environ Model 1:91–102
60. Elvira C, Sampedro L, Benítez E, Nogales R (1998) Vermicomposting of sludges from paper mill and dairy industries with *Eisenia andrei*: a pilot-scale study. Bioresour Technol 63:205–211
61. Farenhorst A, Topp E, Bowman BT, Tomlin AD (2000) Earthworm burrowing and feeding activity and the potential for atrazine transport by preferential flow. Soil Biol Biochem 32:479–488
62. Fracchia L, Dohrmann AB, Martinotti MG, Tebbe CC (2006) Bacterial diversity in a finished compost and vermicompost: differences revealed by cultivation independent analyses of PCR amplified 16S rRNA genes. Appl Microbiol Biotechnol 71:942–952
63. Fialho L, Silva W, Milori D, Simões M, Martin L (2010) Characterization of organic matter from composting of different residues by physicochemical and spectroscopic methods. Bioresour Technol 101:1927–1934
64. Gandhi M, Sangwan V, Kapoor KK, Dilbaghi N (1997) Composting of household wastes with and without earthworms. Environ Ecol 15:432–434
65. Garg P, Gupta A, Satya S (2006) Vermicomposting of different types of waste using *Eisenia foetida*: a comparative study. Bioresour Technol 97:391–395
66. Ghosh M, Chattopadhyay GN, Baral K (1999) Transformation of phosphorus during vermicomposting. Bioresour Technol 69:149–154
67. Göçmez S, Bellitürk K, Görres JH, Turan HS, Üstündağ Ö, Solmaz Y, Adiloğlu A (2019) The effects of the use of vermicompost in olive tree farming on microbiological and biochemical characteristics of the production material. Erwerbs-obstbau 2019:1–8. https://doi.org/10.1007/s10341-019-00437-1
68. Gunadi B, Blount C, Edwards CA (2002) The growth and fecundity of *Eisenia foetida* (Savigny) in cattle solids pre-composted for different periods. Pedobiologia 46:15–23
69. Gupta AK, Pankaj PK, Upadhyava V (2008) Effect of vermicompost, farm yard manure, biofertilizer and chemical fertilizers (N, P, K) on growth, yield and quality of lady's finger (*Abelmoschus esculentus*). Pollut Res 27:65–68
70. Grapelli A, Tomati U, Galli E (1983) Vermicomposting of combined sewage sludge and municipal refuse. In: Proceedings of international syrup on agricultural and environmental prospects in earthworm farming, Rome, Italy, pp 73–80

71. Guerrero RD, Guerrero LA (2006) Response of eggplant (*Solanum melongena*) grown in plastic containers to vermicompost and chemical fertilizer. Asia Life Sci 15:199–204
72. Hasheminajs K, Kalbasi M, Golchin A, Shariatmadari H (2006) Comparison of vermicompost and composts as potting media for growth of tomatoes. J Plant Nutr 27:1107–1123
73. Hawida S, Kapari L, Ossipov V, Ramtala MJ, Ruuhola T, Haukioja E (2007) Foliar phenolics are differently associated with *Epirrita autmnata* growth and immuno competence. J Chem Ecol 33:1013–1023
74. Hijbeek R, Van Ittersum MK, Berge HF, Gort G, Spiegel H, Whitmore AP (2017) Do organic inputs matter. A meta-analysis of additional yield effects for arable crops in Europe. Plant Soil 411:293–303
75. Jadhav AD, Talashilkar SC, Pawar AG (1997) Influence of the conjunctive use of FYM, vermicompost Andurea on growth and nutrient uptake in rice. J Maharashtra Agric Univ 22:249–250
76. Jambhekar H (1992) Use of earthworm as a potential source of decompose organic wastes. In: Proceedings of the national seminar on organic waste utilization, Coimbatore, pp 52–53
77. Jain N (2016) Waste management of temple floral offerings by vermicomposting and its effect on soil and plant growth. Int J Environ Agric Res 2:89–94
78. Julka JM (1986) Earthworms' resources of India. In: Dash RC, Senapathi BK, Mishra PC, (eds), Proceedings of organic waste utilization, vermicomposting, Part B: verms and Vermicomposting, pp 1–7
79. Kale RD, Bano K (1986) Field trials with vermicompost (Vee comp. E. 8. UAS) on organic fertilizers. In: Dass MC, Senapati BK, Mishra PC (eds) Proceedings of the national seminar on organic waste utilization. Sri Artatrana Ront, Burla, pp 151–157
80. Kale RD, Mallesh BC, Bano K, Bagyaray DJ (1992) Influence of vermicompost application on the available macronutrients and selected microbial populations in paddy field. Soil Biol Biochem 24:1317–1320
81. Kalembasa D (1996) The influence of vermicomposts on yield and chemical composition of tomato. Zesz Probl Post Nauk Roln 437:249–252
82. Kang J, Zhang Z, Wang JJ (2011) Influence of humic substances on bioavailability of Cu and Zn during sewage sludge composting. Biores Technol 102:8022–8026
83. Karmegam N, Daniel T (2008) Effect of vermi-compost and chemical fertilizer on growth and yield of hyacinth bean (*Lablab purpureas*). Dyn Soil Dyn Plant Glob Sci Books 2:77–81
84. Kaouachi A, Ibijbijen J, Amane M, Jaafari SE (2013) Management of olive mill waste employing vermicomposting technology. Int J Sci Res 4:486–490
85. Khaleel R, Reddy KR, Overcash MR (1981) Changes in soil physical properties due to organic waste applications: a review. J Environ Qual 10(2):133–141
86. Khalid A, Arshad M, Anjum M, Mahmood T, Dawson L (2011) The anaerobic digestion of solid organic waste. Waste Manage 31:1737–1744
87. Kolodziej M, Kostecka J (1994) Some qualitative features of the cucumbers and carrots cultivated on the vermicompost. Zeszyty Naukowe Akademii Rolniczej W Krakowie 292:89–94
88. Kostecka J, Blazej JB, Kolodziej M (1996) Investigations on application of vermicompost in potatoes farming in second year of experiment. Zeszyty Naukowe Akademii Rolniczej W Krakowie 310:69–77
89. Koul O (2008) Phytochemicals and insect control: an antifeedant approach. Crit Rev Plant Sci 27:1–24
90. Kumar R, Shweta (2011) Removal of pathogens during vermi-stabilization. J Environ Sci Technol 4:621–629
91. Kurowska A, Gora J, Kalemba D (1990) Effects of plant phenols on insects. Pol Wiad Chem 44:399–409
92. Knight C, Kaiser J (1997) Heavy metals in surface water and stream sediments in Jamaica. Env Geochem Health 19:63–66
93. Knuutinen J, Palm H, Hakala H, Haimi J, Huhta V, Salminen J (1990) Polychlorinated phenols and their metabolites in soil and earthworms of a saw mill environment. Chemosphere 20:609–623

94. Lange MG (2005) A comparison analysis of vermicomposting strategies in food substrates with an emphasis on nutrient values and reproduction (Research Advisor: Dr. Christopher Baxter and Dr. Ken Killian). Pioneer Undergraduate Research Fellowship, May 27, pp 1–15
95. Lavelle P, Martin (1992) Small scale and large scale effects of endogeic earthworms on soil organic matter dynamics in soils of the humid tropics. Soil Biol Biochem 24(12):1491–1498
96. Lazcano C, Gomez-Brandon M, Dominguez J (2008) Comparison of the effectiveness of composting and vermicomposting for the biological stabilization of cattle manure. Chemosphere 72:1013–1019
97. Lee KE (1985) Earthworms: their ecology and relationships with soils and land use. Academic Press, Sydney
98. Li L, Wu J, Tian G, Xu Z (2009) Effect of the transit through the gut of earthworm (*Eisenia fetida*) on fractionation of Cu and Zn in pig manure. J Haz Mat 167:634–640
99. Lim SL, Wu TY, Lim PN, Shak LPY (2014) The use of vermicompost in organic farming: overview, effects on soil and economics. J Sci Food Agric 95:1143–1156
100. Lung AJ, Lin C-M, Kim JM, Marshall MR, Nordstedt R, Thompson NP, Wei CI (2001) Destruction of *Escherichia coli* O157:H7 and salmonella enteritidis in cow manure composting. J Food Protect 64:1309–1314
101. Mahanil S, Attajarusit J, Stout MJ, Thipayong P (2008) Over expression of tomato phenol oxidase increases resistance to the common cutworm. Plant Sci 174:456–466
102. Maheswarappa HP, Nanjappa HV, Hegde MR (1999) Influence of organic manures on yield of arrowroot, soil physico-chemical and biological properties when grown as intercrop in coconut garden. Ann Agr Res 20:318–323
103. Maji D, Misra P, Singh S, Kalra A (2017) Humic acid rich vermicompost promotes plant growth by improving microbial community structure of soil as well as root nodulation and mycorrhizal colonization in the roots of *Pisum sativum*. Appl Soil Ecol 110:97–108
104. Malińska K, Golanska M, Caceres R, Rorat A, Weisser P (2017) Biochar amendment for integrated composting and vermicomposting of sewage sludge—the effect of biochar on the activity of *Eisenia fetida* and the obtained vermicompost. Biores Technol 225:206–214
105. Marinari S, Masciandaro G, Ceccanti B, Grego S (2000) Influence of organic and mineral fertilisers on soil biological and physical properties. Bioresour Technol 72:9–17
106. Makulec G (2002) The role of *Lumbricus rubellus* Hoffm. In determining biotic and abiotic properties of peat soils. Pol J Ecol 50:301–339
107. Masullo A (2017) Organic wastes management in a circular economy approach: rebuilding the link between urban and rural areas. Ecol Eng 101:84–90
108. Mbagwu JSC (2003) Aggregate stability and soil degradation in tropics. In: Paper presented at 2003 college of soil physics, ICTP Trieste Italy
109. Mehta N, Karnwal A (2013) Solid waste management with the help of vermicomposting and its applications in crop improvement. J Biol Earth Sci 3:B8–B16
110. Mehta CM, Uma P, Franke-Whittle IH, Sharma AK (2014) Compost: its role, mechanism and impact on reducing soil-borne plant diseases. Waste Manage 34:607–622
111. Mildvan AS (1970) Metal in enzymes catalysis. In: Boyer DD (ed) The enzymes. Academic Press London, vol 11, pp 445–536
112. Mitchell A (1997) Production of *Eisenia fetida* and vermicompost from feed-lot cattle manure. Soil Biol Biochem 29:763–766
113. Mitchell A, Edwards CA (1997) The production of vermicompost using *Eisenia fetida* from cattle manure. Soil Biol Biochem 29:3–4
114. Morgan JE, Morgan AJ (1999) The accumulation of metals (Cd, Cu, Pb, Zn and Ca) by two ecologically contrasting earthworm species (*Lumbricus rubellus* and *Aporrectodea caliginosa*): implications for ecotoxicological testing. Appl Soil Ecol 13:9–20
115. Munnoli PM, Da Silva JAT, Saroj B (2010) Dynamics of the soil-earthworm-plant relationship: a review. Dyn Soil Dyn Plant 1–21
116. Nagavallemma KP, Wani SP, Stephane L, Padmaja VV, Vineela C, Babu Rao M, Sahrawat KL (2004) Vermicomposting: recycling wastes into valuable organic fertilizer. In: Global theme on agrecosystems report no.8. Patancheru 502324. International Crops Research Institute for the Semi-Arid Tropics, Andhra Pradesh, p 20

117. Nakamura Y (1996) Interactions between earthworms and microorganisms in biological control of plant root pathogens. Farming Jpn 30:37–43
118. Nardi S, Dell'Agnola G, Nuti PM (1983) Humus production from farmyard wastes by vermicomposting. In: Proceedings of international symposium on agricultural and environmental prospects in earthworm farming, Rome, pp 87–94
119. Nair J, Sekiozoic V, Anda M (2006) EVect of pre-composting on vermicomposting of kitchen waste. Bioresour Technol 97:2091–2095
120. Nasiru A, Ismail N, Ibrahim MH (2013) Vermicomposting: tool for sustainable ruminant manure management. J Waste Manage 2013:7
121. Nweke IA (2017) Effect of compost and earthworm production on soil properties, growth and dry matter yield of maize in crude oil degraded soil. J Soil Sci Environ Manage 8(1):1–10
122. Okpara M (2023) Vermicompost remediation of contaminated soil: a mini review. Ann Facul Eng Hunedoara-Int J Eng 21(3)
123. Orozco FH, Cegarra J, Trujillo LM, Roig A (1996) Vermicomposting of coffee pulp using the earthworm *Eisenia fetida*: effects on C and N contents and the availability of nutrients. Biol Fertil Soils 22:162–166
124. Park SR, Cho EJ, Yu KH, Kim YS, Suh JJ, Chang CS (1996) Endogenous phenoloxidase from an earthworm lumbricus rubellus. Tongmul Hakoehi 39:36–46
125. Parle JN (1963) A microbiological study of earthworm casts. J Gen Microbiol 31:13–22
126. Pathma J, Sakthivel N (2012) Microbial diversity of vermicompost bacteria that exhibit useful agricultural traits and waste management potential. Springer Plus 1:26
127. Patil SL, Sheelavantar MN (2000) Effect of moisture conservation practices, organic sources and nitrogen levels on yield, water use and root development of rabi sorghum (*Sorghum bicolo*r L.) in the vertisols of semiarid tropics. Ann Agric Res 21:32–36
128. Pedersen JC, Hendrikse NB (1993) Effect of passage through the intestinal tract of detritivore earthworms (*Lumbricus* spp.) on the number of selected gram negative and total bacteria. Biol Fertil Soils 16:227–232
129. Petersen H, Luxton MA (1982) A comparative analysis of soil fauna populations and their role in decomposition process. Oikos 39:287–388
130. Peyvast GH, Olfati JA, Madeni S, Forghani A, Samizadeh H (2008) Vermicompost as a soil supplement to improve growth and yield of parsley. Int J Veg Sci 2:19–27
131. Pimental D, Culliney T, Bashore T (2018) Public health risks associated with pesticides and natural toxins in foods. In: Hutchison B, Radcliffe T (eds) Radcliffe's IPM world textbook. University of Minnesota, Minneapolis
132. Pimonsree S, Vongruang P, Sumitsawan S (2018) Modified biomass burning emission in modeling system with fire radiative power: simulation of particulate matter in mainland southeast Asia during smog episode. Atmos Pollut Res 9:133–145
133. Pizl V, Novokova A (1993) Interactions between microfungi and *Eisenia andrei* (*Oligochaeta*) during cattle manure vermicomposting. Pedobiologia 47:895–899
134. QiTian S (2004) Research on prevention and elimination of agricultural pests by ginkgo phenols phenolic acids. Chem Ind For Prod 24:83
135. Rakkini VM, Vincent S, Kumar AS, Baskar K (2017) An overview: organic waste management by earthworm. J Civil Eng Environ Sci 3(1):013–017
136. Rao KR, Rao PA, Rao KT (2001) Influence of fertilizers and manures on the population of coccinellid beetles and spiders in groundnut ecosystem. Ann Plant Protect Sci 9:43–46
137. Rao KR (2002) Induce host plant resistance in the management sucking pests of groundnut. Ann Plant Protect Sci 10:45–50
138. Rao KR (2003) Influence of host plant nutrition on the incidence of *Spodoptera litura* an helicoverpa armigera on groundnuts. Indian J Entomol 65:386–392
139. Riffaldi R, Levi-Minzi R (1983) Osservazioni Preliminari Sul Ruolo Dell Eisenia foetida Nell'umificazione Del Letame. Agrochimica 27:271–274
140. Rojas RV, Achouri M, Maroulis J, Caon L (2016) Healthy soils: a prerequisite for sustainable food security. Environ Earth Sci 75:180

141. Rasmussen JJ, Reiler EM, Carazo E, Matarrita J, Muñoz A, Cedergreen N (2016) Influence of rice field agrochemicals on the ecological status of a tropical stream. Sci Total Environ 542:12–21
142. Sadowski A, Baer-Nawrocka A (2018) Food and environmental function in world agriculture interdependence or competition? Land Use Policy 71:578–583
143. Sánchez-Monedero MA, Roig A, Paredes C, Bernal MP (2001) Nitrogen transformation during organic waste composting by the Rutgers system and its effects on pH, EC and maturity of the composting mixtures. Bioresour Technol 78:301–308
144. Sainz MJ, Taboada-Castro MT, Vilariño A (1998) Growth, mineral nutrition and mycorrhizal colonization of red clover and cucumber plants grown in a soil amended with composted urban wastes. Plant Soil 205:85–92
145. Sequeira V, Chandrashekar JS (2015) Vermicomposting of biodegradable municipal solid waste using indigenous *Eudrilus* sp. earthworms. Int J Curr Microbiol App Sci 4:356–365
146. Sinha RK, Heart S, Agarwal S, Asadi R, Carretero E (2002) Vermiculture and waste management: study of action of earthworms *Eisenia foetida*, *Eudrilus eugeniae* and *Perionyx excavatus* on biodegradation of some community waste in India and Australia. Environmentalist 22:261–268
147. Sinha RK, Bharambe G, Chaudhari U (2008) Sewage ttreatment by vermifiltration with synchronous treatment of sludge by earthworms: a low-cost sustainable technology over conventional systems with potential for decentralization. Environmentalist 28:409–420
148. Sinha RK, Herat S, Valani D, Chauhan K (2009) Vermiculture and sustainable agriculture. Am-Euras J Agric and Environ Sci 5:1–55
149. Sinha RK, Agarwal S, Chauhan K, Valani D (2010) The wonders of earthworms and its vermicompost in farm production: Charles Darwin's 'friends of farmers', with potential to replace destructive chemical fertilizers from agriculture. Agric Sci 1(2):76–94
150. Singh R, Sharma RR, Kumar S, Gupta RK, Patil RT (2008) Vermicompost substitution influences growth, physiological disorders, fruit yield and quality of Strawberry (*Fragaria x ananassa* Duch.). Bioresour Technol 99:8507–8511
151. Singh RP, Embrandiri A, Ibrahim MH, Esa N (2011) Management of biomass residues generated from palm oil mill: vermicomposting a sustainable option. Res Conser Recyc 55(4):423–434
152. Singleton DR, Hendrixb PF, Colemanb DC, Whitmana WB (2003) Identification of uncultured bacteria tightly associated with the intestine of the earthworm *Lumbricus rubellus* (*Lumbricidae; Oligochaeta*). Soil Biol Biochem 35:1547–1555
153. Scheuerell SJ, Sullivan DM, Mahaffee WF (2005) Suppression of seedling damping-off caused by *Pythium ultimum*, and *Rhizoctonia solani* in container media amended with a diverse range of pacific northwest compost sources. Phytopathology 95:306–315
154. Shamini K, Fauziah SH, Emenike CU (2011) Vermicomposing of spent tea: a sustainable approach for solid waste management. In: Proceeding of the 12th international conference on environmental science and technology, Rhodes, Greece, pp 8–10. A-1695–A-1700
155. Sharma K, Garg V (2018) Comparative analysis of vermicompost quality produced from rice straw and paper waste employing earthworm *Eisenia fetida* (Sav.). Bioresour Technol 250:708–715
156. Shi-wei Z, Fu-zhen H (1991) The nitrogen uptake efficiency from 15N labeled chemical fertilizer in the presence of earthworm manure (cast). In: Veeresh GK, Rajagopal
157. Soobhany N, Mohee R, Vinod Kumar Garg VK (2017) Inactivation of bacterial pathogenic load in compost against vermicompost of organic solid waste aiming to achieve sanitation goals: a review. Waste Manage 64:51–62
158. Subler S, Edwards CA, Metzger PJ (1998) Comparing vermicomposts and composts. Biocycle 39:63–66
159. Suhane RK (2007) Vermicompost. Publication of Rajendra Agriculture University, Pusa, p 88
160. Sumi MG, Vani M, Idicula DV, Mini KD (2014) Solid waste management using vermicomposting and KODMIC? bio-pedestal column and its utility as organic manure. Asian J Microbiol Biotechnol Environ Exp Sci 16:333–338

161. Summers G, Felton GW (1994) Prooxidation effects of phenolic acids on the generalist herbivore helicoverpa zea: potential mode of action of phenolic compounds on plant anti-herbivory chemistry. Insect Biochem Mol Biol 24:943–953
162. Szcech M, Rondomanski W, Brzeski MW, Smolinska U, Kotowski JF (1993) Suppressive effect of a commercial earthworm compost on some root infecting pathogens of cabbage and tomato. Biol Agric and Hortic 10:47–52
163. Tchounwou PB, Yedjou CG, Patloll AK, Sutton DJ (2012) Heavy metals toxicity and the environment. NIH Public Access 101:133–164. https://doi.org/10.1007/978-3-7643-8340-4_6
164. Thomas GV, Mathew AE, Baby G, Mukundan M (2019) Bioconversion of residue biomass from a tropical homestead agro-ecosystem to value added vermicompost by Eudrilus species of earthworm. Waste Biomass Valorization 10:1821–1831
165. Timsina J (2018) Can organic sources of nutrients increase crop yields to meet global food demand? Agronomy 8:214
166. Tiwari SC, Tiwari BK, Mishra RR (1989) Microbial populations, enzyme activities and nitrogen, phosphorous, potassium enrichment in earthworm casts and in the surrounding soil of pine apple plantation. Biol Fertil Soils 8:178–182
167. Ur Rehman S, De Castro F, Aprile A, Benedetti M, Fanizzi FP (2023) Review vermicompost: enhancing plant growth and combating abiotic and biotic stress. Agronomy 13(4):1134
168. Vinken R, Schaeffer A, Ji R (2005) Abiotic association of soil-borne monomeric phenols with humic acids. Org Geochem 36:583–593
169. Webster KA (2005) Vermicompost increases yield of cherries for three years after a single application. EcoResearch, South Australia
170. Westerman PW, Bicudo JR (2005) Management considerations for organic waste use in agriculture. Biores Technol 96:215–221
171. Wilson DP, Carlile WR (1989) Plant growth in potting media containing worm-worked duck waste. Acta Hortic 238:205–220
172. Wong JWC, Selvam A (2006) Speciation of heavy metals during co-composting of sewage sludge with lime. Chemosphere 63:980–986
173. Yuan XZ, Huang HJ, Zeng GM, Li H, Wang JY, Zhou CF, Zhu HN, Pei XK, Liu ZF, Liu ZT (2011) Total concentrations and chemical speciation of heavy metals in liquefaction residues of sewage sludge. Biores Technol 102:4104–4110

Phytoremediation Potential of Bioenergy Plants for Heavy Metals: Supportive Method for Environmental Management

Daka Gopamma, Kolli Suresh Kumar, Parida Tamanna, and Namuduri Srinivas

Abstract This procedure is way more economical and practical than other techniques available in the market, which is rarely explored by the research community. In India, proper waste segregation is very poor, and direct dumping into uncontaminated soil and water is dominant. Phytoremediation can be easily installed in large areas to treat waste, and as its cost is minimal, industries and commercial colonies could approach it. Hyperaccumulators are used for this process as it has a unique defense mechanism to counter heavy metals. It absorbs toxins beyond the limit, reduces toxicity, and accumulates in the root and shoot part. Metals can be claimed after the harvesting period. Introducing bioenergy in the field of phytoremediation is a crucial approach adopted worldwide as it enhances economic values meanwhile sequestrating the heavy metals in a substantial way. It prevents soil erosion, enhancing the soil quality and reducing the carbon load from the environment.

Keywords Heavy metals · Phytoremediation · Phytoextraction · Hyperaccumulator · Chelators

1 Introduction

This era is determined by the globalization of economics with intensifying anthropogenic activities such as industrialization, mining, urbanization, transportation, and agronomic practice but the waste that comes along with the output urges the crucial attention of everyone. Improper segregation of waste puts the planet in a critical situation where animals and plants are witnesses to anthropogenic or natural disasters. Generally, waste is loaded with heavy metals (HMs) such as Cr, Cd, As, Hg, Zn, and Pb and ultimately comes to the soil, and water and gets volatile in the air [57]. HMs are non-biodegradable in nature, and they possess toxicity for a prolonged period so if a natural body gets contaminated by the HMs once, it holds the toxicity for many

D. Gopamma · K. Suresh Kumar · P. Tamanna · N. Srinivas (✉)
Department of Environmental Science, GITAM School of Science, GITAM (Deemed to Be University), Visakhapatnam, Andhra Pradesh, India
e-mail: snamudur@gitam.edu

N. Kumar (ed.), *Heavy Metal Remediation*, Earth and Environmental Sciences Library,
https://doi.org/10.1007/978-3-031-53688-5_5

years [61]. Pb, Cd, and Cr are potentially lethargic to the animal as they straightway attack the endocrine and nervous system even at a minimal concentration [52]. Soil is a major source for the living hood as all depend on it for their basic purpose but direct use of contaminated soil is an obstacle due to its lethal composition where industrial waste plays a vital role. Industrial waste contains potential contaminants in a massive amount which eventually gets deposited on the top layer of soil. It lessens biological activities, strengthens infertility, and reduces yield production which leads to ecological imbalance. It provides a wide platform to get contaminated and introduced into the human food chain [59]. Some metals are essential for plant growth, such as Zn involved in many enzymatic activities like superoxide dismutase whereas Ni and urease enzyme guides the urea metabolism in plants [71]. However, the higher concentration of both non-essential and essential HMs is hazardous to the plant and inhibits the morphological and biological process [73]. The bioavailability of metals influences the ion uptake of roots from the soil which results in the synthesis of biosurfactants and organic acid [3, 77] then mobile through xylem and participates in the growth and storage in the cell organelle [53].

The treatment is mainly categorized into three groups such as physical, chemical, and biological process. Physical and chemical are not gaining popularity due to some major downsides such as the alteration of soil properties, secondary contamination, manpower, high cost, and irreversible changes. A range of machinery has been developed to treat this potential waste. Yet, an effective remediation technology is essential where no residues will be left after post-treatment. A viable solution for it is landfilled; however, this implies risks to water, soil, public health, operational difficulties, and costs. Hence, this is a problematic solution. Phytoremediation falls under biological treatment which is a developing innovative technique for metal-contaminated soil in recent years. The main advantage of phytoremediation is its low expense and can be used for in situ application on large scales. It offers a nature-based treatment of contaminated media. Some special group plants called hyperaccumulators are featured with a specific type of defense mechanism that helps to counter the toxicity of HMs by breaking it and accumulating it in above and below-ground biomass (shoot and root) [28]. It helps to restore the vegetation, and modify the ecosystem [74]. In this way, some native species dominate the field upgrading both soil and air quality and opening up a new field for tourism that helps to explain the importance of ecosystem conservation. More than 101 families together with Asteraceae, Lamiaceae, Caryophyllaceae, Fabacceae, Conouniaceae, Euphorbiaceae, Violaceae, and Brassicaceae claim more than 500 plants for accumulation of HMs. Out of all, Brassica is considered the most effective hyperaccumulator [56]. Some hyperaccumulators produce high biomass without hampering their own metabolism which helps to claim the bioenergy. This way phytoremediation establishes a sustainable eco-friendly method in modern society [32].

2 Role of Hyperaccumulators in Phytoremediation

A hyperaccumulator is a plant species that has the extraordinary ability to accumulate high concentrations of certain heavy metals or metalloids in its tissues without showing significant signs of toxicity or adverse effects. These plants have evolved unique physiological and biochemical mechanisms that enable them to thrive in environments with high levels of metal contamination. Here are some key points about hyperaccumulators [62].

2.1 *Extreme Metal Tolerance*

Hyperaccumulator plants can tolerate and even thrive in soils with metal concentrations that would be toxic to most other plant species. Metal Uptake and Accumulation: Hyperaccumulators actively absorb specific metals from the soil and transport them into their above-ground tissues, such as leaves or stems. This results in the accumulation of high metal concentrations. Metalloids and Heavy Metals: While many hyperaccumulators are known for accumulating heavy metals like cadmium, lead, and nickel, some can also hyperaccumulate metalloids such as selenium and arsenic. Hyperaccumulators are often studied and used for phytoremediation purposes. Phytoremediation is a process where these plants are cultivated in contaminated areas to help remove or reduce metal pollutants from the soil [54].

2.1.1 Ecological Significance

Hyperaccumulators can play a vital role in their ecosystems by influencing soil chemistry, nutrient cycling, and interactions with herbivores and pollinators.

2.1.2 Plant Families

Certain plant families are more likely to include hyperaccumulator species, including Brassicaceae (mustard family) and Asteraceae (daisy family).

2.1.3 Research and Conservation

The study of hyperaccumulators is of significant interest to researchers and environmental scientists due to their potential applications in cleaning up contaminated sites and understanding metal tolerance mechanisms in plants.

2.1.4 Hyperaccumulators

Well-known hyperaccumulator species include *Thlaspi caerulescens* (a zinc and cadmium hyperaccumulator), (a nickel hyperaccumulator), and *Brassica juncea* (Indian mustard, which accumulates heavy metals like lead and cadmium). Hyperaccumulators have applications in phytoremediation projects, where they are used to clean up polluted soils or to extract valuable metals from soils for economic purposes. They continue to be a subject of research to better understand their unique properties and to develop practical strategies for environmental remediation. Some of studies on hyperaccumulator plants is tabulated in Table 1 and benefits and advantages of Phytoremediation is depicted in Fig. 1.

3 Mechanisms and Types of Phytoremediation

The mechanism of sequestrating and accumulating the HMs varies from species to species but the plant has good biomass despite the toxicity of the media considered as the good quality of hyperaccumulator [12, 43]. Generally, the mechanism of phytoremediation is divided into the following parts (i) Phyto stabilisation, (ii) phytoextraction, (iii) phytovolatilization, (iv) rhizodegradation, and (v) phytodegradation. phyto stabilization and phytoextraction are considered to be the most promising mechanisms in phytoremediation [11]. Shoot accumulation is dominant in phytoextraction as the mobility of toxins is continuous from water or soil to shoot and the metabolism of plants isn't affected much by the toxicity of HMs [50]. These plants are categorized with advanced root systems, strong translocation factors with high production of biomass [10]. Metals such as Ni, Zn, Pb, Cd, and Cu are actively absorbed and accumulate in the shoot part through this mechanism [69]. On the other hand, Phyto stabilization claims the root part as metal sequestration is active in below-ground biomass [60]. Adding synthetic chelators such as ethylenediaminetetraacetic acid (EDTA) and biochar accelerates the metal absorption process [8].

Helps to degrade the organic contaminants and mobilize the metals from soil to plant known as phytostimulation or rhizodegradation [27]. Bacteria like *Agrobacterium, Pseudomonas, Arthobacter, Micrococcus, Burkholderia,* and fungi (Glomus) are the main characteristics of rhizodegradation [2]. This mechanism combined with oil crops gives significant results for the removal of contaminants and obtaining a good amount of bioenergy [7].

Phytoremediation is an environmentally friendly and cost-effective approach to clean up contaminated soil, water, and air using plants and their associated microorganisms. There are several types of phytoremediation techniques, each suited to different types of contaminants and environmental conditions. Here are some of the main types of phytoremediation.

Table 1 Researchers study the Hyperaccumulator of plant species

S. No.	Researchers name	Plant name	Plant tissues	Absorption mg/kg	Heavy metal name
1	Kupper et al. [30], Rajput et al. [52]	*Crassula helmsii*	–	9200	Cu
	Ghaderian andRavandi [18], Rajput et al. [52]	*Polypogon fugax*	–	1550–4012	
		Waltheria indica	–	1300–1581	
2	Singh and Fulekar [65], Rajput et al. [52]	*Brassica juncea*	Root	30,550	Zn
	Kupper et al. [32], Rajput et al. [52]	*Arabidopsis halleri*	Shoot	5722	
	Stein et al. [70], Gervais-Bergeron et al. [17]		Forbs	53,900	
	Wang et al. [81]	*Potentilla griffuthii*	Stems leaves	1670	
	Fellet et al. [16], Gervais-Bergeron et al. [17]	*Minuartia verna*	Forbs	18,400	
3	Szczygłowska et al. [72], Rajput et al. [52]	*Brassica juncea*	Root	25,000	Cd
	Liu et al. [38], Rajput et al. [52]	*Phytolacca Americana*	Stem-leaves	10,780	
	Buendı´a-Gonza´lez et al. [10], Rajput et al. [52]	*Prosopis leavigata*	Shoot	8176	
	Likar et al. [35], Rajput et al. [52]	*Thlaspi praecox*	Shoot -seed	5960	
	Kupper et al. [32], Rajput et al. [52]	*Arabidopsis halleri*	Shoot	5722	
	Wang et al. [81], Rajput et al. [52]	*Potentilla griffithii*	Stem-leaves	1670	
	Yang et al. [81]	*Nicotiana tabacum*	Stems and leaves	–	
	Huang et al. [25]	*Zea mays*	–		
4	Adki et al. [1], Rajput et al. [52]	*Opunita cochenillifera*	–	26,000	Cr
	Buendı´a-Gonza´lez et al. [10], Rajput et al. [52]	*Prosopis laevigata*	Shoot	8176	
	Santiago-Cruz et al. [56], Rajput et al. [52]	*Cosmos bipinnatus*	–	4825	
	Liu et al. [36], Rajput et al. [52]	*Lerrsia hexandra*	–	1844	

(continued)

Table 1 (continued)

S. No.	Researchers name	Plant name	Plant tissues	Absorption mg/kg	Heavy metal name
	Mongkhonsin et al. [42], Rajput et al. [52]	*Gynura pseudochina*	–	1611	
5	Singh and Fulekar [65], Rajput et al. [52]	*Brassica juncea*	Root	32,700	Pb
	Ha et al. [23], Rajput et al. [52]	*Potamogeton oxyphyllus*	–	4210	
		Ageratum houstonianum	–	1130	
		Pteris vittata	–	1020	
	Gul et al. [22]	*Mesembryanthemum*	–	–	
6	van der Ent et al. [77], Rajput et al. [52]	*Glochidion cf. sericeum*	–	442–1310	Co
	Hernández et al. [24]	*Lactuca sativa*	–	–	
	Hernández et al. [24]	*Lolium perenne*	–	–	
7	van Der Ent et al. [78], Rajput et al. [52]	*Antidesma*	Foliage	46,500	Mn
		Walsura		45,200	
		Casearia		38,200	
		Beccaria		21,700	
		Aporosa		26,700	
		Mischocarpus		19,400	
		Scolopia		16,600	
		Xylosma		15,600	
		Glochidion		12,600	
		Cleistanthus		11,300	
	Liu et al. [37], Rajput et al. [52]	*Celosia argentea*	–	32,000	
	Liu et al. [38]	*Phytolacca Americana*	Stem-leaves	10,780	
8	Mesjasz-Przybylowicz et al. [39], Rajput et al. [52]	*Phyllanthus balgooyi*	–	1,68,500	Ni
	van Der Ent et al. [78], Rajput et al. [52]	*Dichapetalum*	Foliage	45,600	
		Antidesma		32,700	
		Rinorea		32,200	
		Actephil		14,600	
		Aporosa		10,600	
		Scolopia		9200	
		Walsura		8600	

(continued)

Table 1 (continued)

S. No.	Researchers name	Plant name	Plant tissues	Absorption mg/kg	Heavy metal name
		Beccaria		8000	
		Homalium		5700	
		Hydnocarpus		3200	
		Walsura pinnata	Trees	8000	
	van der Ent et al. [77], Rajput et al. [52]	*Glochidion rubrum*	Foliage	5010	
		Knema matanensis	–	2500–2500	
	Teptin and Paukov [74], Gervais-Bergeron et al. [17]	*Odontarthena tortuosa*	Forbs	1790	
	Hernández et al. [24]	*Lactuca sativa*	–	–	
	Hernández et al. [24]	*Lolium perenne*	–	–	
	Schiavon et al. [59], Rajput et al. [52]	*Stanleya pinnata*	–	14,900	
9	Muller et al. [43], Rajput et al. [52]	*Pteris cretica*	–	1517	Se
	Pfister et al. [48], Gervais-Bergeron et al. [17]	*Symphyotrichum ascendens*	Forbs	4460	
	EI Mehdawi et al. [14], Gervais-Bergeron et al. [17]	*Symphyotrichum ericoides*	Forbs	3000	
	Fehlauer et al. [15]	*Saxifraga paniculata*	Forbs	2320	
10	Hernández et al. [24]	*Lactuca sativa*	–	–	Ba
11	Hernández et al. [24]	*Lolium perenne*	–	–	Fe

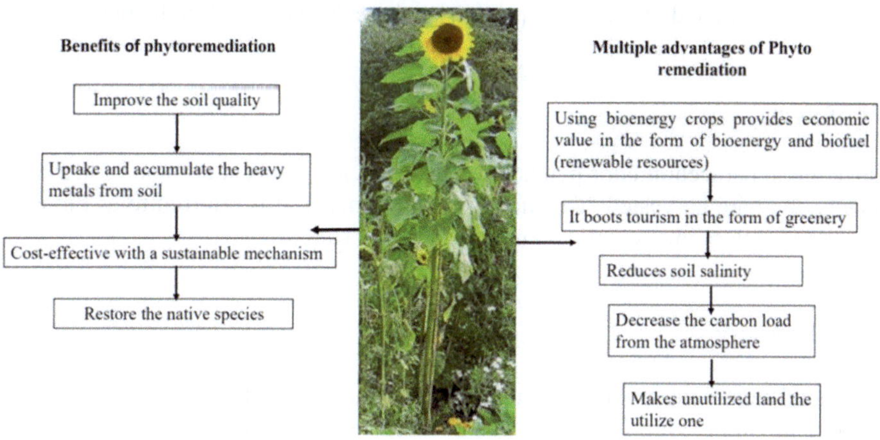

Fig. 1 Benefits and advantages of Phytoremediation

3.1 Phytoextraction

Phytoextraction is a phytoremediation technique that involves using specific plants to remove and accumulate contaminants, particularly heavy metals, from the soil. This process is particularly effective for soils contaminated with heavy metals, such as lead, cadmium, zinc, and copper. Phytoextraction is based on the ability of certain plant species to take up and store these metals in their above-ground tissues without displaying significant toxic effects [18].

3.1.1 How Phytoextraction Works

Plant Selection: Phytoextraction requires the selection of suitable plant species known as hyperaccumulators. Hyperaccumulators are plants that can accumulate large quantities of specific metals in their shoots without experiencing harm. These plants have evolved mechanisms to tolerate and accumulate high metal concentrations.

Planting

Hyperaccumulator plants are cultivated in the contaminated soil, allowing their roots to take up water and nutrients, including the target metals, from the soil.

Metal Uptake

As the hyperaccumulator plants grow, they actively absorb and translocate the target metals from the soil into their above-ground parts, such as leaves or stems. This process is facilitated by transport proteins and mechanisms within the plant.

Harvesting

Once the hyperaccumulator plants have absorbed a significant amount of the target metals, they are harvested. This harvested plant material is referred to as biomass. The metal-rich biomass is then removed from the site.

Metal Recovery

The metal-rich biomass can be further processed to recover the accumulated metals. Depending on the purpose, metals can be extracted and recycled or safely disposed of, thus remediating the contaminated soil.

3.2 Phyto Stabilization

Phyto stabilization aims to immobilize contaminants in the soil, reducing their mobility and bioavailability. Certain plant species are selected for their ability to form a barrier or a "cap" that prevents the spread of contaminants. They can also improve the soil structure and reduce erosion. Phytoextraction is a phytoremediation technique that involves using specific plants to remove and accumulate contaminants, particularly heavy metals, from the soil. This process is particularly effective for soils contaminated with heavy metals, such as lead, cadmium, zinc, and copper. Phytoextraction is based on the ability of certain plant species to take up and store these metals in their above-ground tissues without displaying significant toxic effects [49].

3.2.1 How Phyto Stabilization Works

Plant Selection: Plant species are chosen based on their ability to tolerate and adapt to the contaminated environment and their capacity to accumulate contaminants in their root systems. These plants are often referred to as "stabilizers."

Planting

Stabilizer plants are cultivated in the contaminated soil. As the plants grow, their root systems spread and create a dense network that binds the soil, preventing erosion and the dispersion of contaminants.

Contaminant Immobilization

The root systems of these stabilizer plants secrete substances (such as exudates) that can chemically bind with the contaminants, reducing their mobility in the soil. This process makes the contaminants less available for uptake by other plants or leaching into groundwater.

Reduced Erosion

The dense root systems also help reduce soil erosion, which is particularly important for sites with contaminated sediments near water bodies. The plants stabilize the soil, preventing the contaminants from being transported downstream.

3.3 Phytodegradation

Phytodegradation is a phytoremediation technique that involves using plants and their associated microorganisms to break down or degrade organic contaminants in soil, sediment, or water. This method is particularly effective for addressing organic pollutants, including hydrocarbons, pesticides, and other complex organic compounds. Phytodegradation is often used to remediate contaminated sites and promote the natural breakdown of these substances [66].

3.3.1 How Phytodegradation Works

Plant Selection

Specific plant species are selected based on their ability to tolerate and thrive in the contaminated environment and their potential to enhance the breakdown of organic contaminants.

Planting

These selected plants are cultivated in the contaminated soil, water, or sediment, allowing their root systems to establish and interact with the contaminants.

Contaminant Uptake

The plant roots actively absorb water, nutrients, and organic contaminants from the soil or water. These contaminants may then be transported to various parts of the plant.

Microbial Activity

Many plants form symbiotic relationships with microorganisms, such as mycorrhizal fungi and rhizosphere bacteria. These microorganisms play a crucial role in the degradation process, as they have enzymes and metabolic pathways that can break down organic contaminants.

Contaminant Transformation

Once inside the plant and in close association with microbial communities in the root zone (rhizosphere), the organic contaminants can undergo a series of transformations, including biodegradation, mineralization, and detoxification.

Enhanced Breakdown

Phytodegradation enhances the breakdown of organic contaminants through the combined action of plants and their associated microorganisms. The end products of this process are often less toxic or less persistent than the original contaminants.

3.4 Phytovolatilization

Phytovolatilization is a phytoremediation process that involves the use of plants to extract and release volatile contaminants, particularly metals or metalloids, from the soil into the atmosphere in gaseous form. This method is effective for dealing with contaminants that can be transformed into volatile compounds, making them easier to remove from the environment [66].

3.4.1 How Phytovolatilization Works

Plant Selection

Specific plant species, often referred to as "hyperaccumulators" for the targeted contaminants, are selected based on their ability to tolerate and accumulate volatile substances in their tissues.

Planting

These hyperaccumulator plants are grown in contaminated soil or water, allowing their roots to absorb water, nutrients, and the volatile contaminants present.

Contaminant Uptake

The plant roots actively absorb the volatile contaminants from the soil or water, transferring them to the above-ground parts of the plant. Within the plant, these contaminants can undergo various chemical transformations.

Volatilization

Some of the contaminants that accumulate in the plant's above-ground tissues are transformed into gaseous forms. This volatilization process often involves metabolic or biochemical pathways within the plant.

Release into the Atmosphere

The gaseous contaminants are released into the surrounding atmosphere through the plant's stomata (small openings on leaves and stems) or via evaporation from the plant's surfaces.

Air Dispersal

Once in the atmosphere, the volatile contaminants can disperse, become diluted, and eventually deposit in other areas or be removed from the environment.

3.5 Rhizofiltration

Rhizofiltration is a phytoremediation technique that uses the roots of specially selected plants to filter and remove contaminants, particularly heavy metals, from contaminated water as it passes through the root zone. It's an effective method for treating polluted water and improving water quality [66].

3.5.1 How Rhizofiltration Works

Plant Selection

Certain plant species, known as "accumulator" or "hyperaccumulator" plants, are chosen for their ability to tolerate and accumulate heavy metals and other contaminants in their roots. Common rhizofiltration plants include species like water hyacinth, Indian mustard, and sunflowers.

Planting

The selected plants are grown hydroponically or in wetland areas, allowing their roots to be submerged in the contaminated water.

Water Uptake

The plants actively absorb water, along with dissolved contaminants, through their roots. As the water moves through the root system, contaminants are taken up by the plant.

Contaminant Accumulation

The contaminants are sequestered in the plant's root tissues, and their concentration in the surrounding water is reduced.

Filtering Action

The fine, dense network of roots and associated microorganisms within the rhizosphere (the root zone) acts as a natural filter, capturing and retaining the contaminants. As the water flows through this root zone, contaminants are adsorbed, absorbed, or otherwise retained.

Treatment Efficiency

Over time, the contaminated water passing through the root zone becomes cleaner and less polluted. The extent of treatment depends on factors like the plant species, root structure, and the specific contaminants present.

3.6 *Rhizodegradation*

Rhizodegradation is a process in which the roots of plants, typically certain types of hyperaccumulating plants, help break down and degrade environmental pollutants in the soil. These pollutants can include organic compounds, heavy metals, and other contaminants. Rhizodegradation is a form of phytoremediation, which is the use of plants to clean up or mitigate the effects of soil and water pollution [66].

3.6.1 Root Exudates

Plants release a variety of organic compounds through their roots, such as sugars, organic acids, and enzymes. These exudates can stimulate the growth of beneficial microorganisms in the rhizosphere (the soil region directly influenced by the plant's roots). These microorganisms can break down and metabolize pollutants.

3.6.2 Microbial Activity

The root exudates attract specific types of bacteria and fungi that have the ability to degrade or transform pollutants. These microorganisms may produce enzymes that can break down complex organic compounds or facilitate the immobilization of heavy metals.

3.6.3 Uptake and Translocation

Some plants, known as hyperaccumulators, can take up pollutants from the soil and transport them into their above-ground parts, such as leaves or stems. This process is called phytoextraction. Once the contaminants are in the above-ground parts, they can be harvested and removed from the site, reducing soil pollution.

Rhizodegradation is particularly useful for contaminants that are difficult to remove using traditional remediation methods. It's a sustainable and environmentally friendly approach to dealing with pollution, as it can often be less costly and less disruptive than more aggressive remediation techniques. However, its effectiveness depends on factors such as the type of plant species, the type of contaminants, and the environmental conditions of the site.

4 Bioenergy in the Phytoremediation Field

Biomass represents the organic matter of the plants and produces bioenergy that leads to the different forms of biofuel. It is a potential substitute in many industries, domestic and financial sectors in both developed and developing countries [79]. Bioenergy plants can adapt to the polluted environment with less effort and produce higher biomass that gives high bioenergy. Some promising species such as *Miscanthus sp, Helianthus annuus, Arundo donax, Jatropha curcas, Populus sp., Salix alba, and Ricinus communis* produce high amounts of bioenergy [6]. The toxicity of the shoot is generally below the standard toxicity level in many bioenergy plants [5, 44].

Thus, shoot biomass is considered for bioenergy synthesis in the post-harvesting period while some plants (shoot) absorb above the standard level which is tricky to handle. In such cases, HMs should recover before the bioenergy extraction [29, 40]. Bioenergy extraction from biomass takes place through the thermo-chemical process (gasification, catalytic liquefaction, and pyrolysis) or direct combustion process and biological process (anaerobic/fermentation) [65, 67]. There are many methods available such as electro-kinetic treatment, chemical oxidation or reduction, precipitation, ion exchange, incineration, and excavation reduce the pollutant load but alter the soil properties and sometimes affect the natural habitat [64].

Phytoremediation is based on green technology where plants sequestrate metals from the contaminated media, bio-accumulate them into their roots, and shoot part in a significant way and help to develop the greenery around. In this process, plants may reduce the toxicity of the HMs or stabilize the HMs [20, 26, 48].

5 Bioenergy Crops

Bioenergy crop production enhanced the economic value of phytoremediation which claims less portion of the land and does not affect food production [25]. *Jatropha curcas* is a biodiesel plant that has gained attention in recent years due to its potential feedstock and is widely cultivated in tropical and subtropical regions. It has a high-stress tolerance index as it can survive in harsh climatic conditions, produce a good amount of biogas and biodiesel, remediate contaminated soil widely, improve soil quality, and reduce soil erosion. Some studies state that *J. curcas* sequestrates metals such as Fr, Cr, Pb, Mn, Al, As, Cd, and Zn actively [46, 80]. *Salix alba* is another known bioenergy plant populated for biomass production and remediation potential. It has a profound root-and-shoot system that helps to accumulate metals in a wider way [39]. Adding synthetic chelators such as EDTA, citric acid and introducing endophyte and rhizobacteria is an important strategy adopted by many as it helps to grow plants and accelerate the uptake of toxins in a significant way [4].

6 Benefits of Using Bioenergy Plants

Adopting bioenergy plants in the phytoremediation field gives multiple benefits such as achieving a phytoremediation target, increasing the economic value in terms of bioenergy, bringing unutilized land in a utilization form that prevents hampering food production, and giving a layout for a sustainable approach in low-coast [33]. Bioenergy crops in bioremediation fields give an opportunity to move a step ahead toward renewable energy. It becomes a crucial sector as it remediates and produces a gain amount of bioenergy at the same time [64]. It offers many other advances such as carbon sequestration, biodiversity enrichment, improving water and soil quality, and reducing salinity and soil erosion [19]. Some energy crops such as Ricinus, Miscanthus, Populus, and Jatropha used in phytoremediation can add on profits for carbon credits [45].

7 Conclusions

The phytoremediation potential of bioenergy plants for heavy metals is a highly supportive integrated approach method for environmental management. It offers a sustainable and cost-effective solution, and these plants remove heavy metals and serve as a renewable energy source. Overall, integrating bioenergy plants into environmental management strategies addresses the challenges of heavy metal pollution. Moreover, the by-products of bioenergy production, such as biomass and biochar, can be utilized as valuable resources for other industries, further enhancing the economic benefits. Adopting phytoremediation with bioenergy plants presents a sustainable and multifaceted solution that addresses environmental and economic challenges associated with heavy metal pollution.

Acknowledgements Authors are thankful to the Department of Environmental Science, GITAM School of Science, GITAM (Deemed to be University), Visakhapatnam for encouragement and support.

References

1. Adki VS, Jadhav JP, Bapat VA (2013) Nopalea cochenillifera, a potential chromium(VI) hyperaccumulator plant. Environ Sci Pollut Res 20(2):1173–1180
2. Andrade G, Linderman RG, Bethlenfalvay GJ (1998) Bacterial associations with the mycorrhizosphere and hyphosphere of the arbuscular mycorrhizal fungus Glomus mosseae. Plant Soil 202:79–87
3. Araújo HW, Andrade RF, Montero-Rodríguez D, Rubio-Ribeaux D, Alves da Silva CA, Campos-Takaki GM (2019) Sustainable biosurfactant produced by Serratia marcescens UCP 1549 and its suitability for agricultural and marine bioremediation applications. Microb Cell Fact 18(1):1–13
4. Asad SA, Farooq M, Afzal A, West H (2019) Integrated phytobial heavy metal remediation strategies for a sustainable clean environment-a review. Chemosphere 217:925–941
5. Bauddh K, Singh K, Singh RP (2015) Ricinus communis L.: a value added crop for remediation of cadmium contaminated soil. Bull Environ Contam Toxicol 96:265–269. https://doi.org/10.1007/s00128-015-1669-3
6. Bauddh K, Bhaskar SB, Korstad J (2017) Phytoremediation potential of bioenergy plants. Springer, Singapore. https://doi.org/10.1007/978-981-10-3084-0
7. Bauddh K, Singh K, Singh B, Singh RP (2015) Ricinus communis: a robust plant for bio-energy and phytoremediation of toxic metals from contaminated soil. Ecol Eng 84:640–652
8. Bolan NS, Park JH, Robinson B, Naidu R, Huh KY (2011) Phytostabilization: a green approach to contaminant. Adv Agron 112:145–204. https://doi.org/10.1016/b978-0-12-385538-1.00004-4
9. Buendı´a-Gonza´lez L, Orozco-Villafuerte J, Cruz-Sosa F, Barrera-Dı´az CE, Vernon-Carter EJ (2010) Prosopis laevigata a potential chromium(VI) and cadmium(II) hyperaccumulator desert plant. Bioresour Technol 101(15):5862–5867
10. Chandra R, Cho W, Kang H (2016) Phytoextraction potential of four poplar hybrids under greenhouse conditions. For Sci Technol 12(4):199–206
11. Clemente R, Pardo T, Madejón P, Madejón E, Bernal MP (2015) Food byproducts as amendments in trace elements contaminated soils. Food Res Int 73:176–189
12. Dixit R, Wasiullah X, Malaviya D, Pandiyan K, Singh UB, Sahu A, Shukla R, Singh BP, Rai JP, Sharma PK, Lade H (2015) Bioremediation of heavy metals from soil and aquatic environment: an overview of principles and criteria of fundamental processes. Sustainability 7(2):2189–2212
13. El Mehdawi AF, Reynolds RJB, Prins CN et al (2014) Analysis of selenium accumulation, speciation and tolerance of potential selenium hyperaccumulator Symphyotrichum ericoides. Physiol Plant 152:70–83. https://doi.org/10.1111/ppl.12149

14. Fehlauer T, Collin B, Angeletti B, Negahi MM, Dentant C, Chaurand P, Lallemand C, Levard C, Rose J (2022) Multiscale imaging on Saxifraga paniculata provides new insights into yttrium uptake by plants. Sci Rep 12(1):18268
15. Fellet G, Poščić F, Casolo V, Marchiol L (2012) Metallophytes and thallium hyperaccumulation at the former Raibl lead/zinc mining site (Julian Alps, Italy). Plant Biosystems Int J Dealing Aspects Plant Biol 146(4):1023–1036
16. Gervais-Bergeron B, Paul AL, Chagnon PL, Baker AJ, van der Ent A, Faucon MP, Quintela-Sabarís C, Labrecque M (2023) Trace element hyperaccumulator plant traits: a call for trait data collection. Plant Soil 1–10
17. Ghaderian SM, Ravandi AAG (2012) Accumulation of copper and other heavy metals by plants growing on Sarcheshmeh copper mining area Iran. J Geochem Explor 123:25–32
18. Ghori Z, Iftikhar H, Bhatti MF, Sharma I, Kazi AG, Ahmad P (2016) Phytoextraction: the use of plants to remove heavy metals from soil. In: Plant metal interaction. Elsevier, pp 385–409
19. Gibbs HK, Johnston M, Foley JA, Holloway T, Monfreda C, Ramankutty N, Zaks D (2008) Carbon payback times for crop-based biofuel expansion in the tropics: the effects of changing yield and technology. Environ Res Lett 3(3):034001
20. Glick BR (2010) Using soil bacteria to facilitate phytoremediation. Biotechnol Adv 28:367–374. https://doi.org/10.1016/j.biotechadv.2010.02.001
21. Gul I, Manzoor M, Hashmi I, Bhatti MF, Kallerhoff J, Arshad M (2019) Plant uptake and leaching potential upon application of amendments in soils spiked with heavy metals (Cd and Pb). J Environ Manage 249:109408
22. Ha NTH, Sakakibara M, Sano S, Nhuan MT (2011) Uptake of metals and metalloids by plants growing in a lead–zinc mine area, Northern Vietnam. J Hazard Mater 186(2–3):1384–1391
23. Hernández A, Loera N, Contreras M, Fischer L, Sánchez D (2019) Comparison between Lactuca sativa L. and Lolium perenne: Phytoextraction capacity of Ni, Fe, and Co from Galvanoplastic industry. In: Energy technology 2019: carbon dioxide management and other technologies. Springer International Publishing, pp 137–147
24. Huang X, Luo D, Chen X, Wei L, Liu Y, Wu Q, Xiao T, Mai X, Liu G, Liu L (2019) Insights into heavy metals leakage in chelator-induced phytoextraction of Pb-and Tl-contaminated soil. Int J Environ Res Public Health 16(8):1328
25. Hunce SY, Clemente R, Bernal MP (2019) Energy production potential of phytoremediation plant biomass: helianthus annuus and Silybum marianum. Ind Crops Prod 135:206–216
26. Ibañez S, Talano M, Ontañon O, Suman J, Medina MI, Macek T, Agostini E (2016) Transgenic plants and hairy roots: exploiting the potential of plant species to remediate contaminants. New Biotechnol 33(5):625–635
27. Janeeshma E, Puthur JT (2020) Direct and indirect influence of arbuscular mycorrhizae on enhancing metal tolerance of plants. Arch Microbiol 202:1–16
28. Kafle A, Timilsina A, Gautam A, Adhikari K, Bhattarai A, Aryal N (2022) Phytoremediation: mechanisms, plant selection and enhancement by natural and synthetic agents. Environ Adv 8:100203
29. Kidd PS, Bani A, Benizri E, Gonnelli C, Hazotte C, Kisser J, Konstantinou M, Kuppens T, Kyrkas D, Laubie B, Malina R (2018) Developing sustainable agromining systems in agricultural ultramafic soils for nickel recovery. Front Environ Sci 6:44
30. Kupper H, Götz B, Mijovilovich A, Kupper FC, Meyer-Klaucke W (2009) Copper complexation and toxicity in higher plants. I. Characterization of copper accumulation, speciation, and toxicity in Crassula helmsii as a new copper accumulator. Plant Physiol 151(2):702–14
31. Kupper H, Kochian LV (2010) Transcriptional regulation of metal transport genes and mineral nutrition during acclimatization to cadmium and zinc in the Cd/Zn hyperaccumulator, Thlaspi caerulescens (Ganges population). New Phytol 185(1):114–129
32. Lemus R, Lal R (2005) Bioenergy crops and carbon sequestration. Crit Rev Plant Sci 24:1–21. https://doi.org/10.1080/07352680590910393
33. Licht LA, Isebrands JG (2005) Linking phytoremediated pollutant removal to biomass economic opportunities. Biomass Bioenerg 28(2):203–218

34. Likar M, Pongrac P, Vogel-Mikusˇ K, Regvar M (2010) Molecular diversity and metal accumulation of different Thlaspi praecox populations from Slovenia. Plant Soil 330(1–2):195–205
35. Liu J, Duan C, Zhang X, Zhu Y, Lu X (2011) Potential of Leersia hexandra Swartz for phytoextraction of Cr from soil. J Hazard Mater 188(1–3):85–91
36. Liu J, Shang W, Zhang X, Zhu Y, Yu K (2014) Mn accumulation and tolerance in Celosia argentea Linn.: a new Mn-hyperaccumulating plant species. J Hazard Mater 267:136–141
37. Liu X, Peng K, Wang A, Lian C, Shen Z (2010) Cadmium accumulation and distribution in populations of Phytolacca americana L. and the role of transpiration. Chemosphere 78(9):1136–1141
38. Mesjasz-Przybylowicz J, Przybylowicz W, Barnabas A, Van Der Ent A (2016) Extreme nickel hyperaccumulation in the vascular tracts of the tree Phyllanthus balgooyi from Borneo. New Phytol 209(4):1513–1526
39. Mleczek M, Rutkowski P, Rissmann I, Kaczmarek Z, Golinski P, Szentner K, Strażyńska K, Stachowiak A (2010) Biomass productivity and phytoremediation potential of Salix alba and Salix viminalis. Biomass Bioenerg 34(9):1410–1418
40. Mohanty M (2016) Post-harvest management of phytoremediation technology. J Environ Anal Toxicol 6:398. https://doi.org/10.4172/2161-0525.1000398
41. Mongkhonsin B, Nakbanpote W, Nakai I, Hokura A, Jearanaikoon N (2011) Distribution and speciation of chromium accumulated in Gynura pseudochina (L.) DC. Environ Exp Bot 74:56–64
42. Muller K, Daus B, Mattusch J, Vetterlein D, Merbach I et al (2013) Impact of arsenic on uptake and bioaccumulation of antimony by arsenic hyperaccumulator Pteris vittata. Environ Pollut 174:128–133
43. Oh K, Li T, Cheng HY, Hu XF, Lin Q, Xie YH (2013) A primary study on assessment of phytoremediation potential of biofuel crops in heavy metal contaminated soil. Appl Mech Mater 295:1135–1138
44. Palanivel TM, Pracejus B, Victor R (2020) Phytoremediation potential of castor (Ricinus communis L.) in the soils of the abandoned copper mine in Northern Oman: implications for arid regions. Environ Sci Pollut Res 27:17359–17369
45. Pandey VC, Bajpai O, Singh N (2016) Energy crops in sustainable phytoremediation. Renew Sustain Energy Rev 54:58–73
46. Pandey VC, Singh K, Singh JS, Kumar A, Singh B, Singh RP (2012) Jatropha curcas: a potential biofuel plant for sustainable environmental development. Renew Sustain Energy Rev 16(5):2870–2883
47. Pfister JA, Davis TZ, Hall JO (2013) Effect of selenium concentration on feed preferences by cattle and sheep. J Anim Sci 91:5970–5980. https://doi.org/10.2527/jas.2013-6595
48. Pilon-Smits E (2005) Phytoremediation. Annu Rev Plant Biol 56:15–39
49. Radziemska M, Vaverková MD, Baryła A (2017) Phytostabilization—management strategy for stabilizing trace elements in contaminated soils. Int J Environ Res Public Health 14(9):958
50. Rafati M, Khorasami N, Moattar F, Shivany A, Moraghebi F, Hosseinzadeh S (2011) Phytoremediation potential of Populus alba and Mones alba from cadmium, chromium and nickel adsorption from polluted soil. Int J Environ Res 5:961–970
51. Rajput V, Minkina T, Semenkov I, Klink G, Tarigholizadeh S, Sushkova S (2021) Phylogenetic analysis of hyperaccumulator plant species for heavy metals and polycyclic aromatic hydrocarbons. Environ Geochem Health 43:1629–1654
52. Rizwan M, Ali S, ur Rehman MZ, Rinklebe J, Tsang DC, Bashir A, Maqbool A, Tack FMG, Ok YS (2018) Cadmium phytoremediation potential of Brassica crop species: a review. Sci Total Environ 631:1175–1191
53. Rubio M, Mera MF, Cazón S, Rubio ME, Pérez CA (2020) SR micro-XRF to study Pb diffusion using a one-dimensional geometric model in leaves of Brassica napus for phytoremediation. Radiat Phys Chem 167:108291
54. Sabir M, Waraich EA, Hakeem KR, Öztürk M, Ahmad HR, Shahid M (2014) Phytoremediation: mechanisms and adaptations. Soil Rem Plants Prospects Challenges 85:85–105

55. Santiago-Cruz MA, Villagrán-Vargas E, Velázquez-Rodríguez AS, Vernon-Carter EJ, Cruz-Sosa F, Orozco-Villafuerte J, Buendía-González L (2014) Exploring the Cr (VI) phytoremediation potential of Cosmos bipinnatus. Water Air Soil Pollut 225:1–8
56. Sarma H (2011) Metal hyperaccumulation in plants: a review focusing on phytoremediation technology. J Environ Sci Technol 4(2):118–138
57. Sarwar N, Imran M, Shaheen MR, Ishaque W, Kamran MA, Matloob A, Rehim A, Hussain S (2017) Phytoremediation strategies for soils contaminated with heavy metals: modifications and future perspectives. Chemosphere 171:710–721
58. Schiavon M, Pilon M, Malagoli M, Pilon-Smits EA (2015) Exploring the importance of sulfate transporters and ATP sulphurylases for selenium hyperaccumulation—a comparison of Stanleya pinnata and Brassica juncea (Brassicaceae). Front Plant Sci 6:2
59. Senthilkumar K, Kumar MN (2020) Generation of bioenergy from industrial waste using microbial fuel cell technology for the sustainable future. In: Refining biomass residues for sustainable energy and bioproducts. Academic Press, pp 183–193
60. Shackira AM, Puthur JT (2017) Enhanced phytostabilization of cadmium by a halophyte—Acanthus ilicifolius L. Int J Phytorem 19(4):319–326
61. Shah V, Daverey A (2020) Phytoremediation: a multidisciplinary approach to clean up heavy metal contaminated soil. Environ Technol Innov 18:100774
62. Shrivastava M, Khandelwal A, Srivastava S (2019) Heavy metal hyperaccumulator plants: the resource to understand the extreme adaptations of plants towards heavy metals. Plant-metal interactions, pp 79–97
63. Singh A, Fulekar MH (2012) Phytoremediation of heavy metals by Brassica juncea in aquatic and terrestrial environment. The plant family Brassicaceae: contribution towards phytoremediation, pp 153–169
64. Singh G, Singh P, Guldhe A, Stenström TA, Bux F, Kumari S (2017) Biotechnological intervention to enhance the potential ability of bioenergy plants for phytoremediation. Phytoremediation potential of bioenergy plants, pp 387–408
65. Singh NB, Kumar A, Rai S (2014) Potential production of bioenergy from biomass in an Indian perspective. Renew Sustain Energy Rev 39:65–78
66. Spaczynski M, Seta-Koselska A, Patrzylas P, Betlej AK, Skorzynska-Polit E (2012) Phytodegrdation and biodegradation in rhizosphere as efficient methods of reclamation of soil, contaminated by organic chemicals (a review). Acta Agrophysica 19(1)
67. Srivastava RK (2019) Bio-energy production by contribution of effective and suitable microbial system. Mater Sci Energy Technol 2(2):308–318
68. Stein RJ, Höreth S, de Melo JRF, Syllwasschy L, Lee G, Garbin ML, Clemens S, Krämer U (2017) Relationships between soil and leaf mineral composition are element-specific, environment-dependent and geographically structured in the emerging model Arabidopsis halleri. New Phytol 213(3):1274–1286
69. Swapna KS, Shackira AM, Abdussalam AK, Puthur JT (2014) Accumulation pattern of heavy metals in Chromolaena odorata (L.) King & Robins. grown in nutrient solution and soil. J Stress Physiol Biochem 10(2):297–314
70. Szczygłowska M, Piekarska A, Konieczka P, Namieśnik J (2011) Use of brassica plants in the phytoremediation and biofumigation processes. Int J Mol Sci 12(11):7760–7771
71. Taiz L, Zeiger E, Møller IM, Murphy A (2015) Plant physiology and development, 6th edn. Sinauer Associates Incorporated
72. Teptina AY, Paukov AG (2015) Nickel accumulation by species of Alyssum and Noccaea (Brassicaceae) from ultramafic soils in the Urals Russia. Aust J Bot 63(2):78–84
73. Thakur S, Singh L, Wahid ZA, Siddiqui MF, Atnaw SM, Din MFM (2016) Plant-driven removal of heavy metals from soil: uptake, translocation, tolerance mechanism, challenges, and future perspectives. Environ Monit Assess 188:1–11
74. Tufail MA, Iltaf J, Zaheer T, Tariq L, Amir MB, Fatima R, Asbat A, Kabeer T, Fahad M, Naeem H, Shoukat U (2022) Recent advances in bioremediation of heavy metals and persistent organic pollutants: a review. Sci Total Environ 850:157961

75. van der Ent A, Erskine P, Sumail S (2015) Ecology of nickel hyperaccumulator plants from ultramafic soils in Sabah (Malaysia). Chemoecology 25:243–259
76. van Der Ent A, Ocenar A, Tisserand R, Sugau JB, Echevarria G, Erskine PD (2019) Herbarium X-ray fluorescence screening for nickel, cobalt and manganese hyperaccumulator plants in the flora of Sabah (Malaysia, Borneo Island). J Geochem Explor 202:49–58
77. Wang L, Ji B, Hu Y, Liu R, Sun W (2017) A review on in situ phytoremediation of mine tailings. Chemosphere 184:594–600
78. Wang SL, Liao WB, Yu FQ, Liao B, Shu WS (2009) Hyperaccumulation of lead, zinc, and cadmium in plants growing on a lead/zinc outcrop in Yunnan Province, China. Environ Geol 58:471–476
79. World Energy Council. 2016. World energy resources: Report. https://www.worldenergy.org/
80. Yadav SK, Juwarkar AA, Kumar GP, Thawale PR, Singh SK, Chakrabarti T (2009) Bioaccumulation and phyto-translocation of arsenic, chromium and zinc by Jatropha curcas L.: impact of dairy sludge and biofertilizer. Bioresour Technol 100(20):4616–4622
81. Yang Y, Ge Y, Tu P, Zeng H, Zhou X, Zou D, Wang K, Zeng Q (2019) Phytoextraction of Cd from a contaminated soil by tobacco and safe use of its metal-enriched biomass. J Hazard Mater 363:385–393

Recent Advances Towards Improved Microbial Bioremediation of Heavy Metal Pollution

Farzaneh Dianatdar and Zahra Etemadifar

Abstract Heavy metals refer to metals with a molecular weight of 5 times more than water. Heavy metals enter the environment through mining the metal, metal casting, and smelting industries, and they return to humans through food, meat, vegetables, and contaminated drinking water. Heavy metals are removed by physical remediation, chemical remediation, and bioremediation methods, also bioremediation is cost-effective, has easy availability, and eco-friendliness. The types of mechanism microbial bioremediation of metal are biosorption, bioaccumulation, bioleaching, biomineralization, and biotransformation. Microorganisms provide resistance to heavy metals through various mechanisms, including extracellular or intracellular sequestration, extracellular barriers, reduction, and active transport (efflux). Heavy metals affect the diversity and number of microbes in the environment, which has been studied by metagenomic methods. The consortium produces stromatolites, microbial mats, and biofilms that are resistant and bioremedicated to various metals. Microorganisms also produce nanoparticles by bioremediation of metals. Nanocomposites increase the activities of microbial biostimulation, bioaccumulation, and biotransformation.

Keywords Heavy metal · Microbial bioremediation · Microbial consortia · Biotechnological intervention

1 Introduction

Heavy metals refer to elements with high density and molecular weight (more than 5 times water) amounting to more than 5 cm^{-3}, which are highly toxic in small amounts. Metalloids are compounds between metals and non-metals, the metalloid elements include boron, germanium, silicon, antimony, arsenic, tellurium, and polonium. Arsenic and antimony are toxic metalloids [1]. Today, heavy metals are used in

F. Dianatdar · Z. Etemadifar (✉)
Department of Cell and Molecular Biology and Microbiology, Faculty of Biological Science and Technology, University of Isfahan, Isfahan, Iran
e-mail: z.etemadifar@sci.ui.ac.ir

N. Kumar (ed.), *Heavy Metal Remediation*, Earth and Environmental Sciences Library,
https://doi.org/10.1007/978-3-031-53688-5_6

various industries, agriculture, and medicine, their concentration in the environment has increased and it is a danger to human health. The higher concentration of heavy metals is next to mining the metal, metal casting, and smelting industries [2]. Metals such as Ca, Na, Mg, K, Cu, Fe, Zn, and Cr in low concentrations are necessary for physiological activities for living organisms, and the high concentration of these metals in the environment is toxic [3]. Cd, Hg, Ni, As, and Pb in any amount in the environment are toxic to living organisms [4].

Heavy metals are accumulating in the environment and reach humans through food, meat, vegetables, and contaminated drinking water [2]. The effects of heavy metals on humans depend on the dose, route of exposure, type of heavy metal, as well as genetics, sex, age, and nutritional status of people [5]. Heavy metals destroy cellular functions such as growth, proliferation, differentiation, damage repair processes, and apoptosis, because heavy metals in the body produce ROS, weaken antioxidant defenses, enzyme inactivation, and oxidative stress. Some toxic metals (chromium, cadmium, and arsenic) cause genomic instability by inducing oxidative stress and DNA damage and are known as carcinogens. Toxic metals harm the liver, kidney, brain, heart, and some other body organs [6].

Heavy metals are not decomposed and have toxic effects on the environment, they have been determined by the laws for the entry of fewer metals into the environment and the permissible amounts of metals in drinking water [7]. According to the recommendation of FAO, the permissible amount of metals in Irrigation water includes Al (0.50 mg/l), As (0.10 mg/l), Be (0.10 mg/l), Cd (0.01 mg/l), Co (0.05 mg/l), Cr (0.10 mg/l), Ni (0.20 mg/l), Pd (0.50 mg/l) and Cu (0.20 mg/l) [8].

There are various methods such as coagulation, ion exchange, membrane filtration, chemical precipitation, adsorption, electrochemical treatment, flocculation, and bioremediation to comply with the rules to remove metals in the environment and reach desired standard amounts of heavy metals in the environment [9], but bioremediation is more popular due to its convenience, low cost, environmental friendliness, and absence of side products [10].

The review paper aimed to study different mechanisms of metal removal and resistance to metals in microorganisms. Microorganism cells resist metal by producing biofilm, biosurfactant, slime layer, siderophores, and catecholate in the quorum sensing mechanism (QS). In this, we investigate the effect of heavy metals on microbial communities through metagenomics, morphological, nutritional methods, and modern methods (G+C% DNA, DNA-DNA hybridization, DNA microarray and reverse genome probe, denaturing gradient gel electrophoresis (DGGE), phospholipid fatty acid method (PLFA)). Genetic engineering improves the ability of bioremediated metal. Microorganisms also produce nanoparticles by bioremediation of metals.

Fig. 1 Bioremediation of metal in environment

2 Biological Remediation of Heavy Metals

Heavy metals are removed by physical remediation, chemical remediation, and bioremediation methods. Bioremediation of heavy metals is an effective method to detoxify, degrade, mineralize, neutralize, and transform toxic forms of metals to low toxicity in the environment with biological agents such as bacteria, mold, plants, yeast, and fungi (Fig. 1; Table 1). Bioremediation of heavy metals is a suitable method due to its cost-effectiveness, easy availability, and eco-friendliness. Bacteria and fungi have a potential role in the bioremediation of heavy metals in water resources, soil, and food [11, 12].

3 Type of Mechanism Microbial Bioremediation of Metal

Mechanisms of metal bioremediation include biosorption, bioaccumulation, bioleaching, biomineralization, and biotransformation (Fig. 2) [35, 36]. Biosorption is the reversible binding of water-soluble metal ions (such as Cr, Co, Ni, and Zn ions) on the amine, carboxyl, hydroxyl, sulfhydryl group and phosphate in the cell surface, which is independent of cell metabolism [37]. Mwandira et al., 2020 were biosorbed 97% Pb(II) (initial concentration of 50 mg/L) and 54% Zn(II) (initial concentration of 2 mg/L) by *Oceanobacillus profundus* KBZ 3–2 [38]. Oyewole et al. [39] isolated bacterial (*Pseudomonas, Bacillus, Micrococcus, Escherichia, Streptococcus, Enterobacter*, and *Staphylococcus*) and fungi (*Aspergillus niger, Penicillium notatum,* and *Aspergillus flavus*), that biosorption 89.67% chromium and 90.89% copper by *Pseudomonas aeruginosa*, and *P. notatum* biosorption 77.67% 10 ppm cadmium. *Aspergillus niger* biosorption 81.07% nickel.

Table 1 Examples of bioremediation of metals by bacteria, fungi, and algae

Biological agents	Microorganism	Amount of bioremediation of heavy metals	Reference
Bacteria	*Bacillus thuringiensis* strain "Simi" (KF 916,618.1), and *Bacillus subtilis* strainPSB (KF 279,045.1)	54% Zn	Kumar et al. [13]
	Lactobacillus plantarum MF042018	100% Ni^{2+}, Cr^{2+}, Cd^{2+} and Pb^{2+}	Ameen et al. [14]
	Brevundimonas vesicularis	Ni(II) and Cu(II) 91.6 mg/g, and 129.4 mg/g, respectively	Singh et al. [15]
	Acinetobacter sp	93.7, 55.4, and 68.94% removal of initial 30 mg L^{-1} Cr(VI), 246 mg $^{L-1}$ total Cr, and 51 mg L^{-1} Ni, respectively	Bhattacharya et al. [16]
	Pseudomonas fluorescens SM1	More than 75% Cu^{2+}, Cd^{2+}, Ni^{2+}, Pb^{2+}	Wasi et al. [17]
	Pseudomonas putida, Bacillus licheniformis	*Pseudomonas putida* removal (Co-71%, Ni-51%, Mn-45%, V-83%, Pb-96%, Ti-100% and Cu-49%) and *Bacillus licheniformis* (Al-23% and Zn-53%) and Peranema sp. (Cd-42%)	Kamika et al. [18]
	Rhodopseudomonas palustris TN110, *Rubrivivax gelatinosus* TN414	84% and 55% Cd and Zn, respectively	Sakpirom et al. [19]
	Streptomyces rochei ANH	Ni^{2+} (100%) $\geq Cu^{2+} \geq Mn^{2+} \geq Fe^{2+} > Pb^{2+}$ (95%) $\geq Cd^{2+} > Cr^{6+}$ (86%)	Hamdan et al. [20]
	Bacillus sp. TZ5	48.49% Cd	Wei et al. [21]
	Bacillus megaterium	47.66% Pb, 21.68% Ni, 11.12% Cd	Njoku et al. [22]
Fungi	*Aspergillus flavus* (FS4) and *Aspergillus fumigatus* (FS6)	74% Cr(VI) and cadmium Cd(II)	Talukdar et al. [23]
	Fusarium falciforme (MW898136) and *Fusarium* sp. (MW898137)	82.4% and 76.5% cadmium, respectively	Saad et al. [24]

(continued)

Table 1 (continued)

Biological agents	Microorganism	Amount of bioremediation of heavy metals	Reference
	Fusarium oxysporum NCBT-156 strain	90% for 100 and 200 ppm concentration, up to 80% for 300 and 400 ppm, and up to 65% for 500 ppm to 1000 ppm concentrations	Amatussalam et al. [25]
	Aspergillus lentulus, Aspergillus terreus	100% Cr^{6+} and 81.60% Cu^{2+}	Mishra et al. [26]
	Penicillium coffeae	66.8% arsenate	Bhargavi et al. [27]
	Fomitopsis palustris, Coniophora puteana, and *Laetiporus sulphureus*	100%, 18%, and 85% arsenic, respectively	Kartal et al. [28]
	Rhizopus stolonifer	43.90% Pb, 15.04% Ni, and 8.02% Cd	Njoku et al. [22]
	Aspergillus niger	70% of 250–500 ppm Cr	Srivastava et al. [29]
Algae	*Ulva fasciata* and *Sargassum lacerifolium*	Cd, Cu, Cr, and Ni of 7.45, 1.18, 3.13, and 26.6 mg kg^{-1}, respectively	Ahmed et al. [30]
	Chlorella vulgaris, Chlorella salina	13.61–100% Zn, Cu, Mn, Ni, Co, Fe and Cr	Sheekh et al. [31]
	Porphyra leucosticta	31.45 mg/g Cd(II) and 36.63 mg/g Pb(II)	Ye et al. [32]
	Phormidium sp., Monoraphidium sp., Uronema sp., Chlorella sp. 1, and *Anacystis sp.*	15.93–1.63 Pb(II) (mg/L)	Lakmali et al. [33]
	Fucus vesiculosus	92.76% of 200 mg/L copper and 50.27% of 60 mg/L methyl orange	El-Naggar et al. [34]

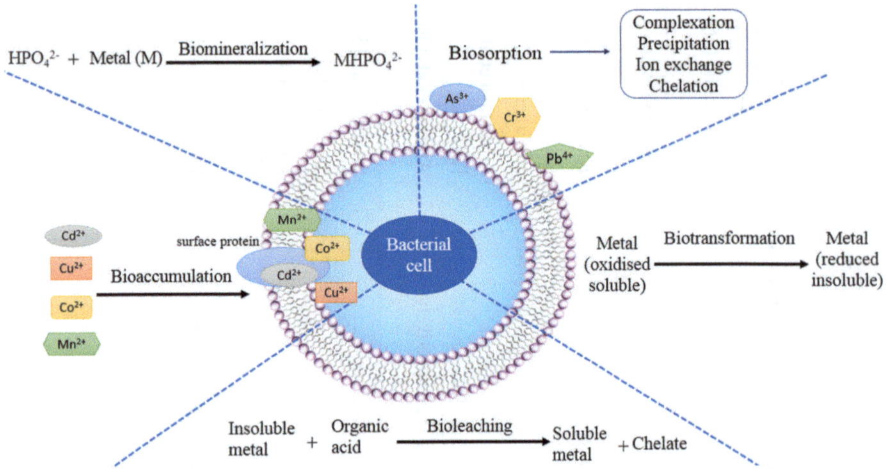

Fig. 2 Mechanisms of microbial bioremediation of metal included biosorption, bioaccumulation, bioleaching, biomineralization, and biotransformation

Bioaccumulation means the absorption of metals with proteins into cells, These metals are used in biological processes such as signaling, catalysis, and stabilization of molecules [40]. Özdemir et al. [41] bioaccumulated Cd^{2+}, Cu^{2+}, Co^{2+}, and Mn^{2+} ions by thermophilic *Geobacillus thermantarcticus* (36.07–12.39 mg/g) and *Anoxybacillus amylolyticus* (28,566–327.3 μg/g dry weight).

Bioleaching means dissolving insoluble metals from solid phases by microorganisms, This method is often used in the extraction of metals from ores [42]. Patel et al. [43] isolated *Aspergillus fumigatus* A2DS from mining industry wastewater, this strain leached 57% copper and 32% nickel.

Biologically induced mineralization means the formation, deposition, and nucleation of minerals through biological processes, in addition, forced biomineralization occurs when extremophiles are deposited and biomineralized in high concentrations of metals (such as pools and mine runoff) [44]. Zhao et al. [45] isolated 4 strains CZW-2, CZW-5, CZW-9, and CZW-12 including *Cupriavidus* sp and *Bacillus* sp, strains biomineralization 53.30%, 27.78%, 42.54% and 53.80% of Cd respectively.

Biotransformation includes precipitation and dissolution of heavy metals by oxidation–reduction [11]. Vogel et al. [46] biotransformed 1–5 mM selenite to selenium particles by *Azospirillum brasilense* (plant growth-promoting bacterium), during this process, selenium nanoparticles are produced and the soil is also removed from this toxic metal.

3.1 Mechanisms of Resistance Against Heavy Metals

Microorganisms provide resistance to heavy metals through various mechanisms, including extracellular or intracellular sequestration, extracellular barriers, reduction, and active transport (efflux) (Table 2) [47].

The external barriers such as the cell wall, membrane, and capsule prevent the entry of heavy metals into the cell, but some bacteria absorb metals due to ionizable groups such as carboxyl, amino, phosphate, and hydroxyl in the cell wall, capsule, extracellular biopolymers, and polysaccharides. Early, metals are quickly and non-specifically absorbed on the surface of the cell, and then the metals slowly enter the cell. This type of absorption is passive because it is also carried out by dead cells [79].

The greatest resistance to metals of microorganisms occurs during the process of active transfer or efflux of metals in the cell. The genes of the metal efflux system are located on the plasmid and chromosome. The efflux of heavy metals is carried out through the mechanisms of magnesium transport system, sulfate transport system, CzcCB2A efflux complex, Proteins families efflux systems ((cation diffusion facilitator), P-type ATPases, CDF and RND (resistance, nodulation, cell division)), and ars system [79, 80].

The Intracellular sequestration method connects metal ions by substances of cytoplasm, In this method two types of peptides, metallothioneins, and phytochelatins, which have a high cysteine amino acid in the cytoplasm (such as phytochelatin-like molecule synthesis in some marine gamma proteobacteria [63], glutathione synthesis by *Rhizobium leguminosarum* cells [64], synthesizing metallothionein by two genes, smtA, and smtB in *Synechococcus* sp. [62]) bind metal ions through sulfhydryl groups [81].

The accumulation and binding of metal ions outside the cell are called extracellular sequestration, For example, copper complex formation in *Pseudomonas syringae* due to the expression of CopA, CopB, and CopC, and blue colony formation [65], *Pseudomonas stutzeri* formed silver sulfide complex in periplasm [82], and binding of SilE periplasmic protein to silver in *Salmonella* sp. [67].

Some bacteria use metals and metallonidines as electron acceptors or donors, because during anaerobic respiration they give electrons to metals such as mercury, chromate, vanadate, molybdate, and chromium, then heavy metals are converted into less dangerous forms [83]. The mer-operon expresses two proteins, MerT and MerA. MerT transport protein enters Hg(II) into the cell, and intracellular MerA reductase protein reductase converts it into Hg° with less toxicity [84]. The first metal resistance genes were observed in the plasmid. Ars operon on the chromosome of *E. coli*, *P. aeruginosa*, and *Bacillus subtilis* is a plasmid-like metal resistance system [85]. The czc operon on *R. metallidurans* CH34 is located on plasmid PMOL30 and causes resistance of bacteria to cadmium, zinc, and cobalt ions. The cop operon causes resistance to cadmium and copper in *E. hirae* and *Pseudomonas* sp. [86].

Table 2 Examples of the method of resistance to metals in microorganisms

Mechanisms of resistance	Microorganisms	Heavy metal	Reference
External barrier	*Pseudomona sputida*	80% Cd(II), Cu(II), Pb(II) and Zn(II)	[48]
	Brevibacterium strain HZM-1	0.1 mM Cu^{2+}, Zn^{2+}	[49]
	Bacillus	0.658 mg Hg(II)	[50]
	Marinobacter	182 nmol Cu^{2+} and 13 nmol Pb^{2+}	[51]
	Pseudomonas aeruginosa	480.54 mg Cu^{2+}	[52]
	Klebsiella and *Pseudomonas*	5–100 mg Cd	[53]
	Enterobacter cloacae	65% cadmium, 20% copper, and 8% cobalt at 100 mg/l heavy metal concentration	[54]
	Mutants of *Escherichia coli*	Hg, and Ag	[55]
	Sphingomonas paucimobilis R40	Cu^{2+}	[56]
Active transport (efflux)	*Ralstonia metallidurans*	cobalt, cadmium, nickel, zinc, and manganese ions	[57]
	Streptococcus mutans JH1005	Cu^{2+}	[58]
	Escherichia coli	As(V)	[59]
	Oryza sativa	Copper, zinc, and cadmium	[60]
	Staphylococcus aureus	Cd^{2+}	[61]
Intracellular sequestration	*Synechococcus* sp. strain PCC 7942	Cadmium and zinc ions	[62]
	P. putida	Copper, cadmium, and zinc ions	[63]
	Rhizobium leguminosarum	Cadmium ions	[64]
Extracellular sequestration	*Pseudomonas syringae*	Cu^{2+}	[65]
	Pseudomonas stutzeri AG259	Silver ion	[66]
	Salmonella sp., *Escherichia coli* K-12, *E. coli* O157:H7	Ag^{+}	[67]
Reduction	*Bacillus cereus* RC607	Hg^{2+}/Hg^{0}	[68]
	Shewanella oneidensis	V^{5+}/V^{4+}	[69]
	Geobacter sulfurreducens	Tc^{7+}/Tc^{4+}	[70]
	Thiobacillus ferrooxidans	Mo^{6+}/Mo^{5+}	[71]
	Geobacter metallireducens	Fe^{3+}/Fe^{2+}	[72]
	Strenotrophomonas sp.	Au^{3+}/Au^{0}	[73]
	Bacillus thermoamylovorans SKC1	Te^{4+}/Te^{0}	[74]

(continued)

Table 2 (continued)

Mechanisms of resistance	Microorganisms	Heavy metal	Reference
	Desulfomicrobium norvegicum	Cr^{6+}/Cr^{3+}	[75]
	Staphylococcus aureus	As^{5+}/As^{3+}	[76]
	Shewanella putrefaciens	Mn^{4+}/Mn^{2+}	[77]
	Ralstonia metallidurans CH34	Se^{4+}/Se^{0}	[78]

3.2 Resistance to Heavy Metals by Microbial Quorum Sensing

Microorganism cells resist stress (such as lack of food, metal, and antibiotics) by quorum sensing mechanism (QS). Autoinducer molecules are secreted during the QS and are effective on the same species of bacteria. Autoinducers molecules include acyl-homoserine lactone (AHL) molecules and peptides [87].

The slime layer is composed of exopolysaccharides (EPS) including carbohydrates, nucleic acids, fatty acids, and polysaccharides [88]. Heavy metals such as lead, cadmium, and uranium binned on negatively charged functional groups (succinyl, phosphate, hydroxyl, and uronic acids) EPS in *Staphylococcus aureus, Micrococcus luteus*, and *Azotobacter* sp, which is effective in immobilizing metals [89, 90].

Siderophores (such as hydroxamate and catecholate) are secondary metabolites and organic molecules with low molecular weight in some bacteria, which send iron from the environment containing low iron ions into the cell. Siderophore is secreted outside the cell to form a soluble complex with ferric ions and to be absorbed by the microorganisms [91]. Cr^{3+}, Al^{3+}, Ga^{3+}, and catecholate are very similar to iron ions size, For example, *Bacillus megaterium* formed Cu^{2+}, Ga^{3+}, Mn^{2+}, Ni^{2+}, and Al^{3+} with siderophore [92, 93]. Pyoverdine is a type of Siderophore in *Pseudomonas aeruginosa* PAOI, which is complex with Ag^{+}, Al^{3+}, Cd^{2+}, Co^{2+}, Cu^{2+}, Fe^{3+}, Ga^{3+}, Hg^{2+}, Mn^{2+}, Ni^{2+} or Zn^{2+} [94].

Catecholate may form complexes with metals other than iron, such as Mn, Co, Cr, B, and Al, then Catecholate of *E. coli* UTI89 binned Cu(II), during this reaction, toxic copper was produced and bacteria produce yersiniabactin and enterobactin to reduce toxicity [95]. The formation of the metal-siderophore complex changes the toxicity and absorption of metals. The molecular mass of the siderophore-metal complex is suitable for absorption by cell porins, and the formation of the siderophore-metal complex reduces the amount of metal dissolved in the environment as a result of metal toxicity. *P. aeruginosa* produced pyoverdine and pyochelin, which are resistant to metals [96].

3.3 Microbial Metagenomics of Heavy Metal

Metagenomics was first used by Handelsman et al. [97]. Metagenomics is a culture-independent method, and this quick method can analyze environmental samples of water and soil. The development of next-generation sequencing (NGS) has revolutionized the genetic and ioinformatics analysis of environmental microbial samples. Metagenomics has two subfields, sequence-based and function-based, which provide information regarding microbial metabolism, key enzymes, genes, diversity of microbial communities, and non-cultivable organisms [98]. A nucleotide database developed with in silico instruments has led to the identification of a new gene responsible for microbial biodegradation. Metagenomics studies provide a good insight into the relationship between environmental microbiota and environmental activities [99]. For example, by analyzing the metagenomics of Greenland ice, strong genes that bioremediation of lead and mercury were discovered [100]. The Metagenomics method studies all cultivable and non-cultivable microbes in biodegradation because DNA was extracted from different environmental niches. The metagenomics method reached a better understanding of the real microbial community, the roles of microbes in biodegradation, and the cooperation and relationships of microbes in a different environment, In other words, the metagenomics method discovered new ecosystems [101]. The metagenomics method has identified important microbes for bioremediation, including *Bacteroidetes*, *Chloroflexi*, *Proteobacteria*, *Firmicutes*, *Actinobacteria*, *Spirochetes*, *Acidobacteria*, and *Patesibacteria* [102]. Muñoz-García et al. [103] investigated metagenomics of metal resistance and tolerance genes in the rhizosphere microbiome of *Avicennia germinans* in a mangrove environment, in this research, 28 metal tolerance genes and 49 metal resistance genes were discovered. The most abundant tolerance genes and resistance genes are related to Zn and Co, respectively. The most abundant genes related to copper metal resistance included cusA and copA genes. They found that *Vibrio fluvialis* strain is tolerant to Cu, Ni, Zn, and Cd. Li et al. [104] studied the metagenomics of the rhizosphere microbiome in metal-contaminated soil, the ecological network of taxa included *Rhizobiaceae*, *Xanthobacteraceae*, *Burkholderiaceae*, and *Actinomycetia.* Drewniak et al. (2016), Jones et al. studied the metagenomics of gold and uranium mine water in Zloty Stok and Kowary mine in Poland to identify useful microbes for bioremediation of mine water, metagenomics analysis showed that two *Methylococcaceae* and *Methylophilaceae* families were the most abundant bacteria in both gold and uranium mines, and the backbone filamentous bacteria *Leptothrix*, *Thiothrix*, and *Beggiatoa.* The bacterial community in the water of Kowary mine was *Helicobacteraceae* family, but the bacterial community in the water of Zloty mine included the Stok community consisting mainly of *Sphingomonadaceae*, *Rhodobacteraceae*, and *Caulobacteraceae* families [105]. Liu et al. [106] was carried out soilless revegetation in gold mine wastewater without soil, and rhizosphere plant bioremediation 44.02% of heavy metal in mine wastewater. Acinetobacter and genes involved in functional genes involved in the carbon, nitrogen, and phosphorus cycle such as pyruvate metabolism, phosphate ester mineralization, phosphate transport, and As and Zn resistance genes were increased.

3.4 Effect of Heavy Metal on Microbial Community

The characteristics of microorganisms such as small size, high abundance, and key role in geochemical cycles are suitable indicators for investigating environmental pollution [107]. The presence of heavy metals in the environment causes a change in the activity of enzymes, the cycle of carbon, nitrogen, and organic substances and reduces the biodiversity of microorganisms and biomass, and also the increase of specific microorganisms resistant to heavy metals in the environment [108–110]. Morphological and nutritional methods for determining the toxicity of metals in the environment include the activity of enzymes, C and N mineralization, and nitrification [111]. Modern methods for determining the microbial community include DNA base ratio (mol% G+C), DNA-DNA hybridization, DNA microarray and reverse genome probe, denaturing gradient gel electrophoresis (DGGE), and phospholipid fatty acid method (PLFA) [112].

Jarosławiecka et al. [113] studied the activity and microbial communities of soil around two Piekary Śląskie and Bukowno mines by total and metal-tolerant culturable bacteria, phospholipid fatty acid method (PLFA), denaturing gradient gel electrophoresis (PCR-DGGE); and metabolic activity using BIOLOG and EcoPlateTM methods. Bacteria resistant to Zn, Ni, Cd, and Cu including *Proteobacteria, Actinobacteria,* and *Bacteroidetes* were isolated and metal-resistant genes such as *czc*A, *cad*A, and *ncc*A were identified. Wang et al. [110] studied the concentrations of nutrients (C, N, P, and S) and concentrations of metals (As, Cd, Cr, Cu, Hg, Ni, Pb, and Zn) in agricultural soils of southwestern China. The concentration of metals, especially Cd, had high levels of risk (RI 898.85, i.e., > 600). Soils with higher contamination of metals had lower microbial diversity. The predominant microbes in the contaminated soil were *Proteobacteria, Acidobacteria,* and *Chloroflexi*, it is assumed that these microorganisms are more resistant to metals. Zhang et al. [114] studied the microbial diversity of three types of soil (highly contaminated, medium contamination, and no metal contamination) by Illumina sequencing of 16s rRNA gene method, and assembly of the microbial community was analyzed by phylogenetic-bin-based null approach (iCAMP) method. The diversity and composition of bacteria and archaea were drastically changed by contamination with heavy metals. The assembly of the microbial community showed that homogeneous selection (deterministic process) increased by 16.2% and drift and dispersal limitation (stochastic process) decreased by 11.4% and 5.4%, respectively. The deterministic proportion of metal-resistant microbes including *Thiobacillus, Euryarchaeota,* and *Crenarchaeota* increased and metal-sensitive microbes decreased, which indicates the change of the microbial community due to the stress of heavy metals.

3.5 Biosorption of Heavy Metals Under Multimetal Stress

Industrial effluents, ore mining sites, and groundwater generally contain a mixture of metals. Since most of the polluted areas are contaminated with several heavy metals, multi-metal-resistant microbes are known to be efficient agents for biosorption. However, the bioabsorption of several metals is influenced by many factors such as the number of metals, their concentration and composition, the order of addition, and contact time in the bioabsorption environment [115].

The presence of a single toxic metal species in a contaminated solution is not common, and metal ions are often present in multiple forms, which causes an interactive effect between them. The use of living and fixed microorganisms for the simultaneous removal of several metals from the metal solution has recently received attention. The presence of other metal ions in the solution leads to competition and changes the overall metal binding ability. Multi-metal biosorption studies can lead to a better understanding of these complex systems and their behavior [115].

4 Recent Advances in Sustainable Microbial Removal of Heavy Metals

4.1 Enzyme Application

Microbial enzymes can biologically transform toxic heavy metals into non-toxic compounds. The oxidases and reductase enzymes have a critical role in metals' biotransformation. Reduction of chromium by chromate reductase in *pseudomonas putida* or oxidation of toxic arsenite to arsenate by arsenite oxidase in *Bacillus cereus* and *Acinetobacter junii* are examples of metal transformation by enzymatic catalysis [116, 117]. Biomineralization of heavy meals such as Cd, Co, Ni, Cu, and Zn is done through urease enzymes produced by *Bacillus* spp. and *Sporosarcina pasteuri.* In this process, hydrolysis of urea to carbonate and NH_3 catalyzed by urease enzyme, which resulted in precipitating heavy metals as carbonate forms [116].

4.2 Biosurfactants

In stress conditions, bacteria secrete biosurfactants, which can bind to metal. *Bacillus* sp. HIP3 produced biosurfactants that bound zinc, cadmium, and lead [118]. Biosurfactants reduce metals and increase the solubility of metals, as a result, biosurfactants are found in environments contaminated with metals [119]. *Citrobacter freundii* MG812314.1 isolated from Al-Rahawy drain sediments, this isolate produce biosurfactant that bound Al^{2+}, Cd^{2+}, Cu^{2+}, Fe^{2+}, Pb^{2+}, Mn^{2+}, and Zn^{2+} [120]. *Bacillus* sp.

MSI has produced lipopeptide biosurfactant and removed the 99.93% Cd, 97.73% Pb, 89.5% Mn, and Hg 75.5% Hg from 1000 ppm metal solutions [121].

4.3 Biomineralization

MICP is considered a novel mechanism for heavy metal biomineralization as metal carbonate precipitation by the urease function of related microorganisms such as *Sporosarcina pasteurii*, *Bacillus pasteurii*, *Kocuria flava* CR1, *B. cereus*, *B. subtilis*, *B. megaterium*, and *B. simplex* [122].

Microbial-induced mineral precipitation (MIMP) as a bio-geochemical process could induce the precipitation of minerals in soil. It is known that several organisms including some species of soil bacteria, photosynthetic microorganisms, and sulfate-reducing bacteria involved in the process of mineral precipitation. The mechanisms of mineralization include anaerobic sulfide oxidation, sulfate reduction, methane oxidation, ammonification, photosynthesis, and urea hydrolysis. One of the most common methods is the use of urease-producing bacteria [123–125]. The most important factor in choosing a bacterial species for the MICP process is its ability to synthesize urease enzymes [126]. Urease enzyme activity in a wide range of microorganisms such as *Sporosarcina pasteuri* (previously called *Bacillus pasteuri*), *Bacillus* spp.

The surface of the bacterial cell acts as a nucleation site for the accumulation of calcium carbonate crystals. Because the bacterial cell wall has functional groups such as carboxyl, phosphate, and amine with a negative charge, which acts as a receptor for divalent cations such as calcium, and binds them to the cell surface at neutral pH and provides ideal nucleation sites for the formation of calcite. This cation (metal ion) reacts with an anion (carbonate) and eventually calcium carbonate is produced in an insoluble form. On the other hand, extracellular polymeric substances (EPS) play an important role in the production of calcium carbonate crystals [127]. Exopolysaccharides, proteins, nucleic acid, and lipids can trap a large amount of divalent cations such as calcium due to their anionic chemical properties. As a result, the concentration of calcium in the place increased, which leads to the precipitation of calcium carbonate crystals in the presence of carbonate. Therefore, bacterial cells are very important for calcium carbonate deposition because they provide nucleation sites for the formation and accumulation of calcium carbonate crystals and also influence the specific type of minerals formed [128, 129].

In a study by Maity et al. [130] the isolated *Bacillus subtilis* from industrial sites could significantly immobilize heavy metals including Cr, Cu, and Zn cations within the same contaminated soil by the MIMP process, simultaneously increased urease and dehydrogenase activities and produced various heavy metals minerals in the soil.

5 Biotechnological Intervention in Microbial Bioremediation)

Both in-situ and ex-situ procedures are used for heavy metal bioremediation by microbial processes. These techniques include bioventing, biosparging, bioaugmentation, land farming, biopiling, composting, bioreactors, and electrodialysis [131]. Progress in genetic engineering and the construction of genetically modified biocatalysts indicated that these organisms bioremediated pollution in the environment more efficiently than natural bacteria [132]. In these techniques, enzymes, gene expression systems, metabolic pathways, and other effective objects on metal removal are modified by genetic engineering [133].

Microbial populations can express various special genes under heavy metals stress such as *szx* for cadmium, *cop* for copper, *mer* for mercury, and *aox* for arsenic, which results in microbial viability in such stress conditions. Overexpression of these genes in genetically engineered microorganisms by metabolic engineering, alteration of enzyme specificity, and the substrate affinity by altering of coding and regulator genes, can accelerate HMs removal [134, 135].

Shehata [136] constructed recombinant *E. coli* capable of displaying metallothionein (SmtB) and lead binding protein (PbrR) onto the outer membrane and removed 100 mg/l Pb^{2+} from the water after 18 h. This recombinant strain was successfully used for the removal of Pb^{2+} from acidic solution that is highly soluble; and remains its metal removal activity after heat treatment, which lost its viability.

Metallothionein (MT) isoforms from rice were expressed by genetically engineered *Escherichia coli* as fusion proteins with glutathione-S-transferase (GST) to remove mercury from the culture medium through metal binding and mercury complex formation [137]. Surface display of metal binding protein by genetic engineering is from other strategies for accelerating its bioremediation.

Microbial membrane transporters in the inner or outer lipid membrane have been genetically modified for more efficient heavy metal removal by bioaccumulation mechanism, which Diept et al. [134] discussed in a review paper.

System biology research in HMs bioremediation that includes the computational biology and multi-omics (genomics, metagenomics, metabolomics, proteomics, and transcriptomics) approach has been performed efficiently in gene mining [133]. In addition, novel HMs biodegrading genes were detected by high-throughput and next-generation sequencing (NGS). CRISPR-Cas is a modern technology for gene editing implemented for the optimization of the bioremediation process by improving strain capable of biodegrading specific pollutants [133, 138].

5.1 Microbial Consortia in Metal Bioremediation

Microorganisms grow together in environmental (such as urban and rural affluent, industrial, mining, and agriculture wastewater) as a consortium, which produces stromatolites, microbial mats, and biofilms and are resistant to various stresses [139, 140]. Microbial consortium forms biofilm, which is this biofilm of exopolysaccharide containing nucleic acid, protein, and carbohydrate. These consortia have a protective barrier against toxins and bioremediation heavy metals [141].

Benazir et al. [142] removed chromium by immobilizing *S.cerevisiae*, *P. aeruginosa*, and B. *subtilis consortia* consortia, which reduced 770 mg/l chromium to 5.2 mg/l. Muneer et al. [143] isolated mercury resistance microorganisms (*Bacillus licheniformis*, *Candida parapsilosis*, and *Tetrahymena ostrate*) from industrial effluents of tanneries, this Consortia reduced 88% of Hg^{2+} after 96 h. Li et al. [144] enriched sulfate-reducing function (included *Clostridium*, *Desulfosporosinus*, and *Desulfovibrio* genera) from smelter soils, in which the consortia removed concentrations of 1181 mg of As/k or 222 mg Pb from water. Shi et al. [145] used a consortium of methanotrophic archaea and bacteria (*Methanosarcina* and *Methylocystis*) to reduce selenite and oxidizing methane.

5.1.1 Efficient Microbial Bioremediation of Heavy Metal via Nanotechnology

Microorganisms also produce nanoparticles by bioremediation of metals, which nanoparticles use in diagnostics, cancer therapy, therapy, biosensors, and drug delivery. Biological nanoparticles have a very small size and a large surface area, as a result of which their chemical activity increases [146]. The surface of biological nanoparticles contains functional groups such as $-NH_2$, –COOH, and –OH that improve their physical and chemical properties [147]. Nanocomposites increase the activities of microbial biostimulation, bioaccumulation, and biotransformation, and as a result, lead to better removal of metals from the environment by microbes [148], for example, *Halomonas* sp is trapped inside the nanocomposite coated with iron nanoparticles and biodegrades Pb(II) and Cd(II) more [149].

Graphene, graphene oxide, activated carbon, and fullerene carbon nanotubes as carbon nanomaterials that are successfully used for heavy metal-contaminated water purification [150]. Constructed bentonite composite zero-valent iron (B-nZVI) by Zarime et al. [151] was used for Pb(II), Co(II), Ni(II), Zn(II), Cd(II), and Cu(II) from wastewater. This composite showed more efficient metal removal compared to nZVI, because of reduced nZVI aggregation, spread in bentonite layers, and provided more sites for heavy metal absorption. Silver mercaptosuccinic acid (Ag-MSA) could be a potential source to remove mercury (II) [152]. Rhamnolipid-modified nZVI (R-nZVI) was effective for heavy metal immobilization on river sediment [153].

Modified carbon black nanoparticle (MCBN) was applied for heavy metal remediation and the result showed that heavy metals availability could significantly decrease by these nanoparticles in Cd and Ni-contaminated soil and simultaneously enhanced plant growth [154].

Magnetic nanoparticles (MNPs) has received attention recently in environmental remediation, because of easily separation using a magnet and special metal-ion adsorption. The application of MNPs for soil and groundwater remediation surveyed in several research. MNPs core–shell $Fe_3O_4@SiO_2$ NPs coated with iminodiacetic acid chelators) experimented by Fan et al. [155] for heavy metals contaminated soil remediation. Heavy metals removed by chelation via these nanoparticles and then separation by magnetic force. The removal rates of Cd and Pb were obtained 84.9% and 72.2%, respectively. Results of this study indicated that heavy metals removal decreased in the presence of organic materials in the soil, whereas the chemical composition of the soil did not change by using the MNPs.

6 Conclusion

Heavy metals cause environmental pollution through industries, factories, and various human sources. Due to the toxic effects of heavy metals on human health and the environment, they must be removed from the environment. Conventional methods of removing metals from the environment should be replaced by biological methods because they are not efficient, cost-effective, and environmentally friendly techniques. New bioremediation methods should be expanded with new methods such as nanocomposites, genetic modification, use of microbial biofilm, and use of microbial consortium, which requires extensive research.

Acknowledgements The authors are thankful to all scientists who have efforts and research on heavy metal bioremediation and progress in this area.

References

1. Duffus JH (2002) "Heavy metals" a meaningless term? (IUPAC Technical Report). Pure Appl Chem 74(5):793–807
2. Briffa J, Sinagra E, Blundell R (2020) Heavy metal pollution in the environment and their toxicological effects on humans. Heliyon 6(9):e04691
3. Fahad S, Hussain S, Saud S, Hassan S, Tanveer M, Ihsan MZ et al (2016) A combined application of biochar and phosphorus alleviates heat-induced adversities on physiological, agronomical and quality attributes of rice. Plant Physiol Biochem 103:191–198
4. De Buyck P-J, Van Hulle SWH, Dumoulin A, Rousseau DPL (2021) Roof runoff contamination: a review on pollutant nature, material leaching and deposition. Rev Environ Sci BioTechnol 20(2):549–606
5. Jaishankar M, Tseten T, Anbalagan N, Mathew BB, Beeregowda KN (2014) Toxicity, mechanism and health effects of some heavy metals. Interdiscip Toxicol 7(2):60–72

6. Balali-Mood M, Naseri K, Tahergorabi Z, Khazdair MR, Sadeghi M (2021) Toxic mechanisms of five heavy metals: Mercury, Lead, Chromium, Cadmium, and Arsenic. Front Pharmacol 12
7. Organization WH (2017) Guidelines for drinking-water quality: first addendum to the fourth edition
8. FAO (2022) Water quality for agriculture 2022 [irrigation and drainage papers]. Available from: https://www.fao.org/3/t0234e/T0234E06.htm
9. Igiri BE, Okoduwa SIR, Idoko GO, Akabuogu EP, Adeyi AO, Ejiogu IK (2018) Toxicity and bioremediation of heavy metals contaminated ecosystem from tannery wastewater: a review. J Toxicol 2018:2568038
10. Yadav M, Gupta R, Sharma RK (2019) Chapter 14—Green and sustainable pathways for wastewater purification. In: Ahuja S (ed) Advances in water purification techniques. Elsevier, pp 355–383
11. Abd Elnabi MK, Elkaliny NE, Elyazied MM, Azab SH, Elkhalifa SA, Elmasry S et al (2023) Toxicity of heavy metals and recent advances in their removal: a review. Toxics 11(7):580
12. Sreedevi P, Suresh K, Jiang G (2022) Bacterial bioremediation of heavy metals in wastewater: a review of processes and applications. J Water Process Eng 48:102884
13. Kumar V, Singh S, Kashyap N, Singla S, Bhadrecha P, Kaur P et al (2015) Bioremediation of heavy metals by employing resistant microbial isolates from agricultural soil irrigated with industrial waste water. Orient J Chem 31(1):357–361
14. Ameen FA, Hamdan AM, El-Naggar MY (2020) Assessment of the heavy metal bioremediation efficiency of the novel marine lactic acid bacterium, Lactobacillus plantarum MF042018. Sci Rep 10(1):314
15. Singh N, Gadi R (2012) Bioremediation of Ni(II) and Cu(II) from wastewater by the nonliving biomass of Brevundimonas vesicularis. J Environ Chem Ecotoxicol 4(8):137–142
16. Bhattacharya A, Gupta A (2013) Evaluation of Acinetobacter sp. B9 for Cr(VI) resistance and detoxification with potential application in bioremediation of heavy-metals-rich industrial wastewater. Environ Sci Pollut Res 20:6628–6637
17. Wasi S, Tabrez S, Ahmad M (2011) Suitability of immobilized Pseudomonas fluorescens SM1 strain for remediation of phenols, heavy metals, and pesticides from water. Water Air Soil Pollut 220:89–99
18. Kamika I, Momba MN (2013) Assessing the resistance and bioremediation ability of selected bacterial and protozoan species to heavy metals in metal-rich industrial wastewater. BMC Microbiol 13(1):1–14
19. Sakpirom J, Kantachote D, Nunkaew T, Khan E (2017) Characterizations of purple non-sulfur bacteria isolated from paddy fields, and identification of strains with potential for plant growth-promotion, greenhouse gas mitigation and heavy metal bioremediation. Res Microbiol 168(3):266–275
20. Hamdan AM, Abd-El-Mageed H, Ghanem N (2021) Biological treatment of hazardous heavy metals by Streptomyces rochei ANH for sustainable water management in agriculture. Sci Rep 11(1):9314
21. Ma H, Wei M, Wang Z, Hou S, Li X, Xu H (2020) Bioremediation of cadmium polluted soil using a novel cadmium immobilizing plant growth promotion strain Bacillus sp. TZ5 loaded on biochar. J Hazard Mater 388:122065
22. Njoku KL, Akinyede OR, Obidi OF (2020) Microbial remediation of heavy metals contaminated media by Bacillus megaterium and Rhizopus stolonifer. Sci Afr 10:e00545
23. Talukdar D, Jasrotia T, Sharma R, Jaglan S, Kumar R, Vats R et al (2020) Evaluation of novel indigenous fungal consortium for enhanced bioremediation of heavy metals from contaminated sites. Environ Technol Innov 20:101050
24. Saad MM, Saad MA, Saad BS, Zakaria FA, Husain A-RA, Abdelgaleil SA (2022) Bioremediation and microbial-assisted phytoremediation of heavy metals by endophytic Fusarium species isolated from Convolvulus arvensis. Bioremediation J 1–11
25. Amatussalam A, Abubacker M, Rajendran RB (2011) In situ Carica papaya stem matrix and Fusarium oxysporum (NCBT-156) mediated bioremediation of chromium

26. Mishra A, Malik A (2014) Novel fungal consortium for bioremediation of metals and dyes from mixed waste stream. Biores Technol 171:217–226
27. Bhargavi SD, Savitha J (2014) Arsenate resistant Penicillium coffeae: a potential fungus for soil bioremediation. Bull Environ Contam Toxicol 92(3):369–373
28. Kartal SN, Munir E, Kakitani T, Imamura Y (2004) Bioremediation of CCA-treated wood by brown-rot fungi Fomitopsis palustris, Coniophora puteana, and Laetiporus sulphureus. J Wood Sci 50:182–188
29. Srivastava S, Thakur IS (2006) Evaluation of bioremediation and detoxification potentiality of Aspergillus niger for removal of hexavalent chromium in soil microcosm. Soil Biol Biochem 38(7):1904–1911
30. Ahmed DAE-A, Gheda SF, Ismail GA (2021) Efficacy of two seaweeds dry mass in bioremediation of heavy metal polluted soil and growth of radish (Raphanus sativus L.) plant. Environ Sci Pollut Res 28:12831–12846
31. El-Sheekh MM, Farghl AA, Galal HR, Bayoumi HS (2016) Bioremediation of different types of polluted water using microalgae. Rendiconti Lincei 27:401–410
32. Ye J, Xiao H, Xiao B, Xu W, Gao L, Lin G (2015) Bioremediation of heavy metal contaminated aqueous solution by using red algae Porphyra leucosticta. Water Sci Technol 72(9):1662–1666
33. Lakmali WGM, Athukorala ADSNP, Jayasundera KB (2022) Investigation of Pb(II) bioremediation potential of algae and cyanobacteria strains isolated from polluted water. Water Sci Eng 15(3):237–246
34. El-Naggar NE-A, Hamouda RA, Saddiq AA, Alkinani MH (2021) Simultaneous bioremediation of cationic copper ions and anionic methyl orange azo dye by brown marine alga Fucus vesiculosus. Sci Rep 11(1):3555
35. Langston WJ, Bebianno MJ (1998) Metal metabolism in aquatic environments. Springer Science & Business Media
36. Rahman Z, Singh VP (2020) Bioremediation of toxic heavy metals (THMs) contaminated sites: concepts, applications and challenges. Environ Sci Pollut Res 27:27563–27581
37. Michalak I, Chojnacka K, Witek-Krowiak A (2013) State of the art for the biosorption process—a review. Appl Biochem Biotechnol 170(6):1389–1416
38. Mwandira W, Nakashima K, Kawasaki S, Arabelo A, Banda K, Nyambe I et al (2020) Biosorption of Pb(II) and Zn(II) from aqueous solution by Oceanobacillus profundus isolated from an abandoned mine. Sci Rep 10(1):21189
39. Oyewole OA, Zobeashia SSL-T, Oladoja EO, Raji RO, Odiniya EE, Musa AM. Biosorption of heavy metal polluted soil using bacteria and fungi isolated from soil. SN Appl Sci 1(8):857
40. Sharma P, Dutta D, Udayan A, Nadda AK, Lam SS, Kumar S (2022) Role of microbes in bioaccumulation of heavy metals in municipal solid waste: Impacts on plant and human being. Environ Pollut 305:119248
41. Özdemir S, Kilinc E, Nicolaus B, Poli A (2013) Resistance and bioaccumulation of Cd^{2+}, Cu^{2+}, Co^{2+} and Mn^{2+} by thermophilic bacteria, Geobacillus thermantarcticus and Anoxybacillus amylolyticus. Ann Microbiol 63(4):1379–1385
42. Vera M, Schippers A, Hedrich S, Sand W (2022) Progress in bioleaching: fundamentals and mechanisms of microbial metal sulfide oxidation—part A. Appl Microbiol Biotechnol 106(21):6933–6952
43. Patel F, Lakshmi B (2021) Bioleaching of copper and nickel from mobile phone printed circuit board using Aspergillus fumigatus A2DS. Braz J Microbiol 52(3):1475–1487
44. Ehrlich H, Bailey E, Wysokowski M, Jesionowski T (2021) Forced biomineralization: a review. Biomimetics 6(3):46
45. Zhao X, Wang M, Wang H, Tang D, Huang J, Sun Y (2019) Study on the remediation of Cd pollution by the biomineralization of urease-producing bacteria. Int J Environ Res Public Health 16(2):268
46. Vogel M, Fischer S, Maffert A, Hübner R, Scheinost AC, Franzen C et al (2018) Biotransformation and detoxification of selenite by microbial biogenesis of selenium-sulfur nanoparticles. J Hazard Mater 344:749–757

47. Ianieva O (2009) Mechanisms of bacteria resistance to heavy metals. Mikrobiolohichnyĭ Zhurnal 71:54–65
48. Pardo R, Herguedas M, Barrado E, Vega M (2003) Biosorption of cadmium, copper, lead and zinc by inactive biomass of Pseudomonas Putida. Anal Bioanal Chem 376(1):26–32
49. Taniguchi J, Hemmi H, Tanahashi K, Amano N, Nakayama T, Nishino T (2000) Zinc biosorption by a zinc-resistant bacterium, Brevibacterium sp. strain HZM-1. Appl Microbiol Biotechnol 54(4):581–588
50. Green-Ruiz C (2006) Mercury(II) removal from aqueous solutions by nonviable Bacillus sp. from a tropical estuary. Bioresour Technol 97(15):1907–1911
51. Bhaskar PV, Bhosle NB (2006) Bacterial extracellular polymeric substance (EPS): a carrier of heavy metals in the marine food-chain. Environ Int 32(2):191–198
52. Kazy S, Sar P, Singh S, Sen A, D'Souza S (2002) Extracellular polysaccharides of a copper-sensitive and a copper-resistant pseudomonas aeruginosa strain: synthesis, chemical nature and copper binding. World J Microbiol Biotechnol 18:583–588
53. Scott JA, Palmer SJ (1990) Sites of cadmium uptake in bacteria used for biosorption. Appl Microbiol Biotechnol 33(2):221–225
54. Iyer A, Mody K, Jha B (2005) Biosorption of heavy metals by a marine bacterium. Mar Pollut Bull 50(3):340–343
55. Li XZ, Nikaido H, Williams KE (1997) Silver-resistant mutants of Escherichia coli display active efflux of Ag+ and are deficient in porins. J Bacteriol 179(19):6127–6132
56. Richau JA, Choquenet D, Fialho AM, Sá-Correia I (1997) Emergence of Cu(++)-tolerant mutants defective in gellan synthesis in Cu(++)-stressed cultures of Sphingomonas paucimobilis. Res Microbiol 148(3):251–261
57. Gonzalez Henao S, Ghneim-Herrera T (2021) Heavy metals in soils and the remediation potential of bacteria associated with the plant microbiome. Front Environ Sci 15
58. Vats N, Lee SF (2001) Characterization of a copper-transport operon, copYAZ, from Streptococcus mutans. Microbiology 147(3):653–662
59. Rensing C, Ghosh M, Rosen BP (1999) Families of soft-metal-ion-transporting ATPases. J Bacteriol 181(19):5891–5897
60. Lee S, Kim Y-Y, Lee Y, An G (2007) Rice P1B-type heavy-metal ATPase, OsHMA9, is a metal efflux protein. Plant Physiol 145(3):831–842
61. Oger C, Mahillon J, Petit F (2003) Distribution and diversity of a cadmium resistance (cadA) determinant and occurrence of IS 257 insertion sequences in Staphylococcal bacteria isolated from a contaminated estuary (Seine, France). FEMS Microbiol Ecol 43(2):173–183
62. Ybarra G, Webb R (1999) Effects of divalent metal cations and resistance mechanisms of the cyanobacterium Synechococcus sp. strain PCC 7942. J Hazard Substance Res 2(1):1
63. Ivanova EP, Kurilenko VV, Kurilenko AV, Gorshkova NM, Shubin FN, Nicolau DV et al (2002) Tolerance to cadmium of free-living and associated with marine animals and eelgrass marine gamma proteobacteria. Curr Microbiol 44:357–362
64. Lima AIG, Corticeiro SC, Figueira EMdAP (2006) Glutathione-mediated cadmium sequestration in Rhizobium leguminosarum. Enzyme Microbial Technol 39(4):763–769
65. Cha J-S, Cooksey DA (1991) Copper resistance in Pseudomonas syringae mediated by periplasmic and outer membrane proteins. Proc Natl Acad Sci 88(20):8915–8919
66. Klaus T, Joerger R, Olsson E, Granqvist C-G (1999) Silver-based crystalline nanoparticles, microbially fabricated. Proc Natl Acad Sci 96(24):13611–13614
67. Silver S (2003) Bacterial silver resistance: molecular biology and uses and misuses of silver compounds. FEMS Microbiol Rev 27(2–3):341–353
68. Gupta A, Phung LT, Chakravarty L, Silver S (1999) Mercury resistance in Bacillus cereus RC607: transcriptional organization and two new open reading frames. J Bacteriol 181(22):7080–7086
69. Carpentier W, Sandra K, De Smet I, Brige A, De Smet L, Van Beeumen J (2003) Microbial reduction and precipitation of vanadium by Shewanella oneidensis. Appl Environ Microbiol 69(6):3636–3639

70. Lloyd JR, Sole V, Van Praagh C, Lovley D (2000) Direct and Fe(II)-mediated reduction of technetium by Fe(III)-reducing bacteria. Appl Environ Microbiol 66(9):3743–3749
71. Sugio T, Tsujita Y, Katagiri T, Inagaki K, Tano T (1988) Reduction of Mo6+ with elemental sulfur by Thiobacillus ferrooxidans. J Bacteriol 170(12):5956–5959
72. Nevin KP, Lovley DR (2000) Lack of production of electron-shuttling compounds or solubilization of Fe(III) during reduction of insoluble Fe (III) oxide by Geobacter metallireducens. Appl Environ Microbiol 66(5):2248–2251
73. Song H-P, Li X-G, Sun J-S, Xu S-M, Han X (2008) Application of a magnetotactic bacterium, Stenotrophomonas sp. to the removal of Au(III) from contaminated wastewater with a magnetic separator. Chemosphere 72(4):616–621
74. Slobodkina G, Bonch-Osmolovskaya E, Slobodkin A (2007) Reduction of chromate, selenite, tellurite, and iron(III) by the moderately thermophilic bacterium Bacillus thermoamylovorans SKC1. Microbiology 76:530–534
75. Michel C, Giudici-Orticoni M-T, Baymann F, Bruschi M (2003) Bioremediation of chromate by sulfate-reducing bacteria, cytochromes c3 and hydrogenases. Water Air Soil Pollut Focus 3:161–169
76. Ji G, Silver S (1992) Reduction of arsenate to arsenite by the ArsC protein of the arsenic resistance operon of Staphylococcus aureus plasmid pI258. Proc Natl Acad Sci 89(20):9474–9478
77. Myers CR, Nealson KH (1988) Bacterial manganese reduction and growth with manganese oxide as the sole electron acceptor. Science 240(4857):1319–1321
78. Roux M, Sarret Gr, Pignot-Paintrand I, Fontecave M, Coves J (2001) Mobilization of selenite by Ralstonia metallidurans CH34. Appl Environ Microbiol 67(2):769–773
79. Joshi S, Gangola S, Bhandari G, Bhandari NS, Nainwal D, Rani A et al (2023) Rhizospheric bacteria: the key to sustainable heavy metal detoxification strategies. Front Microbiol 14
80. Mathivanan K, Chandirika JU, Vinothkanna A, Yin H, Liu X, Meng D (2021) Bacterial adaptive strategies to cope with metal toxicity in the contaminated environment—a review. Ecotoxicol Environ Saf 226:112863
81. Pinto E, Sigaud-kutner TC, Leitao MA, Okamoto OK, Morse D, Colepicolo P (2003) Heavy metal–induced oxidative stress in algae 1. J Phycol 39(6):1008–1018
82. Slawson RM, Trevors JT, Lee H (1992) Silver accumulation and resistance in Pseudomonas stutzeri. Arch Microbiol 158:398–404
83. Smirnova G (2005) Distribution of bacteria resistant to oxygen-containing anions-xenobiotics. Mikrobiolohichnyi Zhurnal 67(5):11–18
84. Brown N, Shih Y-C, Leang C, Glendinning K, Hobman J, Wilson J (2002) Mercury transport and resistance. Biochem Soc Trans 30(4):715–718
85. Mukhopadhyay R, Rosen BP, Phung LT, Silver S (2002) Microbial arsenic: from geocycles to genes and enzymes. FEMS Microbiol Rev 26(3):311–325
86. Bruins MR, Kapil S, Oehme FW (2000) Microbial resistance to metals in the environment. Ecotoxicol Environ Saf 45(3):198–207
87. Roy R, Tiwari M, Donelli G, Tiwari V (2018) Strategies for combating bacterial biofilms: a focus on anti-biofilm agents and their mechanisms of action. Virulence 9(1):522–554
88. Gomathy M, Sabarinathan K (2010) Microbial mechanisms of heavy metal tolerance-a review. Agric Rev 31(2)
89. Igiri BE, Okoduwa SI, Idoko GO, Akabuogu EP, Adeyi AO, Ejiogu IK (2018) Toxicity and bioremediation of heavy metals contaminated ecosystem from tannery wastewater: a review. J Toxicol 2018
90. Escárcega-González CE, Garza-Cervantes JA, Vázquez-Rodríguez A, Morones-Ramírez JR (2018) Bacterial exopolysaccharides as reducing and/or stabilizing agents during synthesis of metal nanoparticles with biomedical applications. Int J Polym Sci 2018
91. Albelda-Berenguer M, Monachon M, Joseph E (2019) Siderophores: from natural roles to potential applications. Adv Appl Microbiol 106:193–225
92. Hu X, Boyer GL (1996) Siderophore-mediated aluminum uptake by Bacillus megaterium ATCC 19213. Appl Environ Microbiol 62(11):4044–4048

93. Schalk IJ, Hannauer M, Braud A (2011) New roles for bacterial siderophores in metal transport and tolerance. Environ Microbiol 13(11):2844–2854
94. Braud A, Hoegy F, Jezequel K, Lebeau T, Schalk IJ (2009) New insights into the metal specificity of the Pseudomonas aeruginosa pyoverdine–iron uptake pathway. Environ Microbiol 11(5):1079–1091
95. Johnstone TC, Nolan EM (2015) Beyond iron: non-classical biological functions of bacterial siderophores. Dalton Trans 44(14):6320–6339
96. Braud A, Geoffroy V, Hoegy F, Mislin G, Schalk I (2010) The siderophores pyoverdine and pyochelin are involved in Pseudomonas aeruginosa resistance against metals: another biological function of these two siderophores. Environ Microbiol Rep 2:419–425
97. Handelsman J, Rondon MR, Brady SF, Clardy J, Goodman RM (1998) Molecular biological access to the chemistry of unknown soil microbes: a new frontier for natural products. Chem Biol 5(10):R245–R249
98. Tripathi M, Singh DN, Vikram S, Singh VS, Kumar S (2018) Metagenomic approach towards bioprospection of novel biomolecule(s) and environmental bioremediation. Annu Res Rev Biol 1–12
99. Bharagava RN, Purchase D, Saxena G, Mulla SI (2019) Chapter 26—Applications of metagenomics in microbial bioremediation of pollutants: from genomics to environmental cleanup. In: Das S, Dash HR (eds) Microbial diversity in the genomic era. Academic Press, pp 459–477
100. Hauptmann AL, Sicheritz-Pontén T, Cameron KA, Bælum J, Plichta DR, Dalgaard M et al (2017) Contamination of the Arctic reflected in microbial metagenomes from the Greenland ice sheet. Environ Res Lett 12(7):074019
101. Datta S, Rajnish KN, Samuel MS, Pugazlendhi A, Selvarajan E (2020) Metagenomic applications in microbial diversity, bioremediation, pollution monitoring, enzyme and drug discovery. A review. Environ Chem Lett 18:1229–1241
102. Sharma P, Kumar S, Pandey A (2021) Bioremediated techniques for remediation of metal pollutants using metagenomics approaches: a review. J Environ Chem Eng 9(4):105684
103. Muñoz-García A, Arbeli Z, Boyacá-Vásquez V, Vanegas J (2022) Metagenomic and genomic characterization of heavy metal tolerance and resistance genes in the rhizosphere microbiome of Avicennia germinans in a semi-arid mangrove forest in the tropics. Mar Pollut Bull 184:114204
104. Li L, Meng D, Yin H, Zhang T, Liu Y (2023) Genome-resolved metagenomics provides insights into the ecological roles of the keystone taxa in heavy-metal-contaminated soils. Front Microbiol 14
105. Drewniak L, Krawczyk PS, Mielnicki S, Adamska D, Sobczak A, Lipinski L et al (2016) Physiological and metagenomic analyses of microbial mats involved in self-purification of mine waters contaminated with heavy metals. Front Microbiol 7
106. Liu C, Li B, Chen X, Dong Y, Lin H (2022) Insight into soilless revegetation of oligotrophic and heavy metal contaminated gold tailing pond by metagenomic analysis. J Hazard Mater 435:128881
107. Bouchez T, Blieux A, Dequiedt S, Domaizon I, Dufresne A, Ferreira S et al (2016) Molecular microbiology methods for environmental diagnosis. Environ Chem Lett 14:423–441
108. Giller KE, Witter E, Mcgrath SP (1998) Toxicity of heavy metals to microorganisms and microbial processes in agricultural soils: a review. Soil Biol Biochem 30(10–11):1389–1414
109. Wang Y, Shi J, Wang H, Lin Q, Chen X, Chen Y (2007) The influence of soil heavy metals pollution on soil microbial biomass, enzyme activity, and community composition near a copper smelter. Ecotoxicol Environ Saf 67(1):75–81
110. Wang X, Gao P, Li D, Liu J, Yang N, Gu W et al (2019) Risk assessment for and microbial community changes in Farmland soil contaminated with heavy metals and metalloids. Ecotoxicol Environ Saf 185:109685
111. Giller KE, Witter E, McGrath SP (1998) Toxicity of heavy metals to microorganisms and microbial processes in agricultural soils: a review. Soil Biol Biochem 30(10):1389–1414
112. Agrawal PK, Agrawal S, Shrivastava R (2015) Modern molecular approaches for analyzing microbial diversity from mushroom compost ecosystem. 3 Biotech 5(6):853–866

113. Jarosławiecka AK, Piotrowska-Seget Z (2022) The Effect of Heavy Metals on Microbial Communities in Industrial Soil in the Area of Piekary Śląskie and Bukowno (Poland). Microbiol Res 13(3):626–642
114. Zhang M, Zhang T, Zhou L, Lou W, Zeng W, Liu T et al (2022) Soil microbial community assembly model in response to heavy metal pollution. Environ Res 213:113576
115. Priyadarshanee M, Das S (2021) Biosorption and removal of toxic heavy metals by metal tolerating bacteria for bioremediation of metal contamination: a comprehensive review. J Environ Chem Eng 9(1):104686
116. Naureen A, Rehman A (2016) Arsenite oxidizing multiple metal resistant bacteria isolated from industrial effluent: their potential use in wastewater treatment. World J Microbiol Biotechnol 32:1–9
117. Tasleem M, El-Sayed A-AA, Hussein WM, Alrehaily A (2022) Bioremediation of chromium-contaminated groundwater using chromate reductase from Pseudomonas putida: an in silico approach. Water 15(1):150
118. Md Badrul Hisham NH, Ibrahim MF, Ramli N, Abd-Aziz S (2019) Production of biosurfactant produced from used cooking oil by Bacillus sp. HIP3 for heavy metals removal. Molecules 24(14):2617
119. Bodour AA, Drees KP, Maier RM (2003) Distribution of biosurfactant-producing bacteria in undisturbed and contaminated arid Southwestern soils. Appl Environ Microbiol 69(6):3280–3287
120. Gomaa EZ, El-Meihy RM (2019) Bacterial biosurfactant from Citrobacter freundii MG812314.1 as a bioremoval tool of heavy metals from wastewater. Bull Natl Res Centre 43(1):69
121. Ravindran A, Sajayan A, Priyadharshini GB, Selvin J, Kiran GS (2020) Revealing the efficacy of thermostable biosurfactant in heavy metal bioremediation and surface treatment in vegetables. Front Microbiol 11:222
122. Prajapati NK, Agnihotri AK, Basak N (2023) Microbial induced calcite precipitation (MICP) a sustainable technique for stabilization of soil: a review. Mater Today: Proc
123. Chaparro-Acuña SP, Becerra-Jiménez ML, Martínez-Zambrano JJ, Rojas-Sarmiento HA (2018) Soil bacteria that precipitate calcium carbonate: mechanism and applications of the process. Acta Agron 67(2):277–288
124. Omoregie AI, Palombo EA, Nissom PM (2021) Bioprecipitation of calcium carbonate mediated by ureolysis: a review. Environ Eng Res 26(6)
125. Omoregie AI, Ngu LH, Ong DEL, Nissom PM (2019) Low-cost cultivation of Sporosarcina pasteurii strain in food-grade yeast extract medium for microbially induced carbonate precipitation (MICP) application. Biocatal Agric Biotechnol 17:247–255
126. Rajasekar A, Moy CK, Wilkinson S (eds) (20017) Stimulation of indigenous carbonate precipitating bacteria for ground improvement. IOP conference series: earth and environmental science. IOP Publishing
127. Görgen S, Benzerara K, Skouri-Panet F, Gugger M, Chauvat F, Cassier-Chauvat C (2021) The diversity of molecular mechanisms of carbonate biomineralization by bacteria. Disc Mater 1:1–20
128. Zhu T, Dittrich M (2016) Carbonate precipitation through microbial activities in natural environment, and their potential in biotechnology: a review. Front Bioeng Biotechnol 4:4
129. Ng W-S, Lee M-L, Hii S-L (2012) An overview of the factors affecting microbial-induced calcite precipitation and its potential application in soil improvement. Int J Civ Environ Eng 6(2):188–194
130. Maity JP, Chen G-S, Huang Y-H, Sun A-C, Chen C-Y (2019) Ecofriendly heavy metal stabilization: microbial induced mineral precipitation (MIMP) and biomineralization for heavy metals within the contaminated soil by indigenous bacteria. Geomicrobiol J 36(7):612–623
131. Gangola S, Sharma A, Bhatt P, Khati P, Chaudhary P (2018) Presence of esterase and laccase in Bacillus subtilis facilitates biodegradation and detoxification of cypermethrin. Sci Rep 8(1):12755

132. Bhatt P, Sethi K, Gangola S, Bhandari G, Verma A, Adnan M et al (2022) Modeling and simulation of atrazine biodegradation in bacteria and its effect in other living systems. J Biomol Struct Dyn 40(7):3285–3295
133. Verma S, Bhatt P, Verma A, Mudila H, Prasher P, Rene ER (2021) Microbial technologies for heavy metal remediation: effect of process conditions and current practices. Clean Technol Environ Policy 1–23
134. Diep P, Mahadevan R, Yakunin AF (2018) Heavy metal removal by bioaccumulation using genetically engineered microorganisms. Front Bioeng Biotechnol 6:157
135. Wu C, Li F, Yi S, Ge F (2021) Genetically engineered microbial remediation of soils co-contaminated by heavy metals and polycyclic aromatic hydrocarbons: advances and ecological risk assessment. J Environ Manage 296:113185
136. Shehata M (2018) Using recombinant E. coli displaying surface heavy metal binding proteins for removal of Pb^{2+} from contaminated water: 北海道大学
137. Shahpiri A, Mohammadzadeh A (2018) Mercury removal by engineered Escherichia coli cells expressing different rice metallothionein isoforms. Ann Microbiol 68(3):145–152
138. Miglani GS (2017) Genome editing in crop improvement: Present scenario and future prospects. J Crop Improv 31(4):453–559
139. Lashani E, Amoozegar MA, Turner RJ, Moghimi H (2023) Use of microbial consortia in bioremediation of metalloid polluted environments. Microorganisms 11(4):891
140. Zhang S, Merino N, Okamoto A, Gedalanga P (2018) Interkingdom microbial consortia mechanisms to guide biotechnological applications. Microb Biotechnol 11(5):833–847
141. Harrison JJ, Ceri H, Stremick CA, Turner RJ (2004) Biofilm susceptibility to metal toxicity. Environ Microbiol 6(12):1220–1227
142. Benazir JF, Suganthi R, Rajvel D, Pooja MP, Mathithumilan B (2010) Bioremediation of chromium in tannery effluent by microbial consortia. Afr J Biotech 9(21):3140–3143
143. Muneer B, JIqbal M, Shakoori F, Shakoori A (2013) Tolerance and biosorption of mercury by microbial consortia: potential use in bioremediation of wastewater. Pak J Zoology 45(1)
144. Li M, Yao J, Sunahara G, Hawari J, Duran R, Liu J et al (2022) Novel microbial consortia facilitate metalliferous immobilization in non-ferrous metal(loid)s contaminated smelter soil: efficiency and mechanisms. Environ Pollut 313:120042
145. Shi L-D, Lv P-L, Wang M, Lai C-Y, Zhao H-P (2020) A mixed consortium of methanotrophic archaea and bacteria boosts methane-dependent selenate reduction. Sci Total Environ 732:139310
146. Pande V, Pandey SC, Sati D, Bhatt P, Samant M (2022) Microbial interventions in bioremediation of heavy metal contaminants in agroecosystem. Front Microbiol 13:824084
147. Arshad F, Selvaraj M, Zain J, Banat F, Haija MA (2019) Polyethylenimine modified graphene oxide hydrogel composite as an efficient adsorbent for heavy metal ions. Sep Purif Technol 209:870–880
148. Tan W, Peralta-Videa JR, Gardea-Torresdey JL (2018) Interaction of titanium dioxide nanoparticles with soil components and plants: current knowledge and future research needs–a critical review. Environ Sci Nano 5(2):257–278
149. Cao X, Alabresm A, Chen YP, Decho AW, Lead J (2020) Improved metal remediation using a combined bacterial and nanoscience approach. Sci Total Environ 704:135378
150. Baby R, Saifullah B, Hussein MZ (2019) Carbon nanomaterials for the treatment of heavy metal-contaminated water and environmental remediation. Nanoscale Res Lett 14:1–17
151. Zarime NA, Yaacob WZW, Jamil H (eds) (2018) Removal of heavy metals using bentonite supported nano-zero valent iron particles. AIP conference proceedings. AIP Publishing
152. Zhang Y, Wu B, Xu H, Liu H, Wang M, He Y et al (2016) Nanomaterials-enabled water and wastewater treatment. NanoImpact 3:22–39
153. Xue W, Huang D, Zeng G, Wan J, Zhang C, Xu R et al (2018) Nanoscale zero-valent iron coated with rhamnolipid as an effective stabilizer for immobilization of Cd and Pb in river sediments. J Hazard Mater 341:381–389

154. Cheng J, Sun Z, Yu Y, Li X, Li T (2019) Effects of modified carbon black nanoparticles on plant-microbe remediation of petroleum and heavy metal co-contaminated soils. Int J Phytorem 21(7):634–642
155. Fan L, Song J, Bai W, Wang S, Zeng M, Li X et al (2016) Chelating capture and magnetic removal of non-magnetic heavy metal substances from soil. Sci Rep 6(1):21027

Biotechnological Strategies for Effective Remediation of Heavy Metals

Muibat Omotola Fashola, Olajide Solomon Anagun, Mautin Lawrence Ogun, Prince Emeka Ndimele, and Olubukola Oluranti Babalola

Abstract Anthropogenic activities resulting from various industrial processes and some natural activities has caused the rise in discharge of heavy metals into the ecosystem. Environmental problems as a result of indiscriminate release of heavy metals is of significance due to the threat pose to all living biota in the environ and their perdurable character which makes them to pile-up in the energy chain and exerts their toxic consequences on all living components of the ecosystem. This necessitates urgent need to prevent their spread and elimination from the environment. The commonly used physicochemical techniques of remediation has many drawbacks that include loss of soil microbial diversity and function, secondary pollution, cost ineffectiveness, alteration of soil properties that render the soil unfit for agricultural purposes and other uses amongst others. Due to these concerns, research into the use of environmentally friendly and sustainable approaches has been on the increase. Novel prospects of employing microorganisms (bioremediation) and plants (phytoremediation, rhizoremediation) to alter or eliminate heavy metals from the environment has been identified. In this section, various plant and microbe-based processes are identified as the biotechnological strategies that are currently adopted for reduction or elimination of heavy metal from polluted areas. This present chapter also discuss the mechanisms responsible for the actualization of the identified biotechnological strategies. It is therefore encouraged to adopt the use of these bio-based approaches in large-scale treatment of heavy metal polluted sites.

M. O. Fashola (✉) · O. S. Anagun
Department of Microbiology, Faculty of Science, Lagos State University, Ojo, Lagos, Nigeria
e-mail: muibat.fashola@lasu.edu.ng

M. O. Fashola · O. O. Babalola
Food Security and Safety Niche Area, Faculty of Natural and Agricultural Science, North West University, Mmabatho, South Africa

M. L. Ogun
Department of Botany, Faculty of Science, Lagos State University, Ojo, Lagos, Nigeria

P. E. Ndimele
Department of Fisheries and Aquatic Biology, Faculty of Science, Lagos State University, Ojo, Lagos, Nigeria

N. Kumar (ed.), *Heavy Metal Remediation*, Earth and Environmental Sciences Library,
https://doi.org/10.1007/978-3-031-53688-5_7

Keywords Bioleaching · Biosorption · Biosurfactant · Phytoremediation · Siderophore

1 Introduction

Heavy metal contamination is becoming a significant universal issue for the ecosystem and the general wellness. Several reports have been made on the excessive concentrations of heavy metals in various microenvironments- air, water-bodies, and soil, affecting underground water and agricultural soils thus elevating the harmful concentration of metal contaminants in consumed plants with consequent effect on human being [64, 104]. The development recorded due to globalization and modernization has intensified heavy metal outflow and contamination from human processes. Human-related activities and sites such as mining, smelting, petroleum refinery and many others have all resulted to heavy metal pollution, harming the well-being ecosystem and its components [11, 32, 81]. Global concern is growing regarding their possible harmful health consequences and prolonged implications on all living biota. Although, minute amount of these metals is present in our surroundings, and food items, when consumed at minute concentrations by organisms, are not harmful because some are essential for growth, such as cobalt, iron, copper, and zinc, but at higher concentrations become toxic [46].

Heavy metals are easily identified because of its larger atomic weight or density. The phrase *heavy metal* refers to metalloids or metallic elements having densities larger than 5 g/cm^3 that are harmful to all forms of life [105]. Metal compounds have the tendency to generate polar bonds, which are necessary for more toxic form known as metalloids. Heavy metals can become lipophilic compounds or ions by covalently attaching to organic groups [42]. Because these metallic compounds have high lipid affinity, they penetrate through the cell membrane and enter the cell. When these metallic compounds interact with cell organelles, they have harmful consequences [21].

Various strategies are available for the reduction or elimination of heavy metals from polluted areas. Physicochemical techniques, such as soil washing, soil flushing, vitrification, encapsulation, chemical precipitation, coagulation, flocculation, electrochemical treatment, adsorption and ion exchange are some of the example of readily used method for treatment of heavy metal polluted environments [57, 120]. These physicochemical approaches though easy to use but are cost intensive with concerns of leading to secondary pollution; in addition, it should be noted that the efficacy of these approaches is contingent upon the presence of elevated levels of heavy metals in the surrounding environment, typically exceeding a concentration of 2 mM [9]. Given the drawbacks associated with conventional approaches, it is imperative to expedite the development of economically viable and environmentally sustainable techniques for the efficient mitigation of heavy metal contamination in the surrounding ecosystem [117].

Biotechnological strategies aimed at reduction or elimination of heavy metals from the ecosystem are usually considered inexpensive and eco-friendly [106]. These methods include plant-based remediation approaches (phytostabilization, phytoextraction, phytoaccumulation, phytovolatilization, phytofiltration, phytotransformation and rhizodegrdation) and microbe-based approaches (biosorption, bioleaching, bioaccumulation, biotransformation, and use of microbial value-added products such as biosurfactants and siderophores). Most recently, research are now designed at harmonizing both bio-based approaches together so as to have higher heavy metal removal efficiency, such methods include plant-growth-enhancing-bacteria aided phytoremediation and arbuscular mycorrhizal fungi aided phytoremediation [14, 73].

This section of the book focuses on the biotechnological strategies adopted in the elimination or reduction of heavy metals from pollutant-altered environment. In this chapter, we also discussed the mechanisms behind each emerging biotechnological approaches for effective elimination or reduction of heavy metals.

2 Heavy Metals and Their Toxicological Effects

Some heavy metals are required for human bioprocesses, but their ingestion into the body can have unanticipated bleak outcomes on health and the physiological system. According to Kim et al. [51], research reveal that, despite their beneficial health advantages, heavy metals are potential candidate of carcinogen. Dissolved heavy metals get into the energy chain as soil pollutants, water pollutants, and air pollutants, eventually accumulating in man, leading to higher severeness on the biological system and opening them to cell outgrowth. Rare metals (As, Cd, Cr) are important carcinogen, according to International Agency for Research on Cancer studies [51].

Asides from the carcinogenic consequence of heavy metals on man, water, soil and floral are also adversely impacted by heavy metals. The direct and indirect transportation of heavy metals into the aquatic ecosystems occurs via water cycle, leaving deposits in water bodies [83]. The presence of heavy metal in different water bodies like streams and many others undergo desorption into the water sediment. Miranda et al. [66] observed that heavy metals present in sediment can move into the hydrosphere, and their mobilities is dependent on a plethora of factors ranging from kind of heavy metal, as well as the physicochemical attributes of both the hydro-based ecosystem and soil-like base of the water, such as acidity, alkalinity, salt concentration and other noticeable water quality parameters.

The stress caused by these contaminants is a key component that has a negative influence on plant agriculture productivity. Heavy metals are commonly recognized as a component of biosphere. Their accumulation in soil and floras, on the other hand, can be harmful to the ecosystem [6]. Heavy metals can move up the energy chain when taken up by flora, eventually concentrating in the bodies of most animates thus, creating biota discomfort [78]. Heavy metals could penetrate underground or aboveground waters, where they can be absorbed by plants or released into the atmosphere as gases. Alternatively, they may develop non-lasting interaction with different soil particulate matter like clay or organic matter, eventually affecting man's wellness. To guarantee sustainability, soil contamination levels must be kept below the permissible toxicity limits [96]. The use of synthetic or inorganic fertilizers and insecticides has been shown to increase the chance of heavy metal saturation of soil, which can then concentrate in the internal organelles of plants grown in such polluted soil [52]. The amount of organic residue and levels acidity and alkalinity have a tangential control on heavy metal mobility in soil. When heavy metals are introduced into soil, their inward migration is restricted by the soil's ability to hold the metals or by the metals' interaction with the neighboring waste cast, which may increase their transportability. Heavy metal level in soil can cause a negative effect on legume development, growth, and nitrogen fixation [71].

3 Traditional Methods of Heavy Metal Removal

Several remediation technologies that involve the use of physicochemical, cellular, conduction and/or thermal procedures to eliminate toxic effects of heavy metals in the ecosystem. This is usually done by containing (e.g., landfilling, encapsulation), immobilizing (e.g., vitrification, solidification) and extracting/removing (e.g., soil flushing, soil washing, electrokinectics) of heavy metal polluted environment [57]. These methods can be in-situ (onsite remediation of polluted soil) or ex-situ (contaminated soils are removed from its natural environment to another location for treatment).

3.1 Soil Washing

This involves the use of physical processes to separate soil particles highly contaminated with heavy metals. These particles then undergo chemical extraction with precise washing agents which release the heavy metals from the soil to the extracting solution. The efficiency of the cleanup process is largely dependent on the extracting reagents used [28, 99]. Soil washing has been identified as one of the methods that can wholly eliminate heavy metals from the soil, however, this technique is limited in durability due to loss of some soil minerals and tier-two pollution [33, 116].

3.2 Soil Flushing

This is an on-site technique in which a solvent is brought into metal-rich soil. These solvents could be an aqueous solution of surfactants, co-solvents, acids, bases, oxidants, chelants, or water introduced into the contaminated soil. Introduction of the flushing agents (solvents) is usually done either within saturated zone, unsaturated zone or both contaminated soils. The mobilization or dissolution of the metals sorbed to the soil cast increases by the flushing solution [113]. Standard ground water extraction wells are usually made to capture and pump the contaminated groundwater and extraction fluids to the surface. Finally, the resulting extraction fluids with the desorbed metals as well as the gaseous release of vaporous pollutants from gained flushing fluids are harvested and treated. The major limitation of this method is the risk of spreading the heavy metals into uncontaminated areas and the effects of flooding the solution into the soil environment [94].

3.3 Vitrification

This is a heat correctional method that uses thermal based sources to alter heavy metal laden soil into glasslike solids. It involves application of intensive energy to metal laden soil to form a high temperature region zone (> 1500 °C) The soil is then melted into molten "lava" and becomes glasslike material after cooling. Captured heavy metals are in the glassy matrix while any other organic pollutants are eliminated. Vitrification technique is destructive, and the processed soil is usually not suitable again for farming purposes due to loss of essential nutrients of the soil [98].

3.4 Encapsulation

This is also known as liner method which is a remedial technique that is used to contain the heavy metal-rich soil in a suitably made physical barrier system comprising of low permeability caps, surrounding underground barriers, and in some cases barrier floors [58]. Isolated of heavy metal rich soil are made for pollutants restriction, removing pollutants through off-site dispersion and bio-exposure of toxicants on-site [47, 65]. Leaching of contaminants into groundwater is prevented by the low permeability caps which is often made of inorganic textile sheets or clay layers. Impermeable barriers underground restrict the pollutant source from horizontal mobility to neighboring areas through potential dispersion and interflow on the subsurface [58].

The physicochemical method of removing heavy metal from contaminated effluents includes ultrafiltration, flocculation, chemical precipitation, adsorption, ion exchange, electrochemical deposition, and membrane filtration.

3.5 *Chemical Precipitation*

Involves adjustment of pH of contaminated water with the utilization of precipitating chemicals like lime and limestones. Heavy metals concentration higher than 1000 mg/L can be precipitated from contaminated water using lime but the major setback of this method is the production of primary sludge that requires additional treatment and final disposal which is expensive. In addition, high amount of chemicals is needed to reduce metal concentration in the water to acceptable level [20].

3.6 *Ion Exchange*

This method encompasses use of a solid cast such as chemical organic ion exchange resins to adsorb, bond and exchange cations or anions from metal laden wastewater. Some of the disadvantages of this method is that it cannot be used for wastewater with elevated levels of heavy metals because the matrix is easily contaminated by other particles in the wastewater. Also, this technique is not selective and is very reactive to changes in pH [120].

3.7 *Adsorption*

Adsorption is the use of agricultural wastes derived adsorbents, a synthetic by-product, natural material, or modified biopolymers to remove heavy metals from water. This usually involve adhesion of the heavy metals onto the surface of solid adsorbent with the aid of chemical bondage and the physical attraction between the contaminant and adsorbent [12].

4 Biotechnological Strategies of Heavy Metals Removal

The traditional methods of heavy metals removal as indicated in the section above are mostly limited in application because of their drawbacks which adversely impacts on the ecosystem, hence, it becomes imperative to source for better ecofriendly strategies with more sustainable practice in the environment. Most of these sustainable approaches are biologically mediated processes aimed at efficient and effective elimination or reduction of heavy metals from contaminated sites. Some of the recent biotechnological approaches and the various strategies involved are highlighted below.

4.1 Phytoremediation and Its Mechanisms

The cleanup of heavy metal laden area with the usage of plants in the technique is known as phytoremediation [69], this word originated from Greek word *phyton* (plant) and the Latin word *remediare* (repair). A low-cost, non-invasive alternative technique to engineering-based remediation approaches. Phytoremediation involve plants, plant products, and their partnership with microbes to detoxify or eliminate heavy metals from contaminated ecosystem [62]. It is an ecofriendly and promising technology for locating, reducing, and eliminating heavy metals. Phytoremediation enables the growth of plants in a polluted site and influence the bio-physicochemical activities occurring in the plants with the aim to significantly contribute to the efficient elimination or reduction of heavy metals from the biosphere. The removal of hazardous trace metals that are toxic to human health and other biological systems can be accomplished utilizing a plethora species of plants [48]. There are more than 400 known ecotypes of plants that are hyperaccumulators of metals (Table 1).

Plants that are hyperaccumulators are those that store large amounts of metals in their internal organelles. The concentration of metals deposited in their tissues is often what sets them apart from other plants [69] and can be hundreds to thousands of numbers higher. If the level of metals in the leaves relative to the dry weight of the leaf tissue is greater than the following thresholds: 100 $\mu g\ g^{-1}$ of Cd, Tl, Se; 300 $\mu g\ g^{-1}$ of Co, Cu, Cr; 1000 $\mu g\ g^{-1}$ of Ni, As, Pb or rare earth elements; 3000 $\mu g\ g^{-1}$ of Zn; 10,000 $\mu g\ g^{-1}$ of Mn for the elements in their natural environment, the plant can be classified as a hyperaccumulator [15, 16, 121].

The uptake of heavy metals from the ecosystem and detoxifying their toxic effects by species of plants is the goal of phytoremediation. This remediation is often done in sustainable manner, which confer on this technology a merit over the traditional techniques, which often has negative side consequences on the biota and environment [44]. Numerous processes or phases are needed to eliminate or reduce heavy metals from water, air and soil leading to the release of nontoxic compounds, thereby bringing about reduction or total reduction of these toxicants from the ecosystem. The various mechanisms involved in phytoremediation of heavy metals include (a) phytostabilization: reduction of heavy metal bioavailability using different species of plants in any ecosystem, (b) phytoextraction: absorption and storage of contaminants-heavy metals within tissues of plants or elimination of heavy metals from soil using floras (c) phytovolatilization: absorption and conversion of harmful heavy metals into non-toxic derivatives and their onward release into the atmosphere using plants (d) phytofiltration: absorption or adsorption of heavy metal ions from waters using hydrophytes, (e) phytotransformation and rhizodegradation: breakdown of contaminants-heavy metals using plant released exudates and microbes [10, 48, 121] and this is illustrated below in Fig. 1.

Table 1 Selected plants with heavy metals phytoremediation potentials

Species	Heavy metals	Habitat types
Brassica juncea	Se, Cd, Pb, Ni	Soil
Eichhornia crissipis	Cd, Cu, Ni, Pb, Zn	Water
Lemna sp.	Cd, Cu, Ni, Pb, Zn	Water
Helianthus annuus	Cd, Zn, Cu	Soil
Populus sp.	Hg	Soil
Zea mays and Orysa sativa	Pb	Soil
Eucalyptus camaldulensis	As, Cd, Cr, Cu, Hg	Soil
Acacia nilotica	As, Cd, Cr, Cu, Hg	Soil
Pistia stratiotes	Cd, Pb, Zn	Water
Amaranthus viridis	Cr	Soil
Arabdopsis thaliana	Cd, As	Soil
Nicotiana tobacum	Cd	Soil
Pteris sp.	As, Cd, Cu	Water/soil
Eleocharis acicularis	As, Cu, Zn	Soil
Phytolacca americana	Cd	Soil
Azolla pinnata	Cd	Water/Soil
Festuca rubra	Hg	Soil/Water
Phyllanthus serpentinus	Ni	Soil
Sorghum bicolar	Cd, Cu, Zn	Soil
Lantana camara	Pb	Soil
Jatropha curcas L	Pb	Soil
Latuca sativa	Cu, Mn, Zn, Ni, Cd, Pb, Co	Soil
Pisum sativum	Pb, Cu, Zn, Ni, Ag, Cr	Soil
Aquilaria malaccensis	Fe	Water
Acacia sp.	Cu, Cd	Soil
Dyera costulata	Cu, Cd	Soil
Hybiscus cannabinus	Pb	Soil

Adapted from [48, 69, 74, 121]

4.1.1 Phytostabilization

The prevention of heavy metals from spreading across the environment and contaminating the food chain, which utilizes of different metal-tolerant plant species to immobilize them underground and reduce their bio-presence is often described as phyto-stabilization. This stops pollutants from spreading from one area to another, toxins in the underground water or soil are sunk in the root or concentrate in the root domain as shown in Fig. 1 [10, 48, 121]. Heavy metals may be phytostabilize

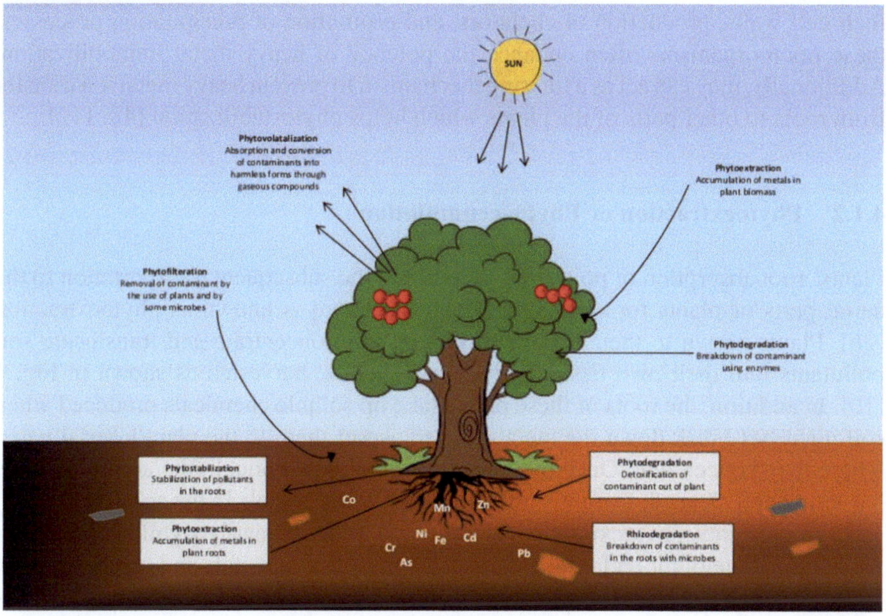

Fig. 1 Plant-based strategies for heavy metal removal

by rhizosphere precipitation, rhizosphere metal valence reduction, root tissue uptake and sequestration, and root cell wall adsorption [36].

Plant growth typically helps conserve soil wellness in heavy metal-rich areas and in addition to eliminating the number of metals that seep into the groundwater. Most developed vegetation cover also stops soil particles from being blown around and dispersing any heavy metals they may contain [48, 53]. Phytostabilization eliminates potentially dangerous biomass of heavy metals. In phytostabilization, choice of floras to be utilized is very important because the plant species must be able to coexist with high heavy metal contents. Plants should have thick rooting systems because they perform a pivotal role in immobilizing heavy metals, soil structure stabilization, with soil erosion prevention [26, 121]. Vegetation cover needs to be established in a site quickly, so plants need to be able to produce a lot of biomasses and develop quickly. It is important that the plant cover is manageable in the field [53, 63]. Several plant species that match the characteristics have been found and employed for phytostabilization of heavy metal-rich soils as documented by Refs. [19, 44, 106]. The effectiveness of phytostabilization could be enhanced by amending the polluted soil with organic or inorganic materials. Soil pH and redox state can be modified by soil amendments [23, 38, 90, 123] to affect metal speciation and decrease heavy metal solubility and bioavailability. Plant colonization and water holding capacity can both be enhanced by amending the soil with organic matter and critical nutrients [90] because of the soil's improved bio-physicochemical qualities. Rhizosphere bacteria play a crucial role in the stabilization of plants. Through adsorption of metals onto their cell walls, production of chelators, and promotion of precipitation processes, these microorganisms often enhance the potency of heavy metal immobilization. Additionally, they can act as a filtering mechanism to prevent heavy metal ion transfer from roots to other parts of the plants which helps phytostabilization [48, 112].

4.1.2 Phytoextraction or Phytoaccumulation

Plants' root absorption of poisonous metals and the subsequent transportation to the aerial parts of plants for storage and later harvested is known as phytoextraction [26]. Plants grown to their maximum height will concentrate and translocate soil pollutants into their own tissues, which can then be harvested as shown in Fig. 1 [10]. In addition, the roots of these plants take up soluble chemicals produced when soil microbes break down the metals and transport them to the plant's aerial parts, where they concentrate. Once the plants have matured enough, the aerial parts are harvested and cleaned, thereby eliminating the heavy metals in the soil [26].

In recent years, phytoextraction has emerged as the primarily significant phytoremediation strategy for reclaiming damaged soil of heavy metals and metalloids [70, 100]. This is because plants can take up heavy metals from soil or water, translocate and concentrate those heavy metals in their uppermost biomass [41, 48]. Unlike phytostabilization, in which plants are grown with heavy metals and the metals are ultimately returned to the soil, phytoextraction permanently removed the heavy metals from the substrate. Heavy metal phytoextraction occurs through several steps, including the transportability of metals in the root domain, plant roots' uptake of metals, the movement of metal ions from the roots to other parts of the plant, the segregation and compartmentalization of metal ions in the internal parts of plant, and finally, the release of metals from the plant. Plant preference, plant yield, heavy metal bio-presence, soil, and root domain characteristics perform a critical role in how effective phytoextraction is [10, 26, 48, 121].

The success of phytoextraction strategies depend on the choice of plant species used. Some of characteristics of good phytoextractors are plants with high tolerance for nutrient-deficient soil, simple cropping and harvesting, and high resistance to metal toxicity and many others. Consequently, a good plant species' phytoextraction capacity is mostly attributed to its metal-accumulating capacities and aboveground biomass. As a result, there are two compelling hypotheses for plant selection: the use of hyperaccumulator plants, which can concentrate heavy metals in their aerial parts to a greater extent, and the use of plants with higher stem and leave biomass production, which may have less metal-accumulating abilities, but overall accumulation of heavy metals is comparable to that of hyperaccumulators [7, 107, 121].

Hyperaccumulators are plant species that can store hazardous amounts of metals in their leaves and stems without showing any signs of damage. At least 45 angiosperm families, including those of the Brassicaceae, Fabaceae, Euphorbiaceae, Asterraceae, Lamiaceae, and Scrophulariaceae [48, 73], have been found to contain metal hyperaccumulators among their members. [48] provide a thorough explanation for how some plants might amass more than two elements (Table 1). However, heavy metals can accumulate in edible sections of plants during phytoremediation, generating concerns about human health if the plants are subsequently consumed by humans or animals. The key to successful and secure heavy metal phytoremediation is, therefore, the careful selection of non-edible hyperaccumulators. A plethora of hyperaccumulators have been documented and utilized in the elimination or reduction of heavy metals, however many of them are transient, producing little biomass at a slow growth rate. Because of its rapid growth, tolerance to toxicant loads, high biomass, and several yield in a single cycle of cultivation, *Trifolium alexandrinum* is chosen as a promising option for phytoextraction of Cd, Pb, Cu, and Zn. Trees and other woody plants are employed in phytoextraction because of their many useful properties [121].

4.1.3 Phytovolatilization

Phytovolatilization, also known as phytoevaporation or phytoxidation, is a mechanism that involves the process of utilizing plants to remove toxicants from the soil, transform them into a non-harmful gaseous state, and then release them into the atmosphere through the transpiration mechanism of the leaves or floral system. The roots take up the pollutants, transport them to the shoot, and then release them into the air via the leaves containing stomata in this process (Fig. 1) [69, 121]. It is also the procedure by which plants absorb contaminants and transform them into less dangerous vaporous forms. Organic pollutants and some heavy metals like selenium, mercury and arsenic are amenable through this strategy [60]. Plants in the family Brassicaceae, such as *Brassica juncea*, are effective selenium vapourizers [121]. For example, selenium is first incorporated into the body as the seleno-amino acids selenocysteine and selenomethionine. Then, biomethylation of selenomethionine yields dimethylselenide (DMSe), a gas that is less hazardous than inorganic selenium and can be released into the atmosphere [119, 122]. Metalloid (heavy metal) pollutants are vaporized and released far from the original site through phytovolatilization, which is preferable to other phytoremediation procedures since it eliminates the requirement for plant harvesting and disposal [50, 69]. However, phytovolatilization as a correctional strategy does not eradicate the contaminants entirely; they are still present in the environment. Toxic volatile substances in the soil are released into the air and contaminate the surrounding area. Thus, prior to its application in the field, ecological risk assessment is required.

4.1.4 Phytofiltration

The use of roots (in rhizofiltration), stems (in caulofiltration), and young plants (in blastofiltration) to filter out contaminants in water encompasses the technique of phytofiltration [88]. This technique is often used because certain plant species' roots can absorb and adsorb rare elements from a wide range of water origin, including underground and top-leveled waters, commercial, rural, and agronomic wastewaters, and water from acid mine sites [69]. Plants employed in rhizofiltration must first be grown in clean water until their root systems are sufficiently developed, and only then are they transferred to contaminated water to acclimatize. This was supported by [118] when they reported that for plants to eliminate or reduce poisonous metals from a polluted area, they must first acclimatize to the environment. Rhizofiltration can lead to the assimilation of poisonous metals by the roots. The acidity of root exudates can change the pH of the root zone, causing heavy metals to precipitate on the plant's roots [43] and thereby eliminating their potential to enter the groundwater supply. Rhizofiltration plants are aquaculturally cultivated in sterile water to produce a dense root structure; later, this water is replaced with dirty water to acclimatise the plants. Rhizofiltration plants, ideally, will have a thick root structure, high yield production, and then poisonous metal tolerance. Rhizofiltration can be done with plants from both the land and the sea. Wetland water is typically remedied with aquatic plants like hyacinth, azolla, duckweed, cattail, and poplar because these plants can accumulate heavy metals efficiently, tolerate toxic conditions well, or grow quickly while producing a lot of biomasses [121]. In comparison to their watery counterparts, the roots of land plants such as sunflower (*H. annuus*) and many others often grew more and hairier. Rhizofiltration studies have found that they are effective at accumulating heavy metals [29, 91]. Water-loving plants such as *Eichhornia crassipes* and *Lemna minor* have been used in various technical variants of rhizofiltration [69, 85]. Noteworthy, in phytofiltration, acclimated plants are then taken to the polluted area where the heavy metals will be extracted immediately roots are formed for absorption of all the toxicants from the water or soil, followed by harvesting and elimination of the plants.

4.1.5 Phytotransformation and Rhizodegradation

In this mechanism, the roots secrete biochemical substances like dehalogenases, nitroreductases, and peroxidases that distintegrate contaminants outside the plant. This process, known as phytotransformation or phytodegradation, is responsible for the breakdown of poisonous metal complexes assimilated by flora through metabolic processes [74, 95] as shown in Fig. 1. With this mechanism, some plants are capable of transforming contaminants-metals into non-toxic derivatives. It has been documented that plants such as *Vetiveria zizanioides* could clean up 97% of trinitrotoluene from the soil. Moreso, *Nicotiana tabacum* have been reported for its breakdown of trinitrotoluene by NfsI nitroreductase enzyme secreted by roots [108]. A great number of plants could break down contaminants through a close association with bacteria and mycorrhizae in best environments. The contents of this breakdown are vaporized or married with the cast of the soil [80]. The release of plant exudates like sugars and organic acids often improve microbial populations [5, 102]. Unlike phytodegradation, rhizodegradation involves the disintegration of pollutants within the root planes using the microbial biota that inhabits the zone [109]. Rhizodegradation often aids favorable soil properties like soil aeration and moisture content which enhanced plants and soil health. However, a study conducted by [82], proved that rhizomes of *Typha latifolia* broke down Terbuthylazine in a wetland ecosystem and this make it a promising species to Terbuthylazine phytodegradation. Also, *Rhizophora mangle* in conjunction with rhizobacteria such as *Pseudomonas aeruginosa* degraded polycyclic aromatic hydrocarbons in polluted sediment [97].

4.2 Phytoremediation: Upskilled Mechanisms for Efficient Remediation

Recently, for easy and efficient phytoremediation process or due to some deficiencies such as entrance of contaminant or their derivatives into the food chain, initiation of mutation, delayed clean-up, absorption of toxic substances, and so on brought about by singleness of phytoremediation, researchers have broadened the effectiveness of plant assisted remediations with use of some modifications. This is because plants do not have the capacity to breakdown all pollutants as submitted by [10]. Thus, it is necessary to explore the merits of the tenacity of plants aided microorganism degradations, and they are as follows.

4.2.1 Plant-Growth Promoting Bacterial (PGPB) Aided Phytoremediation

Plant-growth promoting bacteria are free-living microorganisms usually found around plant roots, that is, the rhizosphere of plants, colonizing it for improved soil enrichment, promoting plant growth, and inhibiting plants from all forms of attacks [77]. PGPB could either be rhizospheric or endophytic in nature. However, several reports have linked the breakdown of toxic pollutants or their reduction to less toxic forms to these bacteria. Many scholars have documented a great number of PGPB with the promotion of phytoremediation of heavy metal contaminants. Most of the bacteria often play a significance in detoxicating metals by the production of chelators like siderophores and some natural acids enabling the bio-presence of heavy metals by providing suitable pH [69, 73]. Also, production of some organic polymers like polysaccharides which enhances heavy metal phytostabilization by reducing metal ions movement have been reported [72, 74, 79, 121]. However, PGPB often aids phytoremediation through higher promotion of plants' decontamination rate, increased enzymatic synthesis for enhanced breakdown of contaminants, and medication of soil pH. [86] documented that *Arthrobacter* inoculated to *Ocimum gratissimum* promotes the roots phyto-removal of Cd. Also, Pb-tolerant *Paxillus involutus* strain have been reported to increase yield and enhanced seedlings of *Populus canescens* to lead tolerance [111]. The outcome of a collaborative association between roots of plants and their allied microbes becomes evident. It has been reported that metabolite production by microbes increased by the root exudates supplied by plants while PGPB fast-track the release of plant-based hormones such as auxin, ethylene and many more and consequently promoting phytoextractions or phytodegradations [8]. Commonly used PGPB in metal extraction with plants are often belonging to the genera like *Agrobacterium*, *Bacillus*, *Burkholderia*, *Pseudomonas*, *Rhizobium* and many others [72].

4.2.2 Arbuscular Mycorrhizal Fungi (AMF) Aided Phytoremediation

Arbuscular mycorrhizal fungi (AMF) are mutualistic fungi relationship with plants' roots to enhance the bio-presence of minerals. Restriction of poisonous metals due to the secretion of chelators and its adherence to the fungal cells wall, and phytoextraction of heavy metals by increasing plant growth with high metals uptake within the domain of the root by changing the physiochemical features of roots exudate are the two major mechanisms employed by AMF aided phytoremediation [72]. Nedjimi [74] in its submission documented *Cassia talic* by AMF characteristically promoted cadmium resilience by inhibiting its mobilization to shoots. Also, relationship between *Festuca arundinacea* and *Glomus m*osseae enhanced both Ni transport and visibility of ABC transporter and metallothionein genes [101]. However, growth, metal pull-up and phosphorus contents in *Zea mays* cultivated in contaminated strontium and cadmium soil have been documented to increased voraciously due to the inoculation with AMF [25]. The use of enzymes produced by AMF aided phytoremediation is ascribed as basis of the advantages because these compounds such as siderophore have a plain chemical model, in addition, the use of these enzymes in the transformation of polluting substances often lead not into the accumulation of harmful derivatives, and the biochemical substances are subsequently used by fungi in the polluted ecosystem.

4.3 *Microbial Remediation of Heavy Metals and Its Mechanisms*

Biotechnological strategies are utilized the removal of poisonous metals from the environment often makes use of microbes-aided techniques which subsequently engenders the remediations of metal polluted sites. Most microbial mediated heavy metals removal involve techniques like biosorption, bioleaching, bioaccumulation, biotransformation, and utilization of microbial products such as siderophores and biosurfactants as illustrated in Fig. 2.

These mechanisms are mostly used because of their low-cost effectiveness, maintenance of soil fertility, and their eco-friendliness.

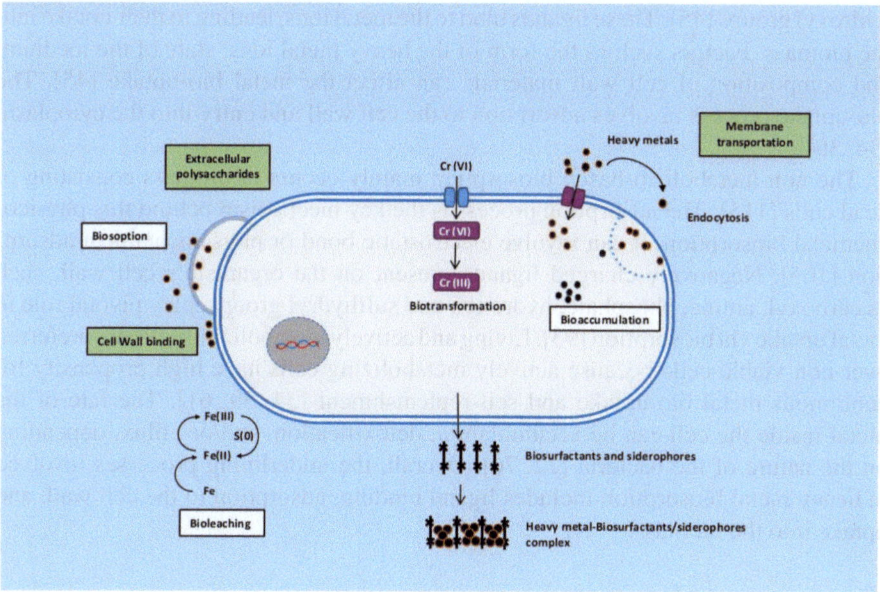

Fig. 2 Microbial-based strategies for heavy metal removal

4.3.1 Biosorption

Biosorption involve the technique where ionic metals are bound to metal binding proteins present on the cell wall of microorganisms or other biosorbent materials [103]. This process is exhibited by numerous organisms like yeasts, bacteria and many others and even agricultural waste materials. One of the advantages of biosorption is that it is a cheaper method compared to traditional metal reduction or elimination methods. The biomass synthesis in bacteria or fungi, is relatively inexpensive. Additionally, biosorption allows for the cleanup extreme quantities of effluents, making it suitable for industrial applications [2]. Also, one other merit of biosorption is that it can remove multiple heavy metals simultaneously. Biosorbents could selectively uptake and remove specific metals, making them highly efficient in metal removal [1]. This selectivity is advantageous in situations where multiple heavy metals are present in contaminated water sources. Furthermore, biosorption offers a more environmentally friendly approach to heavy metal removal. It does not require the addition of chemicals, as the biosorbents have a natural affinity for metal ions [3]. This reduces the need for additional treatment steps and minimizes the generation of chemical waste.

The mechanisms used in biosorption of poisonous metals could be categorized into two main types: metabolism-based biosorption and non-metabolism-based biosorption. Metabolism-based biosorption occurs in living biomass and involves energy from ATP. The metal ions are immobilized by ligands present on the cell wall of biological materials, such as phosphoryl, carboxyl, carbonyl, sulfhydryl, and hydroxyl groups [45]. These ligands bind to the metal ions, leading to their uptake into the biomass. Factors such as the form of the heavy metal ions, state of the medium, and composition of cell wall materials can affect the metal bio-uptake [45]. The bio-uptake process involves adsorption to the cell wall and entry into the cytoplasm [34, 39].

The non-metabolism-based biosorption mainly occurs in biomass consisting of dead cells [115]. The adsorption process is the key mechanism behind this physicochemical biosorption. It can involve electrostatic bond or physicochemical adsorption [103]. Negatively charged ligands present on the organism's cell wall, such as carboxyl, amine, phosphate, hydroxyl, and sulfhydryl groups, play pivotal role in metal uptake via biosorption [93]. Living and actively metabolizing cells are preferred over non-viable cells because actively metabolizing cells have high propensity for continuous metal bio-uptake and self-replenishment [34, 39, 61]. The fate of the metal inside the cell can be accumulation, detoxification, and/or efflux, depending on the nature of the bacteria [22, 76]. Overall, the underlining processes involved in heavy metal biosorption includes ligand binding, adsorption to the cell wall, and uptake into the biomass.

4.3.2 Bioleaching

Bioleaching is a sustainable method used for the removal of heavy metals from various sources such as mine tailings and other metal-laden environments. It involves the use of microorganisms to oxidize or dissolve insoluble components, such as heavy metals, from solid materials [35]. Bioleaching is defined as the solubilization of metals from insoluble solid substrates [30]. In the context of heavy metal removal from the environment, bioleaching has been proven to be an effective and environmentally friendly approach. Indigenous iron-oxidizing microorganisms have been used to bioleach heavy metals from sewage sludge, with ammonium ferrous sulfate and ferrous sulfate as energy sources [84]. Similarly, indigenous bacteria from mine tailings have been found to enhance the mobility of arsenic in response to organic supply, indicating the potential of bioleaching for arsenic removal [55]. Bioleaching can also be combined with other technologies to improve the efficiency of heavy metal removal. For example, mixed moderate thermophiles have been used in the bioleaching of low-grade copper tailings, resulting in increased leaching rates [40]. Additionally, the development of enzymes for bioleaching reactions has allowed for the selective leaching of heavy metals from tailings, improving the leaching rate and economic value while effectively treating metal pollution [19]. This method of microbial remediation of heavy metal, offers a sustainable and efficient technique for the removal of heavy metals from mine tailings especially, contributing to the control of pollution and the recovery of valuable metals [87]. Bioleaching involves the interaction between microorganisms and minerals to extract heavy metals. The mechanisms involved in bioleaching can be categorized into direct and indirect mechanisms [35].

Direct Mechanism

In the direct mechanism, microorganisms directly interact with the mineral surface and facilitate the leaching of heavy metals. This mechanism involves the following processes: *Oxidation*-Acidophilic microorganisms, such as *Acidithiobacillus ferrooxidans* and *Acidithiobacillus thiooxidans*, oxidize ferrous ions (Fe^{2+}) to ferric ions (Fe^{3+}). This oxidation reaction generates protons (H^+) and lowers the pH, creating an acidic environment that promotes metal solubility; *Complexation:* microorganisms produce organic acids, such as gluconic acid and oxalic acid, which can form complexes with heavy metals. These organic acids chelate with metal ions, enhancing their solubility and facilitating their leaching from the mineral matrix; and *Adsorption: m*icroorganisms can attach to the mineral surface, forming a biofilm. This biofilm acts as a barrier between the mineral and the leaching solution, preventing the re-precipitation of dissolved metals and facilitating their release into the solution [18].

Indirect Mechanism

In the indirect mechanism, microorganisms do not directly interact with the mineral surface but produce metabolites that facilitate metal leaching. This mechanism involves the following processes: *Redox Reactions*: microorganisms can catalyze redox reactions by utilizing energy sources such as iron or sulfur. For example, *Acidithiobacillus ferrooxidans* oxidizes ferrous ions to ferric ions, which can then react with sulfide minerals, releasing metal ions into the solution; *Acid Production*: microorganisms produce sulfuric acid through the oxidation of sulfur compounds. This acid production lowers the pH, promoting the dissolution of metal ions from the mineral matrix; and *Biofilm Formation*: microorganisms can form biofilms on the mineral surface, creating a microenvironment with different pH and redox conditions. This biofilm can enhance the leaching efficiency by creating localized conditions favorable for metal solubilization. These mechanisms work in synergy to facilitate the leaching of heavy metals from minerals. The specific mechanisms involved may vary depending on the microorganisms used and the characteristics of the mineral matrix [114].

4.3.3 Bioaccumulation and Biotransformation

Bioaccumulation refers to the process by which heavy metals accumulate and increase in concentration within microorganisms over time. This occurs mainly via direct uptake from the environment [110]. Biotransformation, on the other hand, refers to the chemical alteration or conversion of heavy metals by biological processes. Microorganisms can modify the chemical form of heavy metals, either by reducing their toxicity or by converting them into less mobile or less toxic forms. Biotransformation can involve processes such as reduction, oxidation, methylation, demethylation, and complexation [92].

In the context of heavy metal removal, bioaccumulation and biotransformation play important roles in reducing the concentration and mobility of heavy metals in contaminated environments. Certain microorganisms can accumulate heavy metals within their cells, effectively removing them from the surrounding environment. This process can be enhanced by specific microbial strains or genetic modifications [37, 68]. Biotransformation processes, such as reduction or methylation, can also transform heavy metals into less toxic or less mobile forms. For example, some microorganisms can reduce toxic forms of mercury (e.g., methylmercury) to elemental mercury, which is less harmful. Similarly, certain bacteria can convert toxic forms of arsenic into volatile arsines, which can be released into the atmosphere, reducing their impact on the environment [37, 56]. Overall, bioaccumulation and biotransformation are important strategies adopted by microorganisms in the removal of heavy metals from contaminated environments. They can help reduce the concentration, mobility, and toxicity of heavy metals, making them less harmful to ecosystems and human health. However, the effectiveness of these processes can vary depending on the specific heavy metal, the organisms involved, and the environmental conditions [31, 32].

Microbes use various mechanisms in the bioaccumulation and biotransformation of heavy metals. These mechanisms include active transport, enzymatic detoxification, and reduction reactions.

Active Transport

Microorganisms have transport systems embedded in their cell membranes that allow them to actively uptake heavy metal ions. These transport systems include ion pumps, protein channels, and carrier-mediated transport. Different microbial species have been found to have specific transport systems for different heavy metals, such as *Pseudomonas putida* for cadmium, *Thiobacillus ferrooxidans* for silver, and *Citrobacter* sp. for lead and cadmium [27].

Enzymatic Detoxification

Microorganisms possess enzymatic systems that can transform heavy metals from highly toxic forms to less harmful forms. Enzymes such as oxidoreductases, hydrolases, and methyltransferases are involved in the detoxification of heavy metals. For example, some bacteria such as *Pseudomonas putida* and *Bacillus subtilis* have been shown to be efficient in the enzymatic reduction of toxic form of chromium to less toxic forms (C(VI)) to (Cr(III)) [89].

Reduction Reactions

Microorganisms can also catalyze reduction reactions to transform heavy metals. Reduction reactions involve the transfer of electrons from the microorganism to the heavy metal, resulting in a change in its oxidation state. This can lead to the conversion of highly toxic forms of heavy metals to less toxic or less mobile forms. For instance, certain microorganisms can reduce toxic forms of mercury (e.g., methylmercury) to elemental mercury, which is less harmful.

These mechanisms allow microorganisms to accumulate heavy metals within their cells and transform them into less toxic or less mobile forms, contributing to the removal of heavy metals from contaminated environments.

4.3.4 Biosurfactant and Siderophore

Biosurfactants and siderophores are value-added microbial products that can be used as an alternative heavy metal remediating agent. They are produced by a variety of bacteria and fungi and confers heavy metal tolerance factors on the microorganisms producing them. Both microbial products have received increasing focus over the years due to their low toxicity, high biodegradability, and promising environmental reliability [4, 59]. Biosurfactants are amphiphilic surface-active biomolecules secreted extracellularly on the surface of the producer strains, which can form complex with heavy metals via processes such as ion exchange, precipitation, counter ion binding and electrostatic interaction [13, 54]. Bioremediation of heavy metals by biosurfactants in the environment is enhanced by their ability to form micelles. Nanostructures can be formed through the equilibrium process of micelle production [67]. Micelle production improves the solubility of contaminants, allowing for more efficient removal from the environment [24].

Siderophores, on the other hand, are naturally occurring microorganism-produced iron chelators with the primary purpose of removing environmental iron and re-supplying it to bacteria [17]. Siderophores are not limited to forming stable complexes with iron, they can also do so with other metals of environmental significance, such as aluminum, cadmium, copper, zinc, lead, gallium, indium as well as radionuclides like uranium and neptunium [75]. By increasing the metal's solubility after binding with the siderophore, the metal can be more easily extracted. However, it is common practice to inoculate the rhizosphere of metal-hyperaccumulator plants with bacteria that produce siderophores to increase the effectiveness of metal removal.

5 Conclusion

The continued global environmental pollution arising from heavy metals has called for several remediation approaches. The traditional methods of treatment are costly with residual consequential effects on the environment. The bio-based strategies remain the best option to treat metal laden sites due to the various mechanisms and enzymatic reactions utilized by microbes and plants to efficiently reduce and/or eliminate heavy metals from the contaminated environment. Microbes and plants have intrinsic biological mechanisms that enhance their survival when they are exposed to heavy metal stress and detoxify or eliminate the heavy metals from the environment. However, there is need to improve on the efficiency of the various strategies identified. Genomic and metabolic engineering approaches can be used to enhance tolerance, uptake and hyperaccumulation of heavy metals. The microbes can also be engineered by modification of their metabolic pathways and novel genes and plasmids can be integrated into the genome of promising microbes for improved performance.

References

1. Abbas SH, Ismail IM, Mostafa TM, Sulaymon AH (2014) Biosorption of heavy metals: a review. J Chem Sci Technol 3(4):74–102
2. Abdelfattah I, El-Saied FA, Almedolab AA, El-Shamy AM (2022) Biosorption as a perfect technique for purification of wastewater contaminated with ammonia. Appl Biochem Biotechnol 194(9):4105–4134
3. Abdi O, Kazemi M (2015) A review study of biosorption of heavy metals and comparison between different biosorbents. J Mater Environ Sci 6(5):1386–1399
4. Adamu A, Ijah UJ, Riskuwa ML, Ismail HY, Ibrahim UB (2015) Isolation of biosurfactant producing bacteria from tannery effluents in Sokoto metropolis Nigeria. Int J Innov Sci Eng Technol 2(1):366–373
5. Alawiye TT, Babalola OO (2019) Bacterial diversity and community structure in typical plant rhizosphere. Diversity 11(10):179
6. Alengebawy A, Abdelkhalek ST, Qureshi SR, Wang MQ (2021) Heavy metals and pesticides toxicity in agricultural soil and plants: ecological risks and human health implications. Toxics 9(3):42
7. Ali H, Khan E, Sajad MA (2013) Phytoremediation of heavy metals—concepts and applications. Chemosphere 91(7):869–881
8. Alves AR, Yin Q, Oliveira RS, Silva EF, Novo LA (2022) Plant growth-promoting bacteria in phytoremediation of metal-polluted soils: current knowledge and future directions. Sci Total Environ 838:156435
9. Anirudhan TS, Sreekumari SS (2011) Adsorptive removal of heavy metal ions from industrial effluents using activated carbon derived from waste coconut buttons. J Environ Sci 23(12):1989–1998
10. Aransiola SA, Ijah UJJ, Abioye OP, Bala JD (2019) Microbial-aided phytoremediation of heavy metals contaminated soil: a review. Eur J Biol Res 9(2):104–125. https://doi.org/10.5281/zenodo.3244176
11. Armah FA, Obiri S, Yawson DO, Onumah EE, Yengoh GT, Afrifa EKA, Odoi JO (2010) Anthropogenic sources and environmentally relevant concentrations of heavy metals in surface water of a mining district in Ghana: a multivariate statistical approach. J Environ Sci Health A 45(13):1804–1813
12. Arora R (2019) Adsorption of heavy metals–a review. Mater Today: Proc 18:4745–4750
13. Ayangbenro AS, Babalola OO (2017) A new strategy for heavy metal polluted environments: a review of microbial biosorbents. Int J Environ Res Public Health 14(1):94
14. Ayangbenro AS, Babalola OO (2018) Metal (loid) bioremediation: strategies employed by microbial polymers. Sustainability 10(9):3028

15. Băbău AMC, Micle V, Damian GE, Sur IM (2020) Preliminary investigations regarding the potential of Robinia pseudoacacia L. (Leguminosae) in the phytoremediation of sterile dumps. J Environ Prot Ecol 21:46–55
16. Băbău AMC, Micle V, Damian GE, Sur IM (2021) Sustainable ecological restoration of sterile dumps using Robinia pseudoacacia. Sustainability 13(24):14021
17. Behnsen J, Raffatellu M (2016) Siderophores: more than stealing iron. MBio 7(6):10–1128
18. Bobadilla-Fazzini RA, Poblete-Castro I (2021) Biofilm formation is crucial for efficient copper bioleaching from bornite under mesophilic conditions: unveiling the lifestyle and catalytic role of sulfur-oxidizing bacteria. Front Microbiol 12:761997
19. Borja D, Nguyen KA, Silva RA, Park JH, Gupta V, Han Y, Kim H (2016) Experiences and future challenges of bioleaching research in South Korea. Minerals 6(4):128
20. BrbootI MM, Abid BA, Al-ShuwaikI NM (2011) Removal of heavy metals using chemicals precipitation. Eng Technol J 29(3):595–612
21. Briffa J, Sinagra E, Blundell R (2020) Heavy metal pollution in the environment and their toxicological effects on humans. Heliyon 6(9)
22. Bruins MR, Kapil S, Oehme FW (2000) Microbial resistance to metals in the environment. Ecotoxicol Environ Saf 45(3):198–207
23. Burges A, Alkorta I, Epelde L, Garbisu C (2018) From phytoremediation of soil contaminants to phytomanagement of ecosystem services in metal contaminated sites. Int J Phytoremediation 20(4):384–397
24. Carolin CF, Kumar PS, Chitra B, Jackulin CF, Ramamurthy R (2021) Stimulation of Bacillus sp. by lipopeptide biosurfactant for the degradation of aromatic amine 4-Chloroaniline. J Hazard Mater 415:125716
25. Chang Q, Diao FW, Wang QF, Pan L, Dang ZH, Guo W (2018) Effects of arbuscular mycorrhizal symbiosis on growth, nutrient, and metal uptake by maize seedlings (Zea mays L.) grown in soils spiked with lanthanum and cadmium. Environ Poll 241:607–615
26. Cioica N, Tudora C, Iuga D, Deak G, Matei M, Nagy EM, Gyorgy Z (2019) A review on phytoremediation as an ecological method for in situ cleanup of heavy metals contaminated soils. E3S Web Conf 112:03024
27. de Alencar FLS, Navoni JA, do Amaral VS (2017) The use of bacterial bioremediation of metals in aquatic environments in the twenty-first century: a systematic review. Environ Sci Pollut Res 24:16545–16559
28. Dermont G, Bergeron M, Mercier G, Richer-Laflèche M (2008) Soil washing for metal removal: a review of physical/chemical technologies and field applications. J Hazard Mater 152:1–31. https://doi.org/10.1016/j.jhazmat.2007.10.043
29. Dhanwal P, Kumar A, Dudeja S, Chhokar V, Beniwal V (2017) Recent advances in phytoremediation technology. Adv Environ Biotech 227–241
30. Dusengemungu L, Kasali G, Gwanama C, Mubemba B (2021) Overview of fungal bioleaching of metals. Environ Adv 5:100083
31. Fashola MO, Ngole-Jeme VM, Babalola OO (2016) Heavy metal pollution from gold mines: environmental effects and bacterial strategies for resistance. Int J Environ Res Public Health 13(11):1047
32. Fashola MO, Ngole-Jeme VM, Babalola OO (2020) Physicochemical properties, heavy metals, and metal-tolerant bacteria profiles of abandoned gold mine tailings in Krugersdorp, South Africa. Can J Soil Sci 100(3):217–233
33. Feng W, Zhang S, Zhong Q, Wang G, Pan X, Xu X et al (2020) Soil Washing remediation of heavy metal from contaminated soil with EDTMP and PAA: properties, optimization, and risk assessment. J Hazard Mater 381:120997. https://doi.org/10.1016/j.jhazmat.2019.120997
34. Gadd GM (2009) Biosorption: critical review of scientific rationale, environmental importance and significance for pollution treatment. J Chem Technol Biotechnol 84(1):13–28
35. Gao X, Jiang L, Mao Y, Yao B, Jiang P (2021) Progress, challenges, and perspectives of bioleaching for recovering heavy metals from mine tailings. Adsorp Sci Technol 1–13
36. Gerhardt KE, Gerwing PD, Greenberg BM (2017) Opinion: taking phytoremediation from proven technology to accepted practice. Plant Sci 256:170–185

37. Guo T, Li L, Zhai W, Xu B, Yin X, He Y, Tang X (2019) Distribution of arsenic and its biotransformation genes in sediments from the East China Sea. Environ Pollut 253:949–958
38. Haak MR, Indraratne SP (2023) Soil amendments for vanadium remediation: a review of remediation of vanadium in soil through chemical stabilization and bioremediation. Environ Geochem Health 1–19
39. Hajdu R, Pinheiro JP, Galceran J, Slaveykova VI (2010) Modeling of Cd uptake and efflux kinetics in metal-resistant bacterium Cupriavidus metallidurans. Environ Sci Technol 44(12):4597–4602
40. Hao XD, Liu XD, Qin YANG, Liu HW, Yin HQ, Qiu GZ, Liang YL (2018) Comparative study on bioleaching of two different types of low-grade copper tailings by mixed moderate thermophiles. Trans Nonferrous Mat Soc 28(9):1847–1853
41. Jacobs A, De Brabandere L, Drouet T, Sterckeman T, Noret N (2018) Phytoextraction of Cd and Zn with Noccaea caerulescens for urban soil remediation: influence of nitrogen fertilization and planting density. Ecol Eng 116:178–187
42. Jan AT, Azam M, Siddiqui K, Ali A, Choi I, Haq QMR (2015) Heavy metals and human health: mechanistic insight into toxicity and counter defense system of antioxidants. Int J Mol Sci 16(12):29592–29630
43. Javed MT, Tanwir K, Akram MS, Shahid M, Niazi NK, Lindberg S (2019) Phytoremediation of cadmium-polluted water/sediment by aquatic macrophytes: role of plant-induced pH changes. In: Cadmium toxicity and tolerance in plants. Academic Press, pp 495–529
44. Jeevanantham S, Saravanan A, Hemavathy RV, Kumar PS, Yaashikaa PR, Yuvaraj D (2019) Removal of toxic pollutants from water environment by phytoremediation: a survey on application and future prospects. Environ Technol Innov 13:264–276
45. Joo JH, Hassan SH, Oh SE (2010) Comparative study of biosorption of Zn^{2+} by pseudomonas aeruginosa and Bacillus cereus. Int Biodeterior Biodegradation 64(8):734–741
46. Kabata-Pendias A, Mukherjee AB (2007) Trace elements from soil to human. Springer Science & Business Media
47. Khan FI, Husain T, Hejazi R (2004) An overview and analysis of site remediation 890 technologies. J Environ Manage 71:95–122
48. Khan AU, Khan AN, Waris A, Ilyas M, Zamel D (2022) Phytoremediation of pollutants from wastewater: a concise review. Open Life Sci 17(1):488–496
49. Khan AHA, Kiyani A, Mirza CR, Butt TA, Barros R, Ali B, Yousaf S (2021) Ornamental plants for the phytoremediation of heavy metals: present knowledge and future perspectives. Environ Res 195:110780
50. Khan M, Mir JI (2021) Potential of biotechnology in phytoremediation. Curr World Environ 16(2):622
51. Kim JJ, Kim YS, Kumar V (2019) Heavy metal toxicity: an update of chelating therapeutic strategies. J Trace Elem Med Biol 54:226–231
52. Kladsomboon S, Jaiyen C, Choprathumma C, Tusai T, Apilux A (2020) Heavy metals contamination in soil, surface water, crops, and resident blood in Uthai District, Phra Nakhon Si Ayutthaya, Thailand. Environ Geochem Health 42:545–561
53. Kurade MB, Ha YH, Xiong JQ, Govindwar SP, Jang M, Jeon BH (2021) Phytoremediation as a green biotechnology tool for emerging environmental pollution: a step forward towards sustainable rehabilitation of the environment. Chem Eng J 415:129040
54. Lal S, Ratna S, Said OB, Kumar R (2018) Biosurfactant and exopolysaccharide-assisted rhizobacterial technique for the remediation of heavy metal contaminated soil: an advancement in metal phytoremediation technology. Environ Technol Innov 10:243–263
55. Lee KY, Yoon IH, Lee BT, Kim SO, Kim KW (2009) A novel combination of anaerobic bioleaching and electrokinetics for arsenic removal from mine tailing soil. Environ Sci Technol 43(24):9354–9360
56. Lei M, Dong Z, Jiang Y, Longhurst P, Wan X, Zhou G (2019) Reaction mechanism of arsenic capture by a calcium-based sorbent during the combustion of arsenic-contaminated biomass: a pilot-scale experience. Front Environ Sci Eng 13:1–9

57. Li Q, Wang Y, Li Y, Li L, Tang M, Hu W, ... Ai S (2022) Speciation of heavy metals in soils and their immobilization at micro-scale interfaces among diverse soil components. Sci Total Environ 825:153862
58. Liu L, Li W, Song W, Guo M (2018) Remediation techniques for heavy metal-contaminated soils: principles and applicability. Sci Total Environ 633:206–219
59. Luna JM, Rufino RD, Sarubbo LA (2016) Biosurfactant from Candida sphaerica UCP0995 exhibiting heavy metal remediation properties. Process Saf Environ Prot 102:558–566
60. Mahar A, Wang P, Ali A, Awasthi MK, Lahori AH, Wang Q, Zhang Z (2016) Challenges and opportunities in the phytoremediation of heavy metals contaminated soils: a review. Ecotox Environ Saf 126:111–121
61. Malik A (2004) Metal bioremediation through growing cells. Environ Int 30(2):261–278
62. Mani D, Kumar C (2014) Biotechnological advances in bioremediation of heavy metals contaminated ecosystems: an overview with special reference to phytoremediation. Int J Environ Sci Technol 11:843–872
63. Marques AP, Rangel AO, Castro PM (2009) Remediation of heavy metal contaminated soils: phytoremediation as a potentially promising clean-up technology. Crit Rev Environ Sci Technol 39(8):622–654
64. Masindi V, Muedi KL (2018) Environmental contamination by heavy metals. Heavy Met 10:115–132
65. Meuser H (2013) Soil remediation and rehabilitation: treatment of contaminated and 920 disturbed land. Springer, Dordrecht, The Netherlands
66. Miranda LS, Wijesiri B, Ayoko GA, Egodawatta P, Goonetilleke A (2021) Water-sediment interactions and mobility of heavy metals in aquatic environments. Water Res 202:117386
67. Mishra S, Lin Z, Pang S, Zhang Y, Bhatt P, Chen S (2021) Biosurfactant is a powerful tool for the bioremediation of heavy metals from contaminated soils. J Hazard Mater 418:126253
68. Mishra A, Malik A (2013) Recent advances in microbial metal bioaccumulation. Crit Rev Environ Sci Technol 43(11):1162–1222
69. Mocek-Płóciniak A, Mencel J, Zakrzewski W, Roszkowski S (2023) Phytoremediation as an effective remedy for removing trace elements from ecosystems. Plants 12(8):1653. https://doi.org/10.3390/plants12081653
70. Mohan I, Goria K, Dhar S, Kothari R, Bhau BS, Pathania D (2021) Phytoremediation of heavy metals from the biosphere perspective and solutions. Pollutants and water management: resources, strategies, and scarcity, pp 95–127
71. Mondal NK, Das C, Datta JK (2015) Effect of mercury on seedling growth, nodulation and ultrastructural deformation of Vigna radiata (L) Wilczek. Environ Monit Assess 187:1–14
72. Mosa KA, Saadoun I, Kumar K, Helmy M, Dhankher OP (2016) Potential biotechnological strategies for the cleanup of heavy metals and metalloids. Front Plant Sci 7:303
73. Ndeddy Aka RJ, Babalola OO (2016) Effect of bacterial inoculation of strains of Pseudomonas aeruginosa, Alcaligenes feacalis and Bacillus subtilis on germination, growth, and heavy metal (Cd, Cr, and Ni) uptake of Brassica juncea. Int J Phytoremediation 18(2):200–209
74. Nedjimi B (2021) Phytoremediation: a sustainable environmental technology for heavy metals decontamination. SN Appl Sci 3(3):286
75. Neubauer U, Nowack B, Furrer G, Schulin R (2000) Heavy metal sorption on clay minerals affected by the siderophore desferrioxamine B. Environ Sci Technol 34(13):2749–2755
76. Nies DH (2003) Efflux-mediated heavy metal resistance in prokaryotes. FEMS Microbiol Rev 27(2–3):313–339
77. Ogun ML, Fashola MO, Anagun OS, Adigun SS (2022) Effects of plant growth promoting bacteria on the growth of Capsicum frutescens L. Parana. J Sci Educ 8(9): 31–39. tiny.cc/PJSE24476153v8i9p031-039
78. Ogun ML, Anagun OS, Awote OK, Oluwole SO, Kappo SC, Alonge FO (2023) Abattoirs: the hidden sources of plants' heavy metals and other pollutants in Lagos, Nigeria
79. Ojuederie OB, Babalola OO (2017) Microbial and plant-assisted bioremediation of heavy metal polluted environments: a review. Int J Environ Res Public Health 14(12):1504

80. Oluwole S, Ogun M, Qudus U, Racheal O, Kappo S, Fatimah M (2023) Assessment of the phytoremediation potential of heavy metal contaminated soil using Vigna Unguiculata L. (Walp). Eur Proc Sci Tech Eng Math 22:199–209
81. Oziegbe O, Oluduro AO, Oziegbe EJ, Ahuekwe EF, Olorunsola SJ (2021) Assessment of heavy metal bioremediation potential of bacterial isolates from landfill soils. Saudi J Biol Sci 28(7):3948–3956
82. Papadopoulos N, Zalidis G (2019) The use of Typha Latifolia L. in constructed wetland microcosms for the remediation of herbicide Terbuthylazine. Environ Process 6(4):985–1003
83. Patel P, Raju NJ, Reddy BSR, Suresh U, Sankar DB, Reddy TVK (2018) Heavy metal contamination in river water and sediments of the Swarnamukhi River Basin, India: risk assessment and environmental implications. Environ Geochem Health 40:609–623
84. Pathak A, Dastidar MG, Sreekrishnan TR (2009) Bioleaching of heavy metals from sewage sludge: a review. J Environ Man 90(8):2343–2353
85. Piotrowska-Niczyporuk A, Bajguz A (2013) The effect of natural and synthetic auxins on the growth, metabolite content and antioxidant response of green alga Chlorella vulgaris. Plant Growth Regul 73:57–65
86. Prapagdee B, Khonsue N (2015) Bacterial-assisted cadmium phytoremediation by Ocimum gratissimum L. in polluted agricultural soil: a field trial experiment. Int J Environ Sci Tech 12:3843–3852
87. Qiao D, Wang G, Li X, Wang S, Zhao Y (2020) Pollution, sources and environmental risk assessment of heavy metals in the surface AMD water, sediments and surface soils around unexploited Rona Cu deposit, Tibet, China. Chemosphere 248:125988
88. Raklami A, Meddich A, Oufdou K, Baslam M (2022) Plants—microorganisms-based bioremediation for heavy metal cleanup: recent developments, phytoremediation techniques, regulation mechanisms, and molecular responses. Int J Mol Sci 23(9):5031
89. Ramli NN, Kurniawan SB, Ighalo JO, Mohd Said NS, Marsidi N, Buhari J, Abu Hasan H (2023) A review of the treatment technologies for hexavalent chromium contaminated water. BioMetals 1–31
90. Rashid MS, Liu G, Yousaf B, Hamid Y, Rehman A, Arif M, Ashraf A (2023) Role of biochar-based free radicals in immobilization and speciation of metals in the contaminated soil-plant environment. J Environ Manag 325:116620
91. Rezania S, Taib SM, Din MFM, Dahalan FA, Kamyab H (2016) Comprehensive review on phytotechnology: heavy metals removal by diverse aquatic plants species from wastewater. J Hazard Mater 318:587–599
92. Rinklebe J, Shaheen SM (2017) Redox chemistry of nickel in soils and sediments: a review. Chemosphere 179:265–278
93. Rmaidh AY, Al-Mahdawi AS, Al-Obaidi NS, Al-Obaidi MA (2023) Organic ligands as adsorbent surface of heavy metals and evaluating antibacterial activity of synthesized complexes. Inorg Chem Commun 156:111148
94. Rodríguez Y, Ballester A, Blázquez ML, González F, Muñoz JA (2003) New information on the pyrite bioleaching mechanism at low and high temperature. Hydrometallurgy 71:37–46
95. Rojjanateeranaj P, Sangthong C, Prapagdee B (2017) Enhanced cadmium phytoremediation of Glycine max L. through bioaugmentation of cadmium-resistant bacteria assisted by biostimulation. Chemosphere 185:764–771
96. Sabra M, Aboulnasr A, Franken P, Perreca E, Wright LP, Camehl I (2018) Beneficial root endophytic fungi increase growth and quality parameters of sweet basil in heavy metal contaminated soil. Front Plant Sci 9:1726
97. Sampaio CJ, de Souza JR, Damião AO, Bahiense TC, Roque MR (2019) Biodegradation of polycyclic aromatic hydrocarbons (PAHs) in a diesel oil-contaminated mangrove by plant growth-promoting rhizobacteria. Biotechnology 9:1–10
98. Sanito RC, Bernuy-Zumaeta M, You SJ, Wang YF (2022) A review on vitrification technologies of hazardous waste. J Environ Manag 316:115243
99. Saponaro S, Bonomo L, Barbafieri M, Petruzzelli G (2002) Soil washing feasibility at a manufacturing gas plant site. Soil Sediments Cont. 11(5):751–767. https://doi.org/10.1080/20025891107078

100. Sarwar N, Imran M, Shaheen MR, Ishaque W, Kamran MA, Matloob A, Hussain S (2017) Phytoremediation strategies for soils contaminated with heavy metals: modifications and future perspectives. Chemosphere 171:710–721
101. Shabani L, Sabzalian MR, Pour SM (2016) Arbuscular mycorrhiza affects nickel translocation and expression of ABC transporter and metallothionein genes in Festuca arundinacea. Mycorrhiza 26:67–76
102. Shahzad B, Tanveer M, Hassan W, Shah AN, Anjum SA, Cheema SA, Ali I (2016) Lithium toxicity in plants: reasons, mechanisms, and remediation possibilities–a review. Plant Physiol Biochem 107:104–115
103. Shamim S (2018) Biosorption of heavy metals. Biosorption 2:21–49
104. Siddiquee S, Rovina K, Azad SA, Naher L, Suryani S, Chaikaew PJJMBT (2015) Heavy metal contaminants removal from wastewater using the potential filamentous fungi biomass: a review. J Microb Biochem Technol 7(6):384–395
105. Singh J, Upadhyay SK, Pathak RK, Gupta V (2011) Accumulation of heavy metals in soil and paddy crop (Oryza sativa), irrigated with water of Ramgarh Lake, Gorakhpur, UP, India. Toxicol Environ Chem 93(3):462–473
106. Singh V, Singh N, Rai SN, Kumar A, Singh AK, Singh MP, ... Mishra V (2023) Heavy metal contamination in the aquatic ecosystem: toxicity and its remediation using eco-friendly approaches. Toxics 11(2):147
107. Siyar R, Doulati Ardejani F, Norouzi P, Maghsoudy S, Yavarzadeh M, Taherdangkoo R, Butscher C (2022) Phytoremediation potential of native hyperaccumulator plants growing on heavy metal-contaminated soil of Khatunabad copper smelter and refinery, Iran. Water 14(22):3597
108. Su K, Wu Z, Liu Y, Jiang S, Wang Y, Fu C (2021) Highly efficient detoxification of dinitrotoluene through overexpressing bacterial nitroreductase in switchgrass. Authorea Preprints
109. Sumiahadi A, Acar R (2018) A review of phytoremediation technology: heavy metals uptake by plants. IOP Conf Ser: Earth Environ Sci 142:012023
110. Sun W, Cheng K, Sun KY, Ma X (2021) Microbially mediated remediation of contaminated sediments by heavy metals: a critical review. Curr Pollut Rep 7:201–212
111. Szuba A, Karliński L, Krzesłowska M, Hazubska-Przybył T (2017) Inoculation with a Pb-tolerant strain of Paxillus involutus improves growth and Pb tolerance of Populus × canescens under in vitro conditions. Pl Soil 412:253–266
112. Tak HI, Ahmad F, Babalola OO (2012) Advances in the application of plant growth-promoting rhizobacteria in phytoremediation of heavy metals. Rev Environ Cont Toxicol 223:33–52
113. Trellu C, Mousset E, Pechaud Y, Huguenot D, van Hullebusch ED, Esposito G, Oturan MA (2016) Removal of hydrophobic organic pollutants from soil washing/flushing solutions: a critical review. J Hazard Mater 306:149–174
114. van Alin A, Corbett MK, Fathollahzadeh H, Tjiam MC, Rickard WD, Sun X, Watkin E (2023) Biofilm formation on the surface of monazite and xenotime during bioleaching. Microb Biotechnol
115. Volesky B (1994) Advances in biosorption of metals: selection of biomass types. FEMS Microbiol Rev 14(4):291–302
116. Wei M, Chen J, Wang X (2016) Removal of arsenic and cadmium with sequential soil washing techniques using Na2EDTA, oxalic and phosphoric acid: optimization conditions, removal effectiveness and ecological risks. Chemosphere 156:252–261. https://doi.org/10.1016/j.chemosphere.2016.04.106
117. Wołowiec M, Komorowska-Kaufman M, Pruss A, Rzepa G, Bajda T (2019) Removal of heavy metals and metalloids from water using drinking water treatment residuals as adsorbents: a review. Minerals 9(8):487
118. Wuana RA, Okieimen FE (2011) Heavy metals in contaminated soils: a review of sources, chemistry, risks and best available strategies for remediation. ISRN Ecol 2011:1–20. https://doi.org/10.5402/2011/402647

119. Xu ZN, Lin ZQ, Zhao GS, Guo YB (2023) Biogeochemical behavior of selenium in soil-air-water environment and its effects on human health. Int J Environ Sci Technol 1–22
120. Yadav M, Singh G, Jadeja RN (2021) Physical and chemical methods for heavy metal removal. In: Pollutants and water management: resources, strategies and scarcity, pp 377–397
121. Yan A, Wang Y, Tan SN, Mohd Yusof ML, Ghosh S, Chen Z (2020) Phytoremediation: a promising approach for revegetation of heavy metal-polluted land. Front Plant Sci 11:359
122. Zayed A, Pilon-Smits E, de Souza M, Lin ZQ, Terry N (2020) Remediation of selenium-polluted soils and waters by phytovolatilization. In Phytoremediation of contaminated soil and water. CRC Press, pp 61–83
123. Zhu Y, Wang Y, He X, Li B, Du S (2023) Plant growth-promoting rhizobacteria: a good companion for heavy metal phytoremediation. Chemosphere 139475

Role of Biosurfactants in Remediation of Heavy Metals

Muibat Omotola Fashola, Olajide Solomon Anagun, Ahmeed Olalekan Ashade, and Olubukola Oluranti Babalola

Abstract Environmental contamination by heavy metals is fast becoming a worldwide concern due to their non-degradable nature that enhance their accumulation in the environment. Several conventional methods are in used for treatment of contaminated sites but these methods are usually inefficient financially and negatively impact the qualities of contaminated areas. Therefore, there is a pressing requirement for more economically viable and minimally invasive strategies for detoxifying polluted areas of heavy metals. The exploitation of biosurfactants towards removal of heavy metals from the environment is an exciting new area of research. Micelle production is the fundamental principle of biosurfactants-mediated heavy metal removal. The pace at which metals are removed from contaminated locations is related to the type of biosurfactant generated by the producer strains. Biosurfactant is known to improve heavy metal removal from the environment via a number of mechanisms, the most important of which are complexation, ion exchange, and metal solubilization. Experiments showing that biosurfactants can help remove metals from a laboratory setting pave the way for their potential use on a larger scale. This section describes the properties, classification, and screening procedures of biosurfactants and the mechanisms by which biosurfactants remove heavy metals from soils.

Keywords Biosurfactants · Heavy metal · Complexation · Desorption · Ion-exchange

---

M. O. Fashola (✉) · O. S. Anagun · A. O. Ashade
Department of Microbiology, Faculty of Science, Lagos State University, Ojo, Lagos, Nigeria
e-mail: muibat.fashola@lasu.edu.ng

M. O. Fashola · O. O. Babalola
Food Security and Safety Niche Area, Faculty of Natural and Agricultural Science, North West University, Mmabatho, South Africa

N. Kumar (ed.), *Heavy Metal Remediation*, Earth and Environmental Sciences Library,
https://doi.org/10.1007/978-3-031-53688-5_8

1 Introduction

Increased industrialization and human activities have led to heavy metal pollution of soil and water sources in the modern period. Environmental contamination by heavy metals poses a growing threat to human health, raising alarm among scientists around the world due to its deleterious consequences and toxicological indications [55, 89]. Human-based processes like mining, mineral processing, cowhide curing, burning of natural gas and coal, and electroplating, as well as a wide range of products like paints and colors, additives, composts, and synthetic substances, are the primary sources of heavy metal pollution [15, 43, 51]. There are considerable environmental impacts from these activities and the products they produce due to the ever-increasing amount of toxic heavy metals [8, 91]. Due to their toxicity and chemical configuration, heavy metal contaminants are unable to breakdown easily however, they can be eliminated or transformed to less harmful forms through biochemical activities [32, 60, 61]. Metallic elements and metalloids having densities greater than 6 g/cm^3 are considered heavy metals [5]. According to the United States Environmental Protection Agency's (USEPA) priority pollutants list [20], heavy metals like arsenic, cadmium, chromium, copper, lead, mercury, nickel and zinc are major environmental contaminants.

Heavy metal exposure is becoming linked to congenital malformations, organ malfunctions, and learning impairment [87]. Adsorption of metals [44], precipitation of metals [26], elimination using nanomaterials [97], reverse osmosis [53], exchange of ion [74], and filtration of membrane [76] are just some of the more common physicochemical methods used for decontaminating and restoring heavy metal-laden environments. Nonetheless, these methods are less suited for large scale applications [7, 16, 62]. This is because they are cost and energy intensive and also generate bounteous waste and toxic post-process pollutants. Progress has been achieved in the environmentally friendly elimination of heavy metal pollution utilizing microbial cell factories [39, 80, 98]. This is because of the challenges connected with conventional methods of heavy metal removal.

Biosurfactants are extraordinary microbial products with successful exploitation towards heavy metal elimination and reduction [3, 12, 77]. Biosurfactants, as described by Ayangbenro and Babalola [10], are secondary metabolites released extracellularly by bacterial and fungal cells that are capable of lowering interfacial and surface tension and acting as an excellent emulsifying agent. Heavy metal binding capacities and selectivity is producer-strain dependent. They have low critical micelle concentration (CMC) which promotes their surface tension reduction ability. CMC is defined as lowest amount of microbial surfactant necessary to attain the highest reduction in surface action and induce micelle formation [62]. As indicated by Mishra et al. [62], the lower the CMC value of a biosurfactant, the better its application in heavy metal removal. Typical fields of use include the pharmaceutical, petroleum, cosmetic, detergent, paint, and food industries [58, 68, 96]. The multi-properties of biosurfactants qualifies them as a good candidate for bioremediation agent [3, 30]. In addition, this value-added substance (biosurfactant) provides unique supremacies in heavy metal cleanup based on their compact size, innocuousness, high biodegradability, and activity in extreme environmental conditions [4, 81]. They are more successful in removing contaminants because biosurfactants exhibits a wide array of structural configurations and wide range of metal selectivity and affinity [52, 67]. Since biosurfactants have a solid attraction to metals, a biosurfactant-metal complex is formed. Micelles and reduced surface tension are two of their distinctive qualities that engenders their pivotal role in clean-up of metal-laden environments [30, 56, 82]. *Pseudomonas aeruginosa*, *Bacillus subtilis*, *Burkholderia* sp., *Citrobacter freundii*, *Candida tropicalis*, and *Micrococcus* sp. are among the bacteria that have been identified as possible biosurfactant makers [11, 54, 72].

With their ability to effectively reclaim heavy metal-polluted environments, biosurfactants have piqued the public's interest as a new multifunctional technology in the twenty-first century. Therefore, this chapter provides a thorough overview of the categorization and major properties of biosurfactants, as well as a discussion of their potential application as heavy metal bioremediation agents in polluted environments.

2 Properties of Biosurfactants

The wide availability of substrates and the unique properties of biosurfactants make them competitive with chemically synthesized alternatives. The potential of a surfactant to reduce surface tension, or the attraction force between molecules in liquids, is its most important property [73]. Whereas an interface separates two liquids, a liquid's surface separates it from the air. According to research by Banat et al. [13], the tension between the air/water and oil/water phases is referred to as surface tension and interfacial tension. Tolerance to changes in acidity/alkalinity, temperature, antibacterial activity, biodegradability, emulsifying and demulsifying ability, ionic strength, and low toxicity are further features of microbial surfactants [23].

i. Surface and interface activity: Biosurfactants outperform traditional surfactants in efficiency and effectiveness because of their lower tensio active values at lower concentrations. While the adhesive force between two liquids (interfacial tension-oil/water) and surface tension are roughly 1 and 30 mN/m, respectively, the efficiency of biosurfactants is commonly measured in CMC, which ranges from 1 to 2000 mg/L.
ii. Tolerance to pH and temperature: The economic potential of developing extremophile biosurfactants has piqued the scientific community's interest in this area in recent decades. Most biosurfactants maintain their surface activity in the face of changes in pH and temperature. Singh and Cameotra [87] noted that *Arthrobacter protophormiae* biosurfactant was stable between 30 and 100 °C between 2 and 12 on the pH scale. Nonetheless, new microbial products capable of functioning under temperature, pH, and pressure extremes are required for industrial operations [19].

iii. Biodegradability: Biosurfactants are more suitable to environmental applications like bioremediation of pollutants like hydrocarbons since they degrade quickly [63]. Parthipan et al. [75] harnessed *Bacillus subtilis* A1-biosurfacatant in degrading hydrocarbons consisting of C_{10}–C_{19} fractions. This bacterium showed 87% overall crude oil degradation within 7 days.
iv. Low toxicity: These substances are often regarded to be less or non-toxic, making them suitable for use in pharmaceutical, culinary, and cosmetic contexts. For instance, [22] demonstrated that the biosurfactant sophorolipids from *Candida bombicola* has a low toxicity profile and can be used in the food sector.
v. Emulsification ability: By facilitating the union of otherwise incompatible chemicals, biosurfactants can also serve as emulsifiers. When one immiscible liquid is dispersed in another liquid as droplets larger than 0.1 mm in diameter, the resulting system is called an emulsion.
vi. Antifouling activity: Most bacteria generate biofilm in order to survive in a certain environment. The initial stage in biofilm development is bacterial adhesion to surfaces, which is typically influenced by the surface's hydrophobicity and electrical charge. To reduce microbial adhesion, biosurfactants can be used to alter the surface's hydrophobicity. The colonization of the steel by other thermophilic *Streptococcus* strains, which causes fouling, is slowed by a biosurfactant generated from *Streptococcus thermophilus*. Similarly, *Listeria monocytogenes* adherence to a steel surface was diminished by a biosurfactant from *Pseudomonas fluorescens* [23].

3 Classification

The polar region of chemical surfactants is where the ionic charge is located, hence different types of surfactants have different ionic charges. Depending on electrostatic charges, surfactants can be classified as negatively-charged ions (anionic), positively-charged ions (cationic), lacks any electrical charge (non-anionic), or reacts both as an acid or base (amphoteric) [86]. Biosurfactants are often categorized according to their biochemical make-up or their microbial origin. However, molecular weight is the most used method for classifying biosurfactants [30, 31]. The classification and differentiating features of biosurfactants are shown in Table 1.

4 Methods of Screening for Biosurfactants

The screening methods most commonly used to identify microorganisms that produce biosurfactants are grounded in the observed properties of these substances. The capability of strains to disrupt hydrophobic surfaces can also be tested. The colorimetric cetyltrimethylammonium bromide (CTAB) agar test is one example of an explicit screening approach that can be used, but only with a small number of biosurfactants.

Table 1 Classification of biosurfactants

Class	Sub-class	Feature	References
Low molecular weight biosurfactant	Glycolipids	These biosurfactants have the lowest molecular weight of any commercially produced biosurfactant. They have conjugates of saccharide and lipid moieties. Examples include rhamnolipids, sophorolipids, trehalolipids, and mannosylerythritol lipids	[95, 99]
	Lipopeptides	Hydrophobic fractions of fatty acids are present alongside a hydrophilic peptide sequence. Positively and negatively charged-residues, in addition to amino acids, can be found in the peptide sequence. Many proteins function as surfactants, and some well-known ones include polymyxin, viscosin, serrewettin, fengycin, and iturin	[21, 47]
	Phospholipids and fatty acids	They are responsible for a pivotal role in the membranes of bacteria. In most cases, microorganisms' ability to take in alkanes is linked to their ability to synthesize fatty acids and phospholipids. However, the surface and interfacial activity of fatty acids with chain lengths between 14 and 124 are significantly lower. For instance, *Klebsiella pneumoniae* IVN51 generates phosphatidylethanolamine	[71, 94]
High molecular weight biosurfactant	Polymeric biosurfactants	These are carbohydrate-protein complex. Examples include alasan, biodispersan, and emulsan	[85]
	Particulate biosurfactants	They are naturally occurring surface-active peptides generated mostly by fungus	[62]

i. Du-Nouy-Ring technique: With this method, the amount of energy required to separate a wire ring from a surface is calculated [88]. The separating energy is calculated from the interfacial tension. The ring must be devoid of any foreign substances, and this is typically done by using a platinum wire ring that has been flamed prior to use. Wilhelmy plates, made of platinum, are similar to rings in their functionality [24, 90, 93]. The key benefits of this technology are its accuracy and ease of usage. One drawback is that simultaneous measurements of multiple samples are not possible. The test also has several limitations, such as the small sample size (often just a few milliliters) and the narrow concentration range that can be studied without diluting the sample [18].

ii. Drop collapse test: This technique of biosurfactant screening was developed by Jain et al. [48]. This test is predicated on the ability of biosurfactants to destabilize liquid droplets. As a result, supernatant of cultures or drops of cell suspension are deposited on an oil-coated solid surface. If there are no surfactants in the liquid, the droplets will not collapse because the hydrophobic surface will reject the polar water molecules. Droplets of liquid are dispersed or even collapsed by surfactants because they reduce the interfacial and surface tension between the cell and the oil surface. Surfactant concentration influences drop stability, which is linked to surface and interfacial tension. The flexibility of this test, as described by Maczek et al. [57], to be applied to an automated screening in microplates is particularly noteworthy. Quick and easy to do, the drop collapse test calls for neither specialized tools nor a large volume of material [78]. One major drawback of this method is its low sensitivity,in order for the aqueous droplets on the oil or glass surfaces to collapse, a very high concentration of surface-active chemicals must be present.
iii. Oil spread assay: [64] developed this method for screening of biosurfactant activity. In a Petri dish, 10 μl of crude oil is put to the surface of 40 ml of distilled water to generate a thin oil layer for this test. Then, carefully pour 10 μl of culture suspension on top of the oil layer. When there is a biosurfactant in the cell suspension, displacement of oil occurs and a clearing zone form. Surfactant activity, also known as oil displacement activity, is proportional to the size of this clearing zone on the oil surface. The amount of surfactant produced and diameter of clearing zone observed is influenced by the type of biosurfactant. This method is quick and easy to implement, requiring no specialized equipment and only a small amount of sample [78].
iv. Emulsification index assay: This method provides a quantitative evaluation of the biosurfactant production by a microorganism. Quantification using this method was first introduced by Cooper and Goldenberg [27]. To quantify this quality, kerosene is added to the bacterial sample. The resulting mixture is vortexed at high resolution for 2 min. After 18–24 h, the emulsion layer has stabilized to a certain height. To calculate the emulsion index E_{24}, divide the height of the emulsion layer by the total height of the liquid. E_{24} correlates with the amount of surfactant present. The emulsification index is a quick and easy method for sorting among potential biosurfactant-producing microbes. Therefore, the best this method can do is hint at the presence of biosurfactants.
v. Anthracene solubilization: This screening approach is based on the biosurfactants potential to solubilize anthracene, a very hydrophobic, crystalline molecule. As a result, crystalline anthracene is added to the culture supernatant and placed in rotary shaker incubator at 25 °C for 24 h. The spectrophotometric determination of dissolved hydrophobic anthracene at 354 nm corresponds to biosurfactant formation. This is a simple and quick screening procedure, but it has one big limitation; anthracene toxicity in some bacteria.
vi. CTAB agar plate: Glycolipids and other anionic surfactants can be detected with this semi-quantitative assay. An agar plate containing mineral salt, cetyltrimethylammonium bromide, and methylene blue is used to cultivate the

microorganisms of interest. Anionic surfactants are secreted by the bacteria growing on the cultured plate, as evidenced by the formation of a dark blue and insoluble ion pair between cetyltrimethylammonium bromide and methylene blue.

vii. Hemolytic activity: Biosurfactants can induce erythrocyte lysis. Cultures are injected onto sheep blood agar plates and incubated at 25 °C for 48 h. Positive strains cause blood cell lysis and form a colorless, transparent ring around the colonies. Hemolysis can also be shown using a pure biosurfactant. This procedure is typically used as a preliminary screening for a microorganism's capacity to make biosurfactant. This approach has limitations such as limited specificity because enzymes can be responsible for clear zones and diffusion restriction of the surfactant can prevent the creation of clear zones.

5 Factors Affecting Production of Biosurfactants

Production of biosurfactants is limited by multiple factors. These factors often reduce the activity of the produced bioactive molecule. Asides from the producer strain, these factors are broadly grouped into two main categories, they include the nutritional and environmental factors. The nutritional factors include carbon source, nitrogen source, salt concentration while, the environmental factors include pH, temperature, oxygen availability and metal ion tolerance.

Carbon source: Carbon is a fundamental element for the growth and survival of microorganisms, just as it is for all living organisms. However, the nature and amount of biosurfactant produced are mostly affected and influenced by the type of substrate used [83]. Mostly, hydrocarbon substrates are utilized as carbon source in biosurfactant production but often requires a co-substrate or an inducer usually acetate or glucose before the hydrocarbon can be effectively utilized [46, 49]. Biosurfactants have been reported to grow on non-hydrocarbon, water-soluble carbon sources such as molasses and glycerol [33, 65]. Such biosurfactants, on the other hand, may be less expensive to produce and more beneficial in the food and pharmaceutical sectors since they do not require substantial purification. Ilori [46] was of the opinion that biosurfactant produced from water-soluble carbon sources are less efficient than those produced from hydrocarbon substrates.

Nitrogen source: For any biosurfactant production medium, there must be adequate supply of nitrogen, this is because the element is an essential nutritional factor required for the synthesis of microbial biomolecules such as enzymes, proteins and fatty acids. Organic and inorganic nitrogen are the two forms of nitrogen sources. The unit structure of the two can readily be used in distinguishing them. Examples of organic nitrogen generally include yeast extract, beef extract, tryptone, or peptone which are usually used as nitrogen source in commercially available growth media. On the contrary, examples of inorganic nitrogen generally include ammonium salts which is commonly composed of positive and negative ions like ammonium nitrate (NH_4NO_3) [70]. However, of all, yeast extract is the commonly used nitrogen source for production of biosurfactant [35]. Nonetheless, [38] proposed that, while it can improve the concentration of biomass generated, it has less potential to lower surface tension than other complicated nitrogen sources.

Salt concentration: It is necessary to regulate the concentration of salt in the mineral medium used for biosurfactant production. This is because most cellular operations of microbes are hindered by increasing salt concentrations [70]. The study of [84] further supports this claim, sodium chloride (NaCl) was added to the growth medium in increasing concentrations. The biosurfactant producer strain-*Pseudomonas aeruginosa* RS29 had a lowered production rate when the salt level was increased.

pH: Metabolic processes are pH sensitive and this because pH is a critical factor that influences the chemical reactions of microorganisms. Although the production of biosurfactants varies with microorganisms, it is pH and temperature dependent. However, optimum biosurfactant production is generally near neutral pH, though some reports suggest a slight increase in pH from 7–8.5 [25, 28, 36].

Temperature: A major limiting factor in biosurfactant production is temperature. According to [46], an increase in temperature from 30 to 35 °C had no effect on biosurfactant activity, but a slight increase in activity was detected at 40 °C. However, the biosurfactant's activities decreased at temperatures of 70 and 100 °C. At 100 °C, the biosurfactant retained approximately 77% of its original activity. Although, according to the information released by Maqsood and Jamal [59], the effect of temperature on rhamnolipid production is dependent on the strain of *Pseudomonas aeruginosa* being used. However, in general, rhamnolipid production has been found to increase with temperature from 25 to 30 °C, remain nearly constant from 30 to 37 °C, and decrease slightly when temperature was increased to 42 °C. At 47 °C, *P. aeruginosa* was unable to grow, leading to negligible rhamnolipid production at that temperature.

Oxygen availability: The availability of oxygen is a key component that influences the synthesis of biosurfactant because it allows electron transport from the gas phase to the aqueous phase. Adamczak and Bednarski [1] discovered that the greatest surfactant production value (45.5 g/l) was obtained when the dissolved oxygen content was kept at 50% of saturation.

Metal ion tolerance: Hydrophilic inducers such as metal ions are required for production of biosurfactants. They serve as co-factors that assist cell growth and thereafter, biosurfactants synthesis. Wei et al. [92] increased the synthesis of surfactin by *B. subtilis* ATCC 21332 while growing it on a mineral salt medium, Mg^{2+} (0.6 mM) ion produced the most surfactin (2.23 ± 0.10 gL^{-1}). When 0.6 mM Mg^{2+}, 10 mM K^{+}, 0.1 mM Mn^{2+}, and 1.2 mM Fe^{2+} were employed together, a significant drop in biosurfactant production was found (1.30 ± 0.10 gL^{-1}).

6 Biosurfactants-Mediated Removal of Heavy Metals

By increasing their bioavailability or preparing and eliminating contaminants during a cleaning process through pseudo-solubilization and emulsification, biosurfactants are useful in the remediation of natural compounds like hydrocarbons [17]. Using biosurfactants for inorganic compound recovery speeds up the cleaning process by increasing chelating activity and particle ejection due to interactions between amphipathic chemicals and metal ions [13].

Since surfactant-producing strains do not need to be metal-tolerant, biosurfactants offer indisputable benefits in heavy metals cleanup; nonetheless, biosurfactants require a continual supply of these molecules. Soil can be reused after being treated with biosurfactants, which help in solubilization, dispersion, and desorption of contaminants. Several reports have tested the efficiency of utilizing biosurfactants for clean-up of heavy metal polluted sites [11, 40, 45]. Table 2 however, provides a summarized data on the application of biosurfactants and their corresponding heavy metal removal efficiency rate.

There are two main methods used to clean-up heavy metal polluted sites. The first technique is an ex-situ one, in which a sample of the polluted area (often garbage) is removed, placed in a glass jar, and treated with a biotensioactive solution. The second stage is to use seepage tubes and channels to introduce and sort the biosurfactant solution in order to wash the contaminated sample in situ. Small amounts of blocked soil can be treated with biosurfactants. By blending the soil in a concrete mixer, the biosurfactant metal complex in the soil is extracted. After treating the biosurfactant-metal complex to precipitate the biosurfactant while leaving the metal behind, the soil is returned to the earth. The biosurfactant-metal complex bond is typically strong enough to withstand water washing through the soil.

7 Mechanism of Heavy Metal Removal by Biosurfactants

Microbes can biotransform or bioaccumulate heavy metals into less toxic forms [37], however these processes do not lead to biodegradation. The ubiquitous process of heavy metal complexation with microbial surfactants, which inhibits metal solution phase activity and induces metal desorption from polluted settings [42, 52], is well-documented.

This mechanism is thought to involve complexation reaction between the heavy metal ions and biosurfactants that are adsorbed through chemosorption on the intersection of the metal–surfactant interphase under lowered interfacial activity (see Fig. 1).

Because of their complexation abilities, anionic lipopeptide biosurfactants have high chelating affinity for cationic metal ions like Cu^{2+}, Cd^{2+}, Pb^{2+}, Ni^{2+}, and Zn^{2+} [66, 67]. Similarly, positively charged biosurfactants can compound negatively charged ions. Most importantly, for a biosurfactant to form a complex with heavy

Table 2 Selected studies indicating biosurfactant-mediated removal of heavy metals

S/N	Type of biosurfactant	Producer strain	Source of producer strain	Heavy metals and percentage removed	References
1	Sophorolipid	*Cryptococcus* sp. VITGBN2	Wastewater treatment plant	Zinc (94.34%)	[14]
2	Sophorolipid (SL18)	*Candida bombicola*	Mine tailing	Arsenic (91%)	[6]
3	Rhamnolipid	*Pseudomonas aeruginosa*	Heavy metal contaminated soil	Lead (62%), nickel (56%), cadmium (53%)	[34]
4	Di-rhamnolipid	*Pseudomonas aeruginosa* BS2	Oil sludge	Cadmium (92%), lead (88%)	[50]
5	Sophorolipid	*Candida bombicola* ATCC 22214	American Type Culture Collection	Copper (60%), zinc (25%)	[69]
6	Lipopeptide	*Bacillus cereus* NWUAB01	Mining soil	Lead (69%), cadmium (54%), chromium (43%)	[11]
7	Surfactin	*Bacillus* spp. HIP3	Cooking oil contaminated soil	Copper (13.57%), zinc (12.71%), chromium (1.68%), cadmium (0.7%)	[45]
8	Sophorolipid	*Candida bombicola*	Hydrocarbon polluted soil	Chromium (23.11%), cadmium (15.71%), lead (9.93%), zinc (7.29%), copper (4.96%)	[2]

(continued)

Table 2 (continued)

S/N	Type of biosurfactant	Producer strain	Source of producer strain	Heavy metals and percentage removed	References
9	Unconfirmed	*Citrobacter freundii* MG812314.1	Heavy metal contaminated wastewater	Aluminum (80%), lead (60%), zinc (66%), cadmium (55%)	[40]
10	Unconfirmed	*Candida tropicalis* UCP0996	University culture bank	Zinc (30–80%), copper (30–80%), lead (15%)	[29]

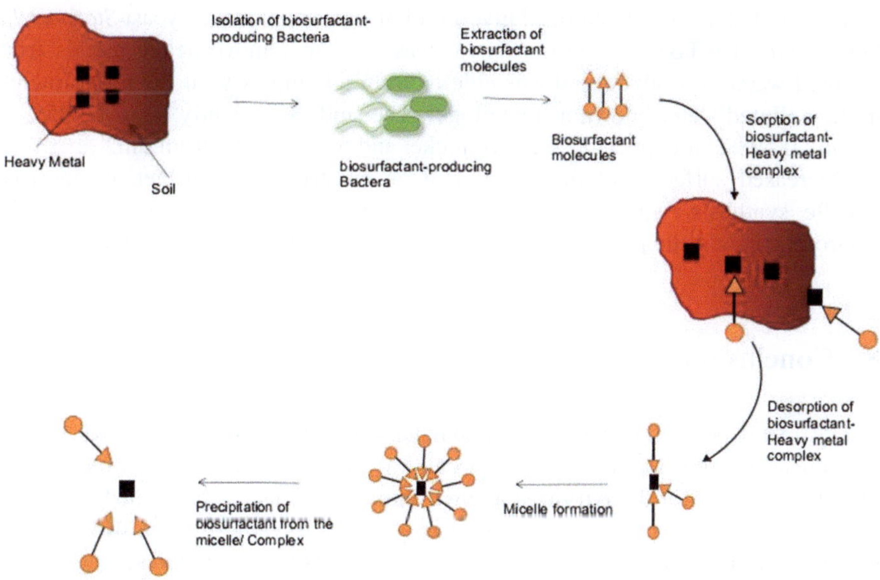

Fig. 1 Mechanism of biosurfactant-mediated removal of heavy metals

metal, two independent processes usually occur. The first process that can lead to formation of heavy metal-biosurfactant complex is via ion-exchange. Ion exchange often uses negatively charged anionic biosurfactants such as rhamnolipids that bind with cationic metals to produce a binding which is under more force than the link formed between soil-metal ion. Biosurfactant gives the cell surface an anionic nature. This procedure promotes electrostatic interaction with metal ions. The polar head moieties of surfactant micelles bond with heavy metal ions, making them water soluble [10]. The surfactant micelles subsequently transport the metal ions into solution, making metal recovery simpler by flushing, washing, or pumping [9]. The heavy metal-biosurfactant combination creates such a strong link that cleaning the soil with water eliminates the complex [73]. Cationic biosurfactants compete with negatively charged surfaces to replenish the same charged ions. The metal ion may then be extracted from the complex and utilized for a variety of purposes [10]. Heavy metal solubilization is another procedure that helps the heavy metal-biosurfactant complex. To limit migration, heavy metals are immobilized on a tightly bonded solid matrix. Heavy metal solubilization improves metal mobility and migration into liquid form, which is frequently accomplished by desorption. This method addresses the pollution problem in the long run. Furthermore, treated soil can be recycled, expanding future land-use options [41]. However, as identified by Ayangbenro and Babalola [10], the composition and cation exchange capacity of the environment, acidity/alkalinity, particulate size and distribution, type and period of contamination, and geological configuration, among other factors, influence the biosurfactant-heavy metal ion complex in polluted environments.

Poirier et al. [79] characterized glycolipid biosurfactant from a yeast *Starmerella bombicola* AUGTB1, this biosurfactant was used to remove heavy metals from polluted water. The glycolipid was able to form a complex with the heavy metals in the polluted water known as metallogel. The study sufficiently removed 95% of cobalt, 78–98% of copper, 77–83% of nickel and 59–65% of chromium.

Decreased surface and interfacial tension created by biosurfactants increase micelle synthesis, which in turn aids the clean-up of heavy metals from the environment. Micelle production is essential for solubilization and mobilization.

8 Conclusion

Heavy metal pollution of the ecosystem has resulted from the accelerated pace of industrialization around the world, calling for the creation of environmentally friendly methods for detoxifying the environment. The use of biosurfactants in the removal of heavy metals is gaining popularity. Micelle synthesis in biosurfactants is essential for the removal or reduction of heavy metals from polluted settings due to their low interfacial and surface activity. Complexation, ion exchange, and solubilization are just few of the biosurfactant-metal binding pathways that contribute to enhanced metal desorption from soil. Thus, biosurfactants are considered a feasible strategy for the elimination of heavy metals from degraded ecosystems. Heavy metal removal efficiency may be affected, however, by the specific biosurfactant used. Also, most studies of biosurfactant-aided heavy metal removal have been undertaken in a controlled laboratory setting. Further field trials are required in addition to the proven laboratory testing in order to build a viable clean-up approach incorporating biosurfactants, due to the increased complexity and instability of field settings. In order to conduct large-scale tests on the in-situ application of biosurfactants in heavy metal removal, a new experimental apparatus should be built.

References

1. Adamczak M, Bednarski W (2000) Properties and yield of synthesis of mannosylerythritol lipids by Candida antarctica. Biotechnol Prog 17:229–234
2. Ahuekwe EF, Okoli BE, Stanley HO, Kinigoma B (2016) Evaluation of hydrocarbon emulsification and heavy metal detoxification potentials of sophorolipid biosurfactants produced from waste substrates using yeast and mushroom. In: African health, safety, security, environment and social responsibility conference and exhibition. Society of Petroleum Engineers, Accra, Ghana, pp 1–16
3. Akbari S, Abdurahman NH, Yunus RM, Fayaz F, Alara OR (2018) Biosurfactants: a new frontier for social and environmental safety: a mini review. Biotechnol Res Innov 2:81–90
4. Akintunde TA, Abioye OP, Oyeleke SB, Boboye BE, Ijah UJJ (2015) Remediation of iron using rhamnolipid-surfactant produced by *Pseudomonas aeruginosa*. Res J Environ Sci 9:169–177
5. Alloway BJ, Ayres DC (1997) Chemical principles of environmental pollution. Blackie Academic and Professional, 2nd edn. CRC Press London
6. Arab F, Mulligan CN (2018) An eco-friendly method for heavy metal removal from mine tailings. Environ Sci Pollut Res 25:16202–16216
7. Arora PK (2020) Bacilli-mediated degradation of xenobiotic compounds and heavy metals. Front Bioeng Biotech 8:570307
8. Arora S, Jain CK, Lokhande RS (2017) Review of heavy metal contamination. Int J Environ Sci Nat Resour 3:139–144
9. Aşçı Y, Nurbaş M, Açıkel YS (2010) Investigation of sorption/desorption equilibria of heavy metal ions on/from quartz using rhamnolipid biosurfactant. J Environ Manage 91(3):724–731
10. Ayangbenro AS, Babalola OO (2018) Metal (loid) bioremediation: strategies employed by microbial polymers. Sustainability 10(9):3028
11. Ayangbenro AS, Babalola OO (2020) Genomic analysis of *Bacillus cereus* NWUAB01 and its heavy metal removal from polluted soils. Sci Rep 10:19660
12. Bachmann RT, Johnson AC, Edyvean RG (2014) Biotechnology in the petroleum industry: an overview. Int Biodeterior Biodegrad 86:225–273
13. Banat IM, Franzetti A, Gandolfi I (2010) Microbial biosurfactants production, applications and future potential. Appl Microbiol Biotechnol 87:427–444
14. Basak G, Das N (2014) Characterization of sophorolipid biosurfactants produced by *Cryptococcus* sp. VITGBN2 and its application on Zn(II) removal from electroplating wastewater. J Environ Biol 35:1087–1094
15. Bharagava RN, Mishra S (2018) Hexavalent chromium reduction potential of *Cellulosimicrobium* sp. isolated from common effluent treatment plant of tannery industries. Ecotoxicol Environ Saf 147:102–109
16. Bhatt P, Gangola S, Bhandari G, Zhang W, Maithani D, Mishra S, Chen S (2021) New insights into the degradation of synthetic pollutants in contaminated environments. Chemosphere 268:128827
17. Bodek I, Lyman WJ, Reehl WF (1998) Environmental inorganic chemistry: properties, processes and estimation methods. Pergamon Press, New York
18. Bodour AA, Miller-Maier RM (1998) Application of a modified drop-collapse technique for surfactant quantitation and screening of biosurfactant-producing microorganisms. J Microbiol Methods 32(3):273–280
19. Cameotra SS, Makkar RS (2004) Recent applications of biosurfactants as biological and immunological molecules. Curr Opin Microbiol 7(3):262–266
20. Cameron RE (1992) Guide to site and soil description for hazardous waste site characterization metals, vol 1. Environmental Protection Agency, Las Vegas, EPA/ 600/4-91/029
21. Carolin F, Kumar SP, Ngueagni PT (2021) A review on new aspect of lipopeptide biosurfactant: types, production, properties and its application in the bioremediation. J Hazard Mater 407:124827

22. Cavalero DA, Cooper DG (2003) The effect of medium composition on the structure and physical state of sophorolipids produced by *Candida bombicola* ATCC 22214. J Biotechnol 103(1):31–41
23. Chakrabarti S (2012) Bacterial biosurfactant: characterization, antimicrobial and metal remediation properties (Doctoral dissertation)
24. Chen CY, Baker SC, Darton RC (2007) The application of a high throughput analysis method for the screening of potential biosurfactants from natural sources. J Microbiol Methods 70(3):503–510
25. Chen J, Huang PT, Zhang KY, Ding FR (2012) Isolation of biosurfactant producers, optimization and properties of biosurfactant produced by Acinetobacter sp. from petroleum-contaminated soil. J Appl Microbiol 112(4):660–671
26. Chen Q, Yao Y, Li X, Lu J, Zhou J, Huang Z (2018) Comparison of heavy metal removals from aqueous solutions by chemical precipitation and characteristics of precipitates. J Water Process Eng. 26:289–300
27. Cooper D, Goldenberg B (1997) Surface-active agents from two Bacillus species. Appl Environ Microbiol 53(2):224–229
28. Czinkóczky R, Németh Á (2020) The effect of pH on biosurfactant production by *Bacillus subtilis* DSM10. Hung J Ind Chem 37–43
29. da Rocha Junior RB, Meira HM, Almeida DG, Rufino RD, Luna JM, Santos VA, Sarubbo LA (2019) Application of a low-cost biosurfactant in heavy metal remediation processes. Int Biodeterior 30:215–233
30. Das AJ, Lal S, Kumar R, Verma C (2017) Bacterial biosurfactants can be an ecofriendly and advanced technology for remediation of heavy metal and co-contaminated soil. Int J Environ Sci Technol 14:1343–1354
31. Dell'Anno F, Sansone C, Ianora A, Dell'Anno A (2018) Biosurfactant-induced remediation of contaminated marine sediments: current knowledge and future perspectives. Mar Environ Res 137:196–205
32. Dhaliwal SS, Can J, Taneja PK, Mandal A (2020) Remediation techniques for removal of heavy metals from soil contaminated through different sources: a review. Environ Sci Pollut Res Int 27:1319–1333
33. Ehinmitola EO, Aransiola EF, Adeagbo OP (2018) Comparative study of various carbon sources on rhamnolipid production. S Afr J Chem Eng 26:42–48
34. Elouzi AA, Akansha AA, Elgerbi AM, El-Baseir M, El Gammudi BA (2012). Removal of heavy metal contamination by Bio-surfactants (rhamnolipids). J Chem Pharm Res 4:4337–4341
35. Fakruddin Md (2012) Biosurfactant: production and application. J Pet Environ Biotechnol 3(4):124
36. Fardami AY, Kawo AH, Yahaya S, Lawal I, Abubakar AS, Maiyadi KA (2022) A review on biosurfactant properties, production and producing microorganisms. J Biochem Microbiol Biotech 10(1):5–12
37. Fashola MO, Ngole-Jeme VM, Babalola OO (2016) Heavy metal pollution from gold mines: environmental effects and bacterial strategies for resistance. Int J Environ Res Public Health 13(11):1047
38. Fooladi T, Hamid ABA, Yusoff WMW, Moazami N, Shafiee Z (2013) Production of biosurfactant by indigenous isolated bacteria in fermentation system. In AIP Conf Proc 1571(1):197–201
39. Gaith ESI, Rizvi S, Namasivayam C, Rahman PKSM (2019) Removal of Cd++ from contaminated water using bio-surfactant modified ground grass as a biosorbent. In: Advances in science and engineering technology international conference (ASET). IEEE, Dubai, United Arab Emirates, pp 1–7
40. Gomaa EZ, El-Meihy RM (2019) Bacterial biosurfactant from *Citrobacter freundii* MG812314.1 as a bioremoval tool of heavy metals from wastewater. Bull Natl Res Cent 43:69
41. Gonzalez Henao S, Ghneim-Herrera T (2021) Heavy metals in soils and the remediation potential of bacteria associated with the plant microbiome. Front Environ Sci 15

42. Gupta P, Kumar V (2017) Value added phytoremediation of metal stressed soils using phosphate solubilizing microbial consortium. World J Microbiol Biotechnol 33:1–15
43. Hanfi MY, Mostafa MYA, Zhukovsky MV (2020) Heavy metal contamination in urban surface sediments: sources, distribution, contamination control, and remediation. Environ Monit Assess 192:32
44. He X, Qiu X, Chen J (2017) Preparation of Fe(II)-Al layered double hydroxide: application to the adsorption/reduction of chromium. Colloids Surf A Physiochem Eng Asp 516:362–374
45. Hisham NHB, Ibrahim MF, Ramil N, Abd-Aziz (2019) Production of biosurfactant produced from used cooking oil by *Bacillus* sp. HIP3 for heavy metals removal. Molecules 24:2617
46. Ilori MO, Amobi CJ, Odocha AC (2005) Factors affecting biosurfactant production by oil degrading Aeromonas spp. isolated from a tropical environment. Chemosphere 61(7):985–992
47. Ines M, Dhouha G (2015) Lipopeptide surfactant: production recovery and pore forming capacity. Peptides 71:100–112
48. Jain DK, Collins-Thompson DL, Lee H, Trevors JT (1991) A drop-collapsing test for screening surfactant-producing microorganisms. J Microbiol Methods 13(4):271–279
49. Joy S, Rahman PK, Sharma S (2017) Biosurfactant production and concomitant hydrocarbon degradation potentials of bacteria isolated from extreme and hydrocarbon contaminated environments. J Chem Eng 317:232–241
50. Juwarkar AA, Nair A, Dubey KV, Singh SK, Devotta S (2007) Biosurfactant technology for remediation of cadmium and lead contaminated soils. Chemosphere 68:1996–2002
51. Kumar M, Gogoi A, Kumari D, Borah R (2017) Review of perspective, problem, challenges and future scenario of metal contamination in the urban environment. J Hazard Toxic Radioact Waste 21:4
52. Lal S, Ratna S, Said OB, Kumar R (2018) Biosurfactant and exopolysaccharide-assisted rhizobacterial technique for the remediation of heavy metal contaminated soil: an advancement in metal phytoremediation technology. Environ Technol Innov 10:243–263
53. Li Q (2017) Rhamnolipid synthesis and production with diverse resources. Front Chem Sci Eng 11:27–37
54. Lopes CSC, Teixeira DB, Braz BF, Santelli RE, de Castilho LVA, Gomez JGC, Castro RPV, Seldin L, Frire DMG (2021) Application of rhamnolipid surfactant for remediation of toxic metals of long and short-term contaminated sites. Int J Environ Sci Technol 18:575–588
55. Lu J, Zhang B, He C, Borthwick AGL (2020) The role of natural Fe(II)-bearing minerals in chemoautotrophic chromium (VI) bio-reduction in ground water. J Hazard Mater 389:121911
56. Luna JM, Leonie DR, Sarubbo A (2016) Biosurfactant from *Candida sphaerica* UCP0995 exhibiting heavy metal remediation properties. Process Saf Environ 102:558–566
57. Maczek J, Junne S, Götz P (2007) Examining biosurfactant producing bacteria—an example for an automated search for natural compounds. Appl Note CyBio AG
58. Mao X, Jiang R, Xiao W, Yu J (2015) Use of surfactants for the remediation of contaminated soils: a review. J Hazard Mater 285:419–435
59. Maqsood MI, Jamal A (2011) Factors affecting the rhamnolipid biosurfactant production. Pakistan J Biotechnol 8(1):1–5
60. Mishra S, Bharagava RN (2016) Toxic and genotoxic effect of hexavalent chromium in environment and its bioremediation strategies. J Environ Sci Health Part C Environ Carcinog Ecotoxicol Rev 34:1–32
61. Mishra S, Bharagava RN, More N, Yadav S, Zainith S, Mani S, Chowdhary P (2019) Heavy metal contamination: an alarming threat to environment and human health. In: Environmental biotechnology for sustainable future. Springer, Singapore, pp 103–125
62. Mishra S, Lin Z, Pang S, Zhang Y, Bhatt P, Chen S (2021) Biosurfactant is a powerful tool for the bioremediation of heavy metals from contaminated soils. J Hazard Mater 418:126253
63. Mohan PK, Nakhla G, Yanful EK (2006) Biodegradability of surfactants under aerobic, anoxic, and anaerobic conditions. Int J Environ Eng 132(2):279–283
64. Morikawa M, Hirata Y, Imanaka TA (2000) A Study on the structure function relationship of lipopeptide bio-surfactants. Biochem Biopsy's Acta. 1488(3):211–218

65. Mouafo TH, Mbawala A, Ndjouenkeu R (2018) Effect of different carbon sources on biosurfactants' production by three strains of Lactobacillus spp. Biomed Res Int
66. Müller MM, Kügler JH, Henkel M, Gerlitzki M, Hörmann B, Pöhnlein M, Syldatk C, Hausmann R (2012) Rhamnolipids—next generation surfactants? J Biotechnol 162(4):366–380
67. Mulligan CN (2021) Sustainable remediation of contaminated soil using biosurfactants. Front Bioeng Biotech 9:635196
68. Mulligan CN, Sharma SK, Mudhoo A (2014) Biosurfactants: research trends and applications. CRC Press, Boca Raton, Florida
69. Mulligan CN, Yong RN, Gibbs BF (2001) Heavy metal removal from sediments by biosurfactants. J Hazard Mater 85:111–125
70. Nurfarahin AH, Mohamed MS, Phang LY (2018) Culture medium development for microbial-derived surfactants production—an overview. Mol 23(5):1049
71. Nwaguma I, Chikere C, Okpokwasili G (2016) Isolation, characterization and application of biosurfactant by *Klebsiella pneumonia* strain IVN51 isolated from hydrocarbon-polluted soil Ogoniland Nigeria. Bioresour Bioprocess 3:40
72. Obayori OS, Fashola MO, Ashade AO, Opere BO, Adeoye SP, Adeyeye MO (2022) Isolation and characterization of biosurfactant producing bacteria from Mile 2 and Ologe Lagoons Nigeria. Malays J Microbiol 18:1
73. Pacwa-Plociniczak M, Plaza GA, Piotrowska-Seget Z (2011) Environmental applications of biosurfactants: recent advances. Int J Mol Sci 12:633–654
74. Pan ZF, An L (2019) Removal of heavy metal from wastewater using ion exchange membranes. In: Ahamed MI, Asiri AM (eds) Application of ion exchange materials in the environment. Springer, pp 25–46
75. Parthipan P, Preetham E, Machuca LL, Rahman PK, Murugan K, Rajasekar A (2017) Biosurfactant and degradative enzymes mediated crude oil degradation by bacterium *Bacillus subtilis* A1. Front Microbiol 8:193
76. Peng H, Guo J (2020) Removal of chromium from wastewater by membrane filtration, chemical precipitation, ion exchange, adsorption, electrochemical reduction, electrodialysis, electrodeionization photocatalysis and nanotechnology: a review. Environ Chem Lett 18:2055–2068
77. Plaza G, Achal V (2020) Biosurfactants: eco-friendly and innovative biocides against biocorrosion. Int J Mol Sci 21:2152
78. Plaza G, Zjawiony I, Banat I (2006) Use of different methods for detection of thermophilic biosurfactants producing bacteria from hydrocarbon contaminated and bio remediated soils. J Petro Science Eng. 50(1):71–77
79. Poirier A, Ozkaya K, Gredziak J, Talbot D, Baccile N (2023) Heavy metal removal from water using the metallogelation properties of a new glycolipid biosurfactant. J Surfactants Deterg 26(2):175–184
80. Qin H, Hu T, Zhai Y, Lu N, Aliyeva J (2020) The improved methods of heavy metal removal by biosorbents: a review. Environ Pollut 258:113777
81. Ravindran A, Sajayan A, Priyadharshini GB, Selvin J, Kiran GS (2020) Revealing the efficacy of thermostable biosurfactant in heavy metal bioremediation and surface treatment in vegetables. Front Microbiol 11:222
82. Sachdev DP, Cameotra SS (2013) Biosurfactants in agriculture. Appl Microbiol Biotechnol 97:1005–1016
83. Sah D, Rai JPN, Ghosh A, Chakraborty M (2022) A review on biosurfactant producing bacteria for remediation of petroleum contaminated soils. 3 Biotech 12(9):218
84. Saikia RR, Deka S, Deka M, Banat IM (2012) Isolation of biosurfactant-producing *Pseudomonas aeruginosa* RS29 from oil-contaminated soil and evaluation of different nitrogen sources in biosurfactant production. Ann Microbiol 62:753–763
85. Shekhar S, Sundaramanickam A, Balasubramaniam T (2014) Biosurfactant producing microbes and its potential applications: a review. Crit Rev Environ Sci Technol 45:1522–1554
86. Silva RL, Farias CBB, Rufino RD (2010) Glycerol as substrate for the production of biosurfactant by *Pseudomonas aeruginosa* UCP0992. Colloids Surf B: Biointerf 79:174–183

87. Singh P, Cameotra SS (2004) Enhancement of metal bioremediation by use of microbial surfactants. Biochem Biophys Res Commun 319(2):291–297
88. Tadros T (2005) Adsorption of surfactants at the air/liquid and liquid/liquid interfaces. Applied surfactants: principles and applications. Wiley VCH, Weinheim, pp 81–82
89. Tang J, Zhnag J, Ren L, Zhou Y, Gao J, Yang Y, Peng Q, Huang H, Chen A (2019) Diagnosis of soil contamination using microbiological indices: a review on heavy metal pollution. J Environ Manag 242:121–130
90. Tuleva B, Christova N, Jordanov B, Nikolova-Damyanova B, Petrov P (2005) Naphthalene degradation and biosurfactant activity by *Bacillus cereus* 28BN. Z Naturforsch C 60(7–8):577–582
91. Wang Z, Zhang B, He C, Shi J, Wu M, Guo J (2021) Sulfur-based microtrophic vanadium (V) bioreduction towards lower organic requirement and sulfate accumulation. Water Res 189:116655
92. Wei YH, Lai CC, Chang JS (2007) Using Taguchi experimental design methods to optimize trace element composition for enhanced surfactin production by *Bacillus subtilis* ATCC 21332. Process Biochem 42(1):40–45
93. Wei YH, Chou CL, Chang JS (2005) Rhamnolipid production by indigenous *Pseudomonas aeruginosa* J4 originating from petrochemical wastewater. Biochem Eng J 27:146–154
94. William CW (2014) Phosphotidylethanolamine and related lipids: structure, occurrence, biochemistry and analysis. The Am Oil Chem Soc Lipid Libr. Lipid Web
95. Yan P, Lu M, Yang Q, Zhang H, Zhang Z, Chen R (2012) Oil recovery from refinery oily sludge using a rhamnolipid biosurfactant-producing Pseudomonas. Bioresour Technol 116:24–28
96. Yin K, Lv M, Wang Q (2016) Simultaneous bioremediation and biodetection of mercury ion through surface display of carboxylesterase E2 from *Pseudomonas aeruginosa* PA1. Water Res 103:383–390
97. Yu G, Wang X, Liu J, Jiang P, You S, Ding N, Guo Q, Lin F (2021) Applications of nanomaterials for heavy metal removal from water and soil: a review. Sustainability 13:713
98. Zhang B, Li Y, Fei Y, Cheng Y (2021) Novel pathway for vanadium (V) biodetoxification by Gram positive Lactococcus raffinolactis. Environ Sci Technol 55:2121–2131
99. Zhao F, Zhou JD, Ma F, Shi RJ, Han SQ, Zhang J, Zhang Y (2016) Simultaneous inhibition of sulfate-reducing bacteria, removal of H2S and production of rhamnolipid by recombinant *Pseudomonas stutzeri* Rhl: applications for microbial enhanced oil recovery. Bioresour Technol 207:24–30

Role of Rhizobacteria in Phytoremediation of Heavy Metal

Zain Mushtaq, Ariba Muzammil, Korkmaz Bellitürk, Waheed Anwar, Adnan Akhter, Hafiz Azhar Ali Khan, Amir Parnian, and Shafeeq Ur Rahman

Abstract Pollution sources in ecosystems are mostly anthropogenic. Heavy metal contamination is one of these concerns, threatening ecosystem health, plants, humans, and animals in different concentrations. Environmental remediation techniques are the probable solution to care for these situations. Every remediation treatment demands resources and some preneeds also have some side effects. Phytoremediation is one of the categories of bioremediations, which is considered a cost-effective and eco-friendly method for techniques categories, which is considered a cost-effective and eco-friendly method for soil and water cleanup. This method could be assisted using rhizobacteria. This chapter gathers and details the different roles of rhizobacteria in the phytoremediation of heavy metals.

Z. Mushtaq (✉)
Department of Soil Science, Faculty of Agricultural Sciences, University of the Punjab, Lahore, Pakistan
e-mail: zain.ss@pu.edu.pk

A. Muzammil
Institute of Soil and Environmental Sciences, University of Agriculture, Faisalabad, Pakistan

K. Bellitürk
Department of Soil Science and Plant Nutrition, Faculty of Agriculture, Tekirdag Namık Kemal University, Suleymanpasa, Tekirdag, Turkey

W. Anwar · A. Akhter
Department of Plant Pathology, Faculty of Agricultural Sciences, University of the Punjab, Lahore, Pakistan

H. A. A. Khan
Department of Entomology, Faculty of Agricultural Sciences, University of the Punjab, Lahore, Pakistan

A. Parnian
National Salinity Research Center (NSRC), Agricultural Research Education and Extension Organization (AREEO), Po. Box. 8917357676, Yazd, Iran

S. U. Rahman
Water Science and Environmental Engineering Research Center, College of Chemical and Environmental Engineering, Shenzhen University, Shenzhen 518060, China

N. Kumar (ed.), *Heavy Metal Remediation*, Earth and Environmental Sciences Library,
https://doi.org/10.1007/978-3-031-53688-5_9

Keywords Phytoremediation · Rhizobacteria · Heavy metals · Environmental remediation · Soil and water cleanup

1 Introduction

Environmental remediation encompasses the utilisation of diverse technological approaches to eliminate pollutants or toxins from distinct components of the environment, including the atmosphere, hydrosphere, and lithosphere. Several technologies commonly employed for environmental remediation include bioremediation, physical remediation, chemical remediation, and thermal treatments. Each of the aforementioned strategies possesses distinct advantages and disadvantages in relation to environmental remediation [80]. Bioremediation is a biotechnological method that uses living organisms, such as bacteria, plants, and algae, to clean up the environment from various pollutants. It can be applied to different environmental settings, such as water and soil, that are contaminated by harmful substances. Bioremediation is an eco-friendly and advanced method for environmental management that can offer a low-cost solution to pollution problems. However, bioremediation has some limitations and challenges, such as the long duration and the specific requirements for the growth of the organisms in the polluted environment [84, 90, 109]. Phytoremediation is a biological process that uses plants and/or algae to clean up and restore polluted environments, such as soil, sediment, air, water, and exhaust. It is a low-cost and environmental-friendly technology that can remove or control contaminants from the ecosystem by using the natural abilities of plants and/or algae [79]. Phytotechnology is a green technology that can address a wide range of pollution problems. It uses plants and/or algae to extract, degrade, contain, and stabilize or immobilize pollutants in different environmental media [78].

The proliferation of agricultural operations, ongoing anthropogenic practices, and global industrialization have resulted in a multitude of environmental issues stemming from the discharge of pollutants, including organic contaminants, smoke and fumes, heavy metals, and hazardous wastes, into the surrounding ecosystem. In the realm of scientific discourse, heavy metals are commonly characterized as elements that possess distinct metallic attributes, such as ligand selectivity, conductivity, ductility, and stability in their cationic form. Furthermore, these elements are identified by having an atomic number greater than 20. The predominant pollutants among heavy metal ions include cadmium, copper, lead, mercury, nickel and chromium [46, 92]. Plants necessitate several heavy metals as micronutrients due to their natural presence in soil. Nevertheless, because of the onset of the industrial revolution, there has been a significant escalation in the contamination of the biosphere with harmful metals. Various anthropogenic activities, including but not limited to mining and metallurgical processes, vehicular emissions, agricultural practices involving fertilizer application, electroplating operations, energy and fuel generation, wastewater discharge, pesticide usage, and the generation of municipal garbage, are significantly contributing to irreversible environmental degradation [121].

Heavy metals can be characterized as elements that possess metallic characteristics such as conductivity, ligand selectivity, stability, and other related properties. Metals naturally occur in soil, as compounds or simple ions, with a significant accumulation of heavy metal ions. These metals serve as necessary micronutrients for plants [72, 97]. The pollution levels experienced a significant increase because of the Industrial Revolution, primarily due to the release of toxic heavy metals resulting from various human activities, including electroplating, mining of metals, and pesticide application [42]. The issue of metal pollution has emerged as a prominent and pressing environmental concern in contemporary times. Rhizobacteria, commonly known as plant growth promoting (PGPR) bacteria, significantly benefit plants by defending against abiotic and biotic stressors. These rhizosphere microbes promote plant growth and soil fertility [55, 103]. The presence of microorganisms in the rhizosphere helps phytoremediation remove heavy metals from contaminated soil. These bacteria emit chelating compounds that change heavy metal mobility, affecting microbial populations and plant availability [73, 109].

Additionally, these bacteria contribute to phytoremediation through acidification, phosphate solubilization, redox shift release, and phytoremediation promotion. The rhizobacteria that are adapted to metal in the phytoremediation process have garnered significant attention [11]. There is a need to enhance our comprehension of the mechanisms implicated in the mobilization and transport of heavy metals within soils that have been poisoned. Soils that have been contaminated with heavy metals give rise to environmental issues due to the very harmful impacts of these metals. There are various methods available for addressing metal pollution in sediments, including ex-situ and in-situ procedures. The objective of in-situ remediation is to improve the retention of metals inside soil particles to promote their maintenance. Other approaches are employed to mitigate the bioavailability or possible mobility of hazardous metals from the environment in plants [32, 76]. The objective of employing ex-situ techniques is to effectively isolate and recover metals from soil by a combination of biological, physical, and chemical processes, facilitated by the utilization of purpose-built reactors.

Hence, it is imperative for individuals to acquire knowledge regarding heavy metals and adopt precautionary steps to mitigate the risks associated with excessive exposure. Heavy metal poisoning is a very infrequent medical illness in most regions within the United States [75, 92]. Nevertheless, its occurrence warrants clinical attention due to its significant impact on health. If toxicity is not acknowledged or treated in an appropriate manner, it can lead to substantial disease and a decrease in the overall quality of life. Testing is crucial for anyone who has concerns about heavy metal toxicity in themselves or someone residing in their household. It may be necessary to consider the pursuit of appropriate conventional and natural medical techniques [27].

Heavy metal pollution of soil is becoming an environmental and public health issue. High soil heavy metal concentrations reduce agricultural yield and microbiological activity. This also puts human health at risk since heavy metals can infiltrate the food chain. Heavy metal buildup harms most plant species [69]. Plant roots quickly absorb heavy metal ions from high environmental concentrations and convey them to the aboveground sections of the plant. This mechanism can slow plant metabolism and growth [35]. Water and soil heavy metal pollution threatens the ecosystem and public health. Polluted soils have higher metal concentrations, which affect soil microbial activity, fertility, and agricultural yields [59]. Non-essential heavy metal cadmium (Cd) is hazardous to plants. It hinders root and shoot growth, disrupts nutrient absorption and balance, and is typically accumulated by agriculturally important crops. Therefore, the consumption of Cd-enriched crop products by animals and humans has the potential to induce many illnesses [24]. Provided that soil cadmium (Cd) contamination exhibits a cumulative nature, characterized by a progressive increase in levels over time, it is plausible that the soil may ultimately become unsuitable for agricultural purposes, specifically crop cultivation. In a similar vein, the presence of cadmium (Cd) in soil can have detrimental effects on both biodiversity and the functioning of soil microbial communities [54]. The investigations conducted by Yang et al. showed that canola seeds exhibited normal development when exposed to nickel chloride concentrations of up to 1 mmol/L. However, the elongation of plant roots and shoots was observed to be hindered when greater levels of nickel chloride were present [112].

2 Remediation Strategies

From a biological standpoint, the destruction of heavy metals is not feasible due to the absence of any degradation or modification in the nuclear structure of these elements. In contrast, heavy metals undergo a process of conversion from one oxidation state or organic complex to another [106]. As a result, the remediation of heavy metal contamination in soil has greater challenges compared to that in aquatic ecosystems.

Thus far, the techniques utilized for the restoration of polluted areas, such as excavation and landfill, thermal treatment, acid leaching, and electro-reclamation, have demonstrated limited feasibility for extensive adoption. The primary reasons for their limited adoption are their high prices, restricted efficacy, substantial impact on soil structure and fertility, and dependence on specific pollutants, soil features, and site conditions [40, 57]. Therefore, it is imperative to establish phytoremediation techniques for soils that are contaminated with heavy metals.

Phytoremediation technology uses plants and algae to clean up different types of aquatic environments, such as groundwater, seawater, and wastewater. Some of these plants and algae are native to aquatic habitats, while others are adapted from land habitats. Algae and aquatic plants are natural remediators of water and air. For phytoremediation applications, plant species that can grow in polluted water are preferred. Phytoremediation is a feasible option for contaminated environments if the plants can survive in those conditions or if the pollutants can be rooted in the culture medium. The following factors should be considered for selecting suitable plants and algae for phytoremediation [79, 115] (Fig. 1):

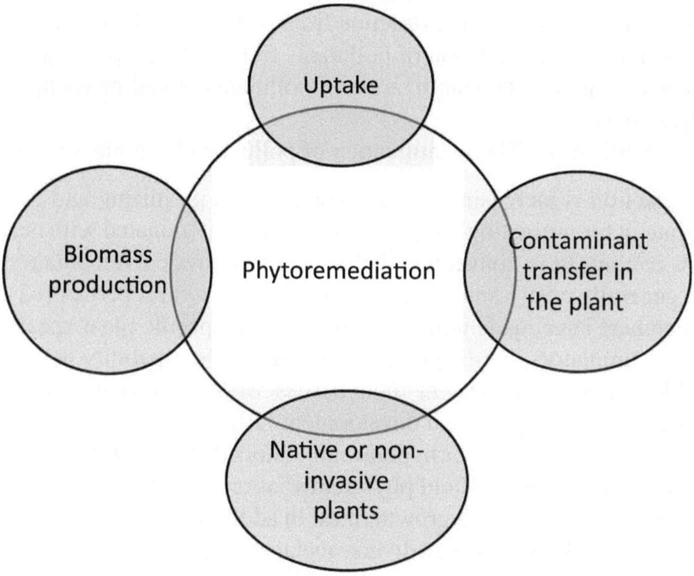

Fig. 1 Factors for selecting suitable plants/algae for phytoremediation

I. High contaminant uptake: The ability to absorb and store large amounts of pollutants in their tissues.
II. High biomass production: The ability to grow fast and produce biomass that can be harvested and disposed of safely.
III. High ability to transfer contaminants from absorbing to removable organs or parts: The ability to transport pollutants from their roots to their shoots or leaves that can be easily removed from the environment.
IV. Ideally native or non-invasive species that can quickly adapt to the climatic conditions: The preference for using local or harmless species that can cope with the environmental changes and reduce the economic costs and ecological impacts.

Plants and algae can clean up pollutants from the environment through two pathways: the foliar surface and the root system. The foliar surface is the active part of the plant in phytoremediation that directly contacts the air or water, while the root system is the plant's active part that contacts the water, soil or sediment in some remediation situation. In aquatic medium remediation, some plants and algae use their entire organs for phytoremediation. The larger active parts promotes and advertise the greater remediation potential, as it provides more space for absorption, uptake, and host for beneficial microbes. Phytoremediation involves different mechanisms, such as [79, 47]:

(a) Phytoaccumulation: The accumulation of pollutants in plant or algae tissues.
(b) Rhizodegradation: The degradation of pollutants by microbes in the root zone.
(c) Phytodegradation: The degradation of pollutants by plant or algae enzymes.

(d) Phytoextraction: Pollutants extraction from soil or water by plant or algae roots.
(e) Rhizofiltration: The filtration of pollutants from water by plant or algae roots.
(f) Phytostabilization: The stabilization of pollutants in soil or sediment by plant or algae roots.
(g) Phytovolatilization: The volatilization of pollutants from plant or algae leaves.

Phytoextraction is increasingly acknowledged as a promising and economically feasible strategy for mitigating the problem of soils contaminated with heavy metals. This stands in contrast to conventional chemical and physical remediation methods, which are generally costly and can negatively impact soil properties [62]. In recent times, researchers have made notable findings about specific plant species, referred to as hyperaccumulators, which possess the remarkable capability to store substantial quantities of heavy metals in their tissues without encountering detrimental consequences on their growth and development [29].

However, the vast majority of hyperaccumulators discovered thus far are deemed unsuitable for practical use in field phytoremediation endeavors, mostly due to their restricted biomass and sluggish growth rates. In addition, it is important to acknowledge that a considerable quantity of diverse metals is absorbed or occluded by carbonates, organic compounds, iron-manganese oxides, and primary or secondary minerals [91]. The effectiveness of phytoremediation may be constrained by the diminished bioavailability of heavy metals within the contaminated ecosystem.

3 Phytoremediation in Assistance with Rhizobacteria in Porous Medium

Phytoremediation is a technique that utilizes plants to remove, isolate, and/or neutralize pollutants through physio-chemical, and biological processes. This approach has been documented as a successful, on-site, non-invasive, cost-effective, visually appealing, environmentally friendly, and socially accepted method for remediating contaminated soils [48, 61]. Various studies have reported the efficacy of phytoremediation, while other research has highlighted its benefits. Furthermore, it is worth noting that this approach not only aids in the prevention of landscape destruction but also contributes to the improvement of soil microorganism activity and diversity, thereby promoting the maintenance of healthy ecosystems [35]. Consequently, it is regarded as a more appealing alternative to the conventional methods currently employed for addressing the issue of heavy metal contamination [18].

Remediating heavy metals with plants involves several methods, such as phytoextraction, rhizofiltration, phytostabilization, and phytovolatilization. Plants are used in phytoextraction, rhizofiltration, phytostabilization, and phytovolatilization to alter metal concentration and mobility [65, 114]. Phytoextraction entails the accumulation of metals in the roots and shoots of plants, effectively extracting them from the soil. Rhizofiltration, on the other hand, involves the absorption, concentration, or precipitation of metals from effluents through the roots of plants. Phytostabilization employs plants to mitigate the movement of heavy metals by means of absorption and precipitation, thereby diminishing their bioavailability [5]. Lastly, phytovolatilization encompasses the uptake and subsequent release of volatile substances, such as compounds containing mercury or arsenic, into the atmosphere.

The optimal plant species for phytoextraction should exhibit rapid growth, possess a high capacity for biomass production, and demonstrate the ability to withstand and accumulate elevated levels of metals in its shoots [104]. Most well-known heavy metal accumulators are classified within the Brassicaceae family. While hyperaccumulator plants possess a remarkable ability to accumulate metals, it is worth noting that many of these plants have a sluggish development rate and tend to yield restricted biomass in situations where the concentration of accessible metals in the contaminated soil is exceedingly high [39, 120]. One potential approach involves utilizing species that have a reduced capacity for metal accumulation but exhibit higher rates of growth, such as *Brassica juncea*, commonly known as Indian mustard. Another option is to introduce associated plant growth-promoting rhizobacteria, which are recognized as a significant element of phytoremediation technology [111]. The rhizosphere is known to harbor a substantial microbial community that exhibits greater metabolic activity in comparison to the surrounding bulk soil.

The impact of microbial populations on the mobility and availability of heavy metals to plants is well-documented in scientific literature. This influence is mostly attributed to the microbial release of chelating agents, acidification, phosphate solubilization, and redox shifts. Certain bacteria that promote plant growth and are found in association with plant roots can have positive effects on plant growth and nutrition [53]. The utilization of rhizobacteria in conjunction with plants is anticipated to yield significant efficacy in the field of phytoremediation [31].

Phytoremediation uses phytoextraction, phytostabilization, phytovolatilization, rhizo filtration, and phyto filtering to clean up polluted metal-lands, water, and groundwater.

Many phytoremediation methods breakdown organic and inorganic pollutants by rhizode grading and phytodegradation. We focus on the most common phytoextraction, phytostabilization, phytoremediation, phytovolatilization, and phytofiltration approaches for metal-contaminated soil and water. Soils and aquatic resources are heavily polluted by metal ions, harming land, surface, and groundwater ecosystems, food safety, and human health. These anthropogenic practises seem to be increasing polluted places that can be amended to prevent chemical transfer to terrestrial and aquatic ecosystems [5]. Therefore, there has been a new focus on exploring how rhizobacteria can enhance phytoremediation to address heavy metal soil contamination.

4 Phytoremediation in Assistance with Rhizobacteria in Aquatic Medium

Phytoremediation is a bio-technique, which can apply plants and/or algae to clean up pollutants from the aquatic environment. Rhizobacteria are bacteria that live in the root zone of plants and alongside algae-associated bacteria can help them to fix, decontaminated, remove or degrade contaminants. Phytoremediation in assistance with rhizobacteria is a promising technique for treating polluted aquatic environments, such as water and sediment [45]. Phytoremediation in assistance with rhizobacteria is a cost-effective and eco-friendly pathway for cleaning up contaminated aquatic environments e.g. water bodies, wastewater, drainage, etc. However, there are many challenges and limitations that need to be attended by method planners and managers, such as (a) the selection of suitable plants and/or algae species, (b) the optimization of rhizobacterial inoculation/colonization, (c) the evaluation of phytoremediation efficiency, feedbacks and safety, and (d) the management of phytoremediation residues/waste/ by-products [60, 88].

The addition of the rhizobacteria consortium could enhance both the growth of aquatic plants and the metal accumulation in constructed wetlands systems. Therefore, aquatic medium remediation with hydrophytes regarding many good rhizobacteria effects on the system performance could be a prosperous method for future remediation research and development projects.

5 Interaction in Rhizosphere

The efficacy of phytoremediation is contingent upon the interplay of soil composition, heavy metal presence, bacterial activity, and plant physiology. As depicted in Fig. 2, the intricate interplay of these interactions is influenced by a multitude of factors, including the attributes and behavior of both plants and rhizobacteria, meteorological circumstances, and soil parameters, among others.

5.1 Plants and Bacteria

The study conducted by Tarekegn et al. [107] emphasizes the importance of the interactions between plant roots and diverse microorganisms in shaping the efficacy of phytoremediation. The effectiveness of plant-bacterial symbioses in soil contaminated with heavy metals can be influenced by the microorganisms engaged in the symbiotic relationship (specifically, bacteria associated with plants) as well as the plant species hosting the symbiosis. Soil microbes play significant roles in various ecological processes, including plant nutrition cycling, soil structure preservation, detoxification of detrimental compounds, and regulation of plant pests and growth

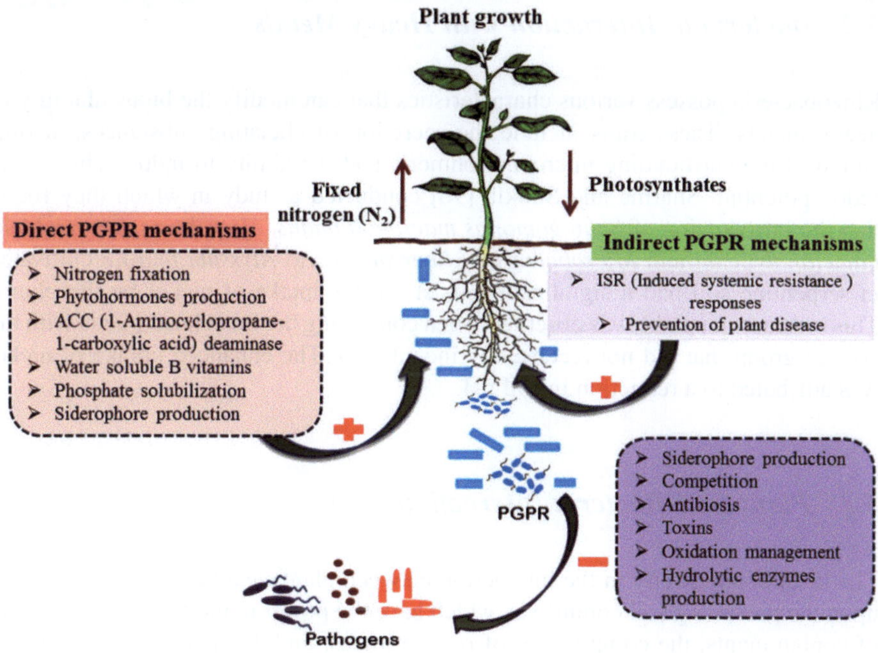

Fig. 2 Mechanism of PGPR [15]

[102]. Therefore, microorganisms have the potential to enhance the ability of plants to remediate contaminated soil or mitigate the toxic effects on plant growth. Moreover, there exists a symbiotic relationship between plants and bacteria, wherein the plant supplies the bacteria with a particular carbon source that stimulates the bacteria to mitigate the phytotoxic effects of the polluted soil [1]. In an alternative scenario, it is possible for plants and bacteria to establish nonspecific connections. In such associations, the regular physiological activities of plants can stimulate the microbial population. Consequently, the microbial community, through its inherent metabolic activity, is capable of degrading pollutants present in the soil [31].

The roots of plants can release root exudates, which in turn can enhance the solubility of ions. The biochemical pathways described herein serve to enhance the remediation activity of bacteria that are symbiotically connected with plant roots [38]. In conclusion, the capacity for adaptation exhibited by both participants in the associative symbiotic relationship, along with the bioremediation potential of the microsymbiont, play a crucial role in mitigating the adverse impacts of heavy-metal contamination [28]

5.2 Bacteria in Interaction with Heavy Metals

Rhizobacteria possess various characteristics that can modify the bioavailability of heavy metals. These traits include the secretion of chelating substances, acidification of the surrounding microenvironment, and the ability to induce changes in redox potential. Sharma and Shukla [96] conducted a study in which they found that the introduction of *Sphingomonas macrogoltabidus, Microbacterium arabinogalactanolyticum* and *Microbacterium liquefaciens,* to *Alyssum murale* cultivated in serpentine soil had a significant impact on the uptake of nickel by the plants. This increase in uptake was observed when comparing the inoculated plants with the control group that did not receive any inoculation. The enhanced uptake of nickel was attributed to a reduction in soil pH.

5.3 Plant-Soil-Bacteria Interaction

The level of specificity in the interaction between plants and bacteria is contingent upon the prevailing soil conditions, which have the power to modify the availability of contaminants, the composition of root exudates, and the quantities of nutrients [16] Furthermore, the metabolic demands associated with the remediation of heavy metals may also influence the nature of the interaction between plants and bacteria, determining whether it is selective or general [17, 49].

5.4 PGPR Interaction for Removal of Heavy Metals

By causing changes in redox potential and the release of chelating compounds from bacteria, the PGPR interacts with heavy metals in a way that can alter their availability to plants in the soil environment. This results in an acidified microenvironment.

The kind of interaction that rhizobacteria, or PGPR, would have with the plant depends on several factors, including the type of plant, the quantity and kind of heavy metal contaminants in the soil, the consistency and makeup of root exudates, and the amount of nutrients present [120]. Because of the metabolic activities of plants, heavy metals must be removed from the soil environment. However, this can also reveal the type of bacterial interaction, which may be particular or non-specific.

The following circumstances lessen the interaction between bacteria and plants:

- inadequate water supply
- poor soil fertility and unfavorable weather
- inadequate dietary intake or impractical nutrient exchanges

Rhizobacteria can block plant heavy metal supply, making it a good bioremediation option. The PGPR can interact with plants at the cellular level to help them endure severe heavy metal toxicity [110].

6 Mechanisms of Rhizobacteria Influencing Heavy Metal Accumulation

6.1 Rhizobacterial Secretions

Rhizobacteria secretions play a big role in the process of phytoremediation. Rhizobacteria help by fixing nitrogen and making siderophores, which dissolve iron in soil and repair it. Plants can grow faster and better dissolve minerals, especially phosphorus, when phytohormones like auxins and cytokinin are made. In a roundabout way, rhizobacteria stop phytopathogens that hurt plant growth and development [85].

Rhizobacteria make siderophores, which are metal-chelating substances. Additionally, these organic substances help plants grow in soil that is metal-contaminated or iron-deficient by increasing the bioavailability of iron that is linked to other substances [51] Microbes called siderophores control how much iron is available in the rhizosphere of plants. Concentration of different types of siderophore, exchange rates, and the supply of Fe-complexes to both microbes and plants are some of the most important factors that affect this process. It is not as strong for phytosiderophores to bind iron as it is for bacteria siderophores.

Several types of PGPR have an enzyme called 1-aminocyclopropane-1-carboxylate (ACC) deaminase that breaks down ACC and lowers its level. Plant hormone ethylene is made up of ACC and ethylene precursors. ACC lowers the production of ethylene by plants [50]. In some plants, ACC is taken from the roots or seeds and broken down by ACC deaminase into ammonia and α-Keto butyrate. The bacteria get nitrogen from this ammonia, which lowers the amount of ACC in the plant. The plant must make a lot of ACC to keep the levels of ACC inside and outside of it in balance. As the amount of ACC in the plant drops, the amount of ethylene in the plant also drops. This model can be used to show how bacteria that help plants grow control the amount of stress ethylene that plants have. PGPR makes indole acetic acid (IAA), which can help plant roots grow. Rhizobacteria make IAA. Low levels of IAA help primary roots grow longer, while high levels of IAA help side and adventitious roots form and stop primary root growth. So, bacteria that help plants grow can do this by changing the mix of hormones inside the plant [89]. In the same way, an ethylene shock can help break the dormancy of seeds, but too much ethylene can stop seeds from germinating. When the amount of ethylene rises, phytopathogens can infect plants. Plant growth-promoting bacteria (PGPB) make ACC deaminase, which makes up for this [41]. Therefore, bacteria that help plants grow change the amount of ethylene they produce and lessen some of the damage that heavy metals seem to do to plants.

6.2 Plant Pathogen Inhibition

Plant growth-promoting rhizobacteria (PGPR) kill plant diseases in numerous ways. These strategies include competition for nutrition and space, antibody production, and siderophores that limit iron, a pathogen-vital element [22]. Additionally, chitinases and β-1,3-glucanases are synthesized to degrade chitin and glucan in fungal cell walls [66].

6.3 Transformation of Heavy Metals

Bacterial microorganisms possess the ability to transform hazardous heavy metals into readily absorbable forms for plant roots. An instance of enhancing selenium (Se) accumulation in plants is by the conversion of selenate to organic forms of Se. Specifically, the observed concentration difference between the bulk soil and rhizosphere indicated that the relative changes in the levels of organic bound Cu, Zn, and Pb were + 5%, + 23%, and + 3%, respectively, in the rhizosphere affected by infection. In contrast, the non-infected rhizosphere exhibited relative changes of 0.8%, − 3%, and − 2% for the same elements [52, 126]. Significant quantities of copper (Cu), zinc (Zn), and lead (Pb) were sequestered by organic matter within the rhizosphere affected by the infection.

Soil rhizobacteria have the capacity to modulate the bioavailability of metals through alterations in their chemical characteristics, including pH, organic matter content, redox state, and other related factors. This process facilitates the leaching of pollutants from soil. An instance of *Pseudomonas maltophilia* was seen to facilitate the conversion of the mobile and harmful Cr^{6+} into the harmless and immobile Cr^{3+}, while also reducing the ambient mobility of other dangerous ions, including Hg^{2+}, Pb^{2+}, and Cd^{2+}.

6.4 Stimulating Transporting Protein

The ability of bacteria to survive and proliferate in their surroundings and within different organisms is contingent upon the acquisition and containment of transition metals, including manganese, zinc, and iron [82]. As an illustration, cells possess the ability to tightly regulate the quantity of zinc inside their intracellular environment, owing to the detrimental effects that elevated levels of zinc can have on many cellular processes. Bacteria have the potential to induce the activation of the sulphate transport protein, which is situated in the plasma membrane of the root [7]. This protein is responsible for facilitating the transportation of selenate as well.

6.5 *Heavy Metal Speciation Versus Bioavailability in Soils*

The bioavailability of heavy metals is influenced by several bacterial features, including alterations in redox potential, acidity of the microenvironment, and the production of chelating agents. While an elevated concentration of heavy metals can have detrimental effects on microbial physiology, it is important to note that small quantities of heavy metals are necessary for the normal growth of microorganisms [85]. These trace amounts play a crucial role in facilitating redox reduction and numerous cellular functions. The way bacteria interact with heavy metals is contingent upon the speciation of the metal and the forms of bioavailability.

Bacterial activity has a significant impact on the bioavailability of metals by modifying the speciation of heavy metals within the soil. To mitigate the adverse effects of heavy metal toxicity on plants, a mechanism is employed wherein the speciation of metals in the soil is modified, transforming them from a bioavailable state to a non-bioavailable state [37, 44, 102].

7 Rhizobacterial Impact on Phytoremediation

7.1 *Crop-Plant Growth*

Plant growth-promoting rhizobacteria (PGPR) are tightly related to roots. Rhizobacteria, a diverse group of free-living soil bacteria, reduce heavy metal damage and boost plant growth in heavy metal-contaminated soils (Table 1).

The presence of elevated levels of heavy metals in the surrounding environment poses a potential risk to plants that can accumulate and tolerate these metals. According to [71], plant species have the potential to experience adverse effects due to a deficiency of iron in soil that is contaminated with heavy metals. The presence of elevated heavy metal concentrations in plants can result in reduced iron levels, which can subsequently lead to chlorosis. This occurrence arises from the inhibitory effect of iron deficiency on chloroplast development and the manufacture of chlorophyll. Plants can uptake microbial iron-siderophore complexes, so acquiring a source of iron. To mitigate the occurrence of chlorotic plants in situations with high concentrations of heavy metals, the implementation of siderophore-producing bacteria has been proposed. According to a study conducted by DalCorso et al. [20], it has been observed that some bacteria with the ability to promote plant growth can enhance the growth of plants in settings contaminated with heavy metals. This bacterial activity results in the development of longer roots and improved establishment of plants

Table 1 Illustrates PGPR variant effects on plants

PGPR	Crop	Impact
Delftia tsuruhatensis strain HR4,	Cucumber, Rice	Inhibits the growth of plant pathogens
Bacillus licheniformis, Serratia liquefaciens 2–68,	Corn, Mung bean, Pea	Enhancement of plant growth and optimization of nutrient absorption
Achromobacter piechaudii ARV8	Tomatoes	Exhibiting resistance to water and salt-induced stress
Kluyvera ascorbata SUD165	Canola, Grasses	The plant seedlings exhibited resistance to Cd, Zn, Cu, Ni, Co, Cr, and Pb, as well as a stimulatory effect on root elongation
Pseudomonas corrugata 13 or aureofaciens	Rice	Inhibits the growth of plant pathogens

during their first growth stages. Following the establishment of seedlings, the bacteria facilitate the process of acquiring iron to promote optimal growth of plants. In their study, [14] discovered that pseudomonads resistant to chromium, derived from effluents of the paint industry, had the ability to enhance the germination and growth of Triticum aestivus seeds when exposed to potassium bichromate. The facilitation of seedling development by bacteria led to a reduction in the absorption of chromium. The experiment investigated the effects of the introduction of K. ascorbata SUD165, a bacterium known for boosting plant growth, to canola or tomato seeds prior to germination. This was done in conjunction with the application of inhibitory doses of Ni^{2+}. The results of the trials demonstrated that the inclusion of K. ascorbata SUD165 led to a significant decrease in the toxicity of nickel across all concentrations (ranging from 1 to 6 mmol/L Ni^{2+}). This reduction was observed in both low-level and high-level bacterial cell treatments, with the former using a concentration of 0.025 and the latter using a concentration of 0.50. The effectiveness of K. ascorbata SUD165 was evident in several components of canola and tomato plants, including roots, shoots, pouches, and pots.

The accumulation of potentially dangerous trace elements by plant tissues is facilitated by the presence of rhizosphere microorganisms. The saltmarsh bulrush plants that were inoculated with rhizosphere bacteria exhibited elevated selenium (Se) concentrations in both their roots ($70 \pm 80\%$) and shoots ($40 \pm 60\%$) in comparison to plants grown in a sterile environment. Four out of the six bacterial strains that were examined showed a substantial increase in selenium accumulation in the roots and shoots of plants when provided as pure cultures. This finding was statistically significant with a p-value of less than 0.05, and the experiment was conducted with a sample size of three. The accumulation of selenium in axenic bulrush plants was enhanced using a combination of six bacterial strains, as demonstrated in previous studies conducted by Cailliatte et al. [10] and Bastow et al. [8]. Several studies have indicated that the presence of ectomycorrhizal or vesicular–arbuscular fungi on the roots of plants can potentially lead to a reduction in metal uptake while simultaneously promoting an increase in biomass. Certain plant species have been observed to employ growth-promoting bacteria or mycorrhizal fungi in order to alleviate the adverse impacts caused by heavy metal contamination.

7.2 *Rhizosphere-Microorganisms*

Considerable scholarly investigation has been devoted to the rhizobacteria that are linked to plants that accumulate or hyperaccumulate metals. This research has shed light on the crucial role that these bacteria play in enabling the plants to tolerate the absorption of heavy metals. Previous studies have indicated that a significant number of rhizobacteria exhibit tolerance towards heavy metals and fulfil crucial functions in the process of mobilizing or immobilizing these metals [125] (Fig. 3).

The population of rhizobacteria has been shown to be substantially greater than that of the bulk soil. The size, composition, and general activity of the soil microbial communities are all significantly impacted by the high concentrations of heavy metals present in these soils. According to [108], the experimental findings revealed that the D. fusca rhizosphere's bacterial population had grown to 1.0 × 107 colony forming

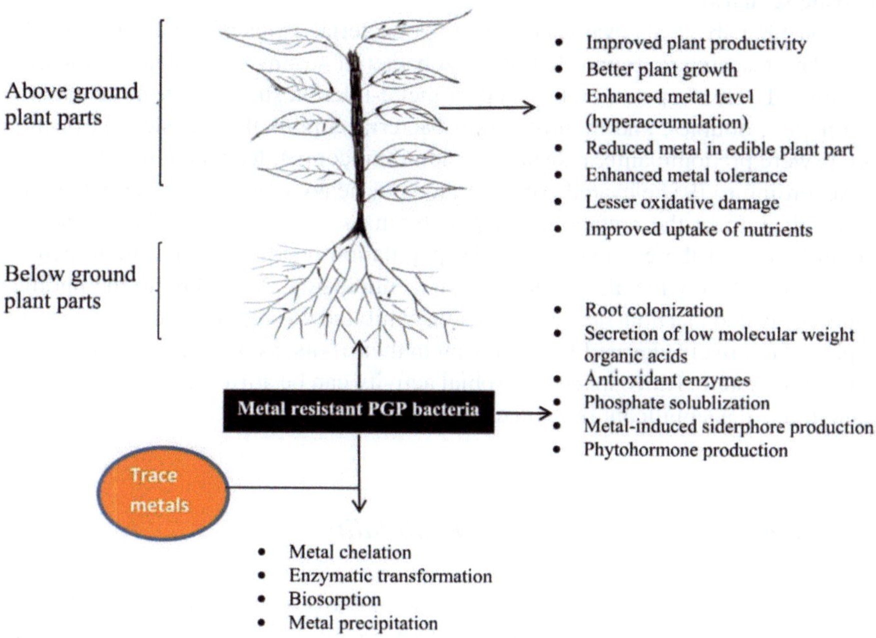

Fig. 3 Association between plant roots and metal tolerant PGP bacteria and their subsequent effects [70]

units per gramme (CFU/g). The higher concentrations of heavy metals in the soil—39 mg Co/kg, 3 mg Cd/kg, 79 mg Ni/kg, 30 mg Cu/kg, 4834 mg Zn/kg, 123 mg Cr/kg, and 114 mg Pb/kg—can be used to explain the observed decline in the bacterial population [30]. Manara [63] found that when the concentrations of Cd in the soil treatments were more than 7 mg/kg, the populations of rhizobium were reduced. Prior studies on soils contaminated with metals have repeatedly demonstrated that elevated metal concentrations can cause microbial communities to shrink.

Moreover, the composition of the microbial community in the rhizosphere has a significant role in the facilitation of plant growth. This phenomenon can be primarily ascribed to the discovery that microbial populations frequently develop a form of mutually beneficial collaboration with the host plant system [9]. An instance of soil pollution involving heavy metals has the potential to result in the emergence of rhizobacteria that are resistant to heavy metals in the soil found in industrial areas. The presence of a significant number of bacteria that are resistant to metal was discovered in the rhizosphere of hyperaccumulators *Thalaspi caerulescens* [26]. These hyperaccumulators were cultivated in soil that was contaminated with zinc (Zn) and nickel (Ni), or solely with nickel, respectively. Previous studies have demonstrated that the introduction of rhizobacteria has led to elevated levels of zinc (Zn) in *T. caerulescens*, nickel (Ni) in *A. murale,* and selenium (Se) in *B. juncea* [12]. The prevalence of multiple metal-resistance (MMR) in bacteria appears to be more common than the opposite scenario.

Abou-Shanab et al. evaluated 107 rhizobacterial isolates' heavy metal tolerance. The researchers found that all rhizobacterial strains were metal ion-tolerant at 1 mmol/L. Hexa-, penta-, tetra-, and tri-metal-tolerant strains were more common than hepta-, double-, and mono-tolerant bacteria. Strains that resisted six or more metals were predominantly resistant to cadmium, copper, lead, and nickel.

According to [95] elevated concentrations of heavy metals have the potential to negatively impact the metabolic activity, biomass, and diversity of rhizobacteria. The influence of the extensive bacterial population residing in the rhizosphere on the uptake of heavy metals by plants is also anticipated. According to the findings, plants cultivated in a non-sterile soil system exhibited no signs of iron deficiency and displayed relatively elevated levels of iron in their roots, as opposed to plants grown in a sterile system. Rhizosphere microbial activity can be attributed to, as it has been found to have a significant influence in iron accumulation.

7.3 *Heavy Metal Toxicant Bioavailability*

The presence of soil rhizobacteria has the potential to directly impact the solubility of metals by altering the chemical forms in which heavy metals are present in the rhizosphere. The investigation into the functions of mycorrhiza in the process of metal speciation within the rhizosphere, as well as its impact on enhancing the ability of host plants to tolerate excessive levels of heavy metals in soil, revealed notable alterations in the speciation of copper (Cu), zinc (Zn), and lead (Pb) within the rhizosphere of maize plants infected and not infected with arbuscular mycorrhiza (AM), in comparison to the surrounding bulk soil [96]. Particularly, the most significant change was observed in the exchangeable form of copper, which exhibited a 26% increase in the non-infected rhizosphere and a 43% increase in the AM-infected rhizosphere, as compared to the bulk soil [16].

Copper was stable in the rhizosphere of both AM and non-AM treatments, apart from copper that is organically bonded in arbuscular mycorrhizal (AM) fungi. Radziemska et al. [83] showed that it is possible to activate Cu via inducing rhizobacteria. The levels of organically bound zinc (Zn) and lead (Pb) were significantly higher in the rhizosphere than in the bulk soil. However, no major changes were found in the zinc carbonate or lead Fe–Mn oxides. Mycorrhiza may protect host plants from copper, zinc, and lead poisoning by changing the chemical form of these elements from a bioavailable state to a non-bioavailable state, as suggested by the results. Andrade Junior et al., Shakira and Puthur [4] found that the levels of copper and zinc accumulation in the roots and shoots of mycorrhiza-infected plants were significantly lower compared to those of non-infected plants. This discovery is suggestive of a significant role for mycorrhiza in preventing the host plants from taking in an excessive amount of copper and zinc.

7.4 *Soil Properties*

Numerous studies have demonstrated that the chemical characteristics of the rhizosphere exhibit notable distinctions from those of the surrounding bulk soil, primarily as a result of several processes instigated by plant roots and/or rhizobacteria [118]. Plant-bacteria interactions may increase the generation of chemicals that modify soil characteristics and increase heavy metal buildup in plants. Rue et al. [90] discovered that soil acidification in *Thalaspi caerulescens* rhizosphere enhances metal ion uptake by increasing mobility around roots. Hg accumulation increased with decreased pH in the culture media. Additionally, microorganisms can reduce certain metals to a lower redox state, removing them from the environment [25]. There exist a multitude of bacterial species that facilitate various biochemical reactions by employing metals as terminal electron acceptors during anaerobic respiration. Dissimilatory metal-reducing bacteria, which exhibit a wide range of phylogenetic and physiological characteristics, typically employ Fe^{3+} and S0 as their preferred terminal electron acceptors. The process of microbial reduction, namely the conversion of Cr^{6+} to Cr^{3+}, is a widely investigated approach in the field of metal bioremediation [74]. This activity is performed by a multitude of heterotrophic organisms, and it can take place either in the absence or presence of oxygen.

7.5 Mechanisms to Overcome Heavy Metal Stress in PGPR

Heavy metal cations play a crucial role as trace elements in facilitating several biochemical events within microbial cell metabolism. Nevertheless, an excessive concentration of these cations can have detrimental effects on cellular function. Bacterial biological activities are negatively impacted by the presence of heavy metals, including Hg^{2+}, Cd^{2+}, and Ag^{+} [6]. When the concentration of metals surpasses the biological need, it has been seen that bacterial growth is inhibited. Alternatively, bacteria have been found to exhibit different resistance mechanisms in response to elevated levels of metals. It was tested in living things how sensitive Rhizobium, *Bradyrhizobium, Pseudomonas*, and other bacteria that help plants grow are to metal ions like Cu^{2+}, Zn^{2+}, Co^{2+}, Mn^{2+}, Mo^{2+}, and Fe^{2+} [2, 93]. The findings showed that Rhizobium *leguminosarum* strains were the most sensitive to Cu^{2+}, Zn^{2+}, and Co^{2+}. On the other hand, *Bradyrhizobium* and Pseudomonas isolates were able to handle the highest amount (10 μg/ml) of these metals without any problems. The extended duration of heavy metal exposure to microorganisms elicits a selective pressure, hence facilitating the proliferation of bacteria that possess a higher degree of tolerance or resistance to metal-induced stress. The investigation of this adaptive mechanism of metal resistance has involved the analysis of habitats that have been subjected to either anthropogenic or natural metal contamination for a significant duration. Additionally, researchers have conducted experiments where heavy metals were intentionally introduced to samples, and subsequent changes were observed over periods ranging from several months to a few years [94]. Therefore, the initial requirement for the manifestation of metal toxicity is the introduction of metal into the bacterial cell. To demonstrate metal toxicity, it is necessary for heavy metal entry to occur within a bacterial cell. Typically, bacterial cells absorb heavy metal cations that possess comparable size, structure, and valency by the same method. Bacteria employ two distinct mechanisms for the absorption of heavy metal ions. The initial approach, which is propelled by the chemiosmotic gradient across the cytoplasmic membrane, is characterized by its rapidity and lack of specificity (Hu et al. 2017). Another approach is substrate-specific, characterized by a sluggish rate of reaction, and requires ATP hydrolysis Feasibility of Pb phytoextraction using nanomaterials assisted ryegrass: Results of a one-year field-scale experiment. The main mechanisms encompass physical sequestration, exclusion, and complexation [3, 121, 123].

The process of heavy metal attachment to extracellular material can immobilize the metal, hence impeding its uptake by bacterial cells. Numerous metallic elements exhibit the ability to form chemical bonds with negatively charged functional groups present on the surfaces of cells [23, 123, 126]. Heavy metals can bind to extracellular polymers, including polysaccharides, proteins, and humic substances. This binding process serves to detoxify the metals by either generating complexes or establishing an efficient barrier that surrounds the cell. Nevertheless, siderophores can decrease the bioavailability of metals, so mitigating their harmful effects [67]. The generation of metabolites leads to the formation of crystals or the precipitation of heavy metals. Various bacterial species exhibit efflux transporters that possess a notable affinity for substrates, resulting in the expulsion of toxic metals from the cell at high concentrations [113]. These efflux systems, which are encoded by plasmids and rely on energy, involve ATPase and chemiosmotic ion/proton pumps, and have been observed in the uptake of chromium and cadmium.

An alternative approach exists for mitigating heavy metal toxicity, whereby ions can be transformed into a harmless state upon entering a bacterial cell. The phrase used to describe this mechanism is cytosolic sequestration mechanism [20]. This process facilitates the absorption of elevated concentrations of heavy metals. One illustrative instance pertains to the synthesis of metallothionein's. These proteins possess a low molecular weight and are characterized by their cysteine-rich nature, enabling them to attach to metals such as cadmium, copper, silver, and mercury with a greater affinity [68]. The synthesis of these metal-binding proteins is controlled by the presence of the respective metals. Certain bacteria employ methylation as an alternate mechanism, wherein methyl groups are transferred to metals and metalloids. However, it is important to note that this process is limited to a select few metals that can undergo methylation. An alternative approach to soil cleaning is the mitigation of heavy metal concentrations [21]. The bacterial species implicated in this phenomenon is commonly known as dissimilatory metal-reducing bacteria. An illustration of the process of chromium detoxification entails the enzymatic reduction of hexavalent chromium (Cr (VI)) to trivalent chromium (Cr (III) by bacterial strains. These bacterial strains are known to possess a significant role in the sequestration of various heavy metals [56, 86].

8 The Role of Rhizobacteria in Phytoremediation

8.1 *PGPR as a Bio-Remediating Agent*

Elevated concentrations of heavy metals have been found to impede microbial activity within soil ecosystems, consequently exerting a detrimental impact on agricultural productivity. This is mostly attributed to the accumulation of heavy metals within various plant organs [19]. Metal ions that are found in the soil are initially taken up by the roots and afterward transferred to other organs. The proteins and enzymes found in plant cells exhibit a propensity for binding with heavy metals, resulting in their inactivation and subsequent loss of functionality [64]. Upon exposure to heavy metals, an alteration in the protein conformation ensues, leading to a subsequent impact on plant development through the deactivation of the photosystem [77]. The biochemical and physiological activities of plants are negatively impacted by the oxidative action generated by free radicals. Growth and development are hindered because of disruptions in respiratory and photosynthetic processes.

The PGPR have implemented many strategies to mitigate the effects of metallic stress, which include:

- The process of transporting substances through the plasma membrane

- The process of biosorption involves the binding of substances to the cell walls of organisms.
- The process of encapsulating metal within extracellular capsules
- The process of heavy metal precipitation
- Oxidation–reduction, commonly referred to as redox reactions, is a fundamental

Singh et al. [99] and Genchi et al. [33] state that the capacity of microbes to efficiently react to heavy metals is a critical determinant in their potential application for the remediation of perilous metal accumulation in plants. In addition to possessing unique metal detoxification properties, the PGBP enzyme is also capable of utilising growth-promoting substances in agronomic applications. These agents play a vital role in the phytoremediation process by acting as bioinoculants and facilitating the restoration of contaminated soils. A researcher examined the efficacy of Microbacterium *arabinogalactanolyticum, Sphingomonas macrogoltabidus*, and *Microbacterium liquefaciens* in treating *A. murale* plants in a prior investigation [84]. The findings of this study demonstrated notable improvements in the uptake of Ni by the plants treated with these bacterial strains, in comparison to the untreated control group. The observed phenomenon can be attributed to the process of root-to-shoot absorption in the seedlings of these plant species. In their investigation [124]. Moreover, the siderophores, which are actively released by PGPB, have been demonstrated to effectively reduce the quantities of free radicals generated by heavy metals upon their binding to the roots [64]. This method has the potential to save microbial auxins from oxidative damage caused by free radicals, hence facilitating the enhancement of plant growth.

In a similar vein, the development of Phaseolus vulgaris was significantly enhanced through the inoculation of *Pseudomonas putida* KNP9, a strain known for its resistance to lead and cadmium. This protective effect was observed as it shielded the plant from the harmful effects of lead and cadmium [120]. Several further instances of PGPB exist, including *Pseudomonas* sp. Ps29C and *Bacillus megaterium* Bm4C. These bacteria have been isolated from soils contaminated with nickel (Ni) and have demonstrated a significant reduction in the harmful effects of Ni on *Brassica juncea*, as well as an improvement in plant development [64]. Plant growth increased due to phytohormones, siderophores, and 1-aminocyclopropane-1-carboxylic acid deaminase production and release. These studies have shed light on how plant growth-promoting bacteria (PGPB) boost plant biomass under heavy metal stress [13]. Thus, remediation using PGPB, and other plant growth-promoting substances is effective.

8.2 Uptake and Transformation

- Microbes play a significant role in influencing the mobility and availability of metals in soil.

- The process of acidification refers to the increase in acidity levels, typically in a liquid or
- Redox reactions refer to chemical processes involving the transfer of electrons between species, resulting in changes in their oxidation states.
- The manufacturing of iron chelators and siderophores, which play a crucial role in maintaining iron accessibility.
- The process of mobilizing metal phosphates.

Typically, heavy metals found in soil are typically associated with organic and inorganic compounds or exist as insoluble precipitates, rendering them inaccessible for root uptake [116]. However, bacterial siderophores produced by plant growth-promoting rhizobacteria (PGPR) can solubilize heavy metals containing iron, thereby transforming them into a readily available form that can be taken up by plant roots. Bacterial plant growth-promoting rhizobacteria (PGPRs) aid in mitigating the toxicity of metals through their biosorption mechanism, wherein bacterial cells have an enhanced capacity to absorb significant quantities of heavy metals [59, 104, 124]. The determinants contributing to enhanced plant development in soils polluted with metals include:

- The phenomenon of biosorption and bioaccumulation
- The synthesis of ACC deaminase
- Phytohormones, also known as plant hormones, are naturally occurring chemical compounds.

8.3 *The Utilization of Genetic Engineering Techniques*

A substantial assemblage of proteins possesses the capability to selectively bind a diverse array of heavy metals with heightened affinity. Metal binding proteins are in the outer membranes of plants and microorganisms. They engage with metal ions found in the surrounding environment, facilitating the transportation of these metals to the cytosol. Subsequently, metallochaperones transfer these metals to the appropriate receptor proteins [21]. Plants have a physiological response to elevated concentrations of heavy metals by engaging in the process of synthesis.

- Metallothionein's are a class of small, cysteine-rich proteins.
- Phytochelatins are small peptides that are synthesized by plants
- The repair of metal-contaminated soil through phytoremediation can be achieved by employing several genetically modified rhizobacteria.

9 Discussion

Despite its ecofriendly and cost-effective nature, phytoremediation is not widely recognized as a viable alternative by numerous regulatory bodies. Consequently, its utilization is restricted due to the extended duration required to attain clean-up objectives. Another constraint of the procedure is the toxicity of chemicals towards the plants in question, as well as the ambiguity surrounding the eventual destiny of numerous related compounds [101]. Given the involvement of plants in the process, there exists a significant likelihood of disruption to the food chain. Moreover, it is widely acknowledged that plants possess a remarkable ability to withstand larger concentrations of harmful substances compared to most microorganisms commonly employed in phytoremediation processes [43].

A potential application of the approach could be its integration with other remediation techniques. In contemporary society, it has become customary to engage in the remediation of affected regions after incidents of oil spills [36]. Given the extensive impact of oil spills on impacted areas, phytoremediation remains the preferred method of remediation.

10 Conclusion

Upon assessing the involvement of rhizobacteria in the phytoremediation process of toxic metals in soil that is contaminated, it becomes evident that individual bacteria have devised distinct tactics for the purpose of bioremediation. In addition to rhizobacteria, there exist alternative approaches for remediating harmful metals from contaminated environments, such as enhancing the presence of metal-accumulating plant species or employing mechanical agitation of plant biomass. Phytoremediation by rhizobacteria is the preferred and implemented method for remediating severely contaminated soils that include harmful metal concentrations above plant tolerance levels. This approach not only enhances the plant's ability to withstand metals such as arsenic, mercury, and sulphur, but also promotes the stable growth of plants involved in the rehabilitation of environments contaminated with metals. This approach has successfully restored the environment productivity that was previously compromised by the deposition of heavy metals. Scientists are aware of the significance of plant growth-promoting rhizobacteria, although there is still much to be comprehended regarding the intricate mechanisms underlying phytoremediation. There are a limited number of enquiries that require responses, which include:

Further examination and research are required to examine the effects of microorganisms in the rhizosphere on the buildup of metals and environmental pollutants. Further investigation is necessary to quantify the impact of the phytoremediation procedure on the Phyto availability of hazardous metals. Enhancing the efficiency of phytoremediation can be achieved by augmenting the comprehension of the phytoextraction capacity exhibited by plant growth-promoting rhizobacteria. The significance of investigating rhizosphere-microbial ecology should be duly acknowledged. Despite the presence of several challenges, the efficacy of the phytoremediation process has been demonstrated to surpass that of alternative approaches. The conventional methods not only incurred significant time and financial expenses, but also resulted in the emission of additional harmful by-products that exhibited higher toxicity levels compared to the primary pollutants. In contrast, the utilization of rhizobacteria for enhancing bioremediation is a cost-effective and environmentally sustainable approach.

References

1. Abioye OP, Oyewole OA, Oyeleke SB, Adeyemi MO, Orukotan AA (2018) Biosorption of lead, chromium and cadmium in tannery effluent using indigenous microorganisms. Braz J Biol Sci 5(9):25–32
2. Ali H, Khan E, Ilahi I (2019) Environmental chemistry and ecotoxicology of hazardous heavy metals: environmental persistence, toxicity, and bioaccumulation. J Chem 2019:1–14
3. Ali S, Rizwan M, Zaid A et al (2018) 5-Aminolevulinic Acid-induced heavy metal stress tolerance and underlying mechanisms in plants. J Plant Growth Regul 37:1423–1436
4. Ali H, Khan E, Sajad MA (2013) Phytoremediation of heavy metals-concepts and applications. Chemosphere 91(7):869–881
5. Ashraf S, Ali Q, Zahir ZA, Ashraf S, Asghar HN (2019) Phytoremediation: environmentally sustainable way for the reclamation of heavy metal polluted soils. Ecotoxicol Environ Saf 174:714–727
6. Baleta J, Mikulcic H, Klemes JJ, Urbaniec K, Dui´c N (2019) Integration of energy, water, and environmental systems for sustainable development. J Cleaner Prod 215:1424–1436
7. Balzano S, Sardo A, Blasio M, Chahine TB, Dell'Anno F, Sansone C, Brunet C (2020) Microalgal metallothioneins and phytochelatins and their potential use in bioremediation. Front Microbiol 11:517
8. Bastow EL, De La Torre VSG, Maclean AE, Green RT, Merlot S, Thomine S, Balk J (2018) Vacuolar iron stores gated by NRAMP3 and NRAMP4 are the primary source of iron in germinating seeds. Plant Physiol 177(3):1267–1276
9. Bennett AE, Evans DM, Powell JR (2019) Potentials and pitfalls in the analysis of bipartite networks to understand plant–microbe interactions in changing environments. Funct Ecol 33(1):107–117
10. Cailliatte R, Schikora A, Briat JF, Mari S, Curie C (2010) High-affinity manganese uptake by the metal transporter NRAMP1 is essential for Arabidopsis growth in low manganese conditions. Plant Cell 22(3):904–917
11. Cameselle C, Gouveia S (2019) Phytoremediation of mixed contaminated soil enhanced with electric current. J Hazard Mater 361:95–102
12. Cameselle C, Gouveia S, Urrejola S (2019) Benefits of phytoremediation amended with DC electric field. Application to soils contaminated with heavy metals. Chemosphere 229:481–488
13. Cetinkaya G, Sozen N (2011) Plant species potentially useful in the Phytostabilization process for the abandoned CMC mining site in northern Cyprus. Int J Phytoremediation 13:681–691
14. Chaabene Z, Rorat A, Hakim IR, Bernard F, Douglas GC, Elleuch A, Vandenbulcke F, Mejdoub H (2018) Insight into the expression variation of metalresponsive genes in the seedling of date palm (Phoenix dactylifera). Chemosphere 197:123–134

15. Chandran H, Meena M, Swapnil P (2021) Plant growth-promoting rhizobacteria as a green alternative for sustainable agriculture. Sustain Sci 13:109–186
16. Chen H, Zhou W, Zhu S, Liu F, Qin L, Xu C, Wang Z (2020) Biological nitrogen and phosphorus removal by a phosphorus-accumulating bacteria Acinetobacter sp. strain C-13 with the ability of heterotrophic nitrification–aerobic denitrification. Bioresour Technol 124507
17. Chen L, Luo S, Li X, Wan Y, Chen J, Liu C (2014) Interaction of Cdhyperaccumulator Solanum nigrum L. and functional endophyte Pseudomonas sp. Lk9 on soil heavy metals uptake. Soil Biol Biochem 68:300–308
18. Chibuike GU, Obiora SC (2014) Heavy metal polluted soils: effect on plants and bioremediation methods. Appl Environ Soil Sci 2014:1–12
19. Cristaldi A, Conti GO, Jho EH, Zuccarello P, Grasso A, Copat C, Ferrante M (2017) Phytoremediation of contaminated soils by heavy metals and PAHs. A brief review. Environ Technol Innov 8:309–326
20. DalCorso G, Fasani E, Manara A, Visioli G, Furini A (2019) Heavy metal pollutions: state of the art and innovation in phytoremediation. Int J Mol Sci 20:3412
21. Deb VK, Rabbani A, Upadhyay S, et al (2020) Microbe-assisted phytoremediation in reinstating heavy metal-contaminated sites: concepts, mechanisms, challenges, and future perspectives. In: Microbial technology for health and environment, pp 161–189
22. Dixit P, Mukherjee PK, Sherkhane PD, Kale SP, Eapen S (2011) Enhanced tolerance and remediation of anthracene by transgenic tobacco plants expressing a fungal glutathione transferase gene. J Hazard Mater 192:270–276
23. Dutta S, Datta A, Zaid A, Bhat JA (2020) Metalloids and their impact on the environment. In: Metalloids in plants. Wiley, pp 19–26
24. Ebbs SD, Bradfield SJ, Kumar P, White JC, Ma X (2016) Projected dietary intake of zinc, copper, and cerium from consumption of carrot (Daucus carota) exposed to metal oxide nanoparticles or metal ions. Front Plant Sci 7:188
25. Eissa MA, Negim OE (2018) Heavy metals uptake and translocation by lettuce and spinach grown on a metal-contaminated soil. J Soil Sci Plant Nutr 18:1097–1107
26. Ekta P, Modi NR (2018) A review of phytoremediation. J. Pharmacogn. Phytochem. 7(4):1485–1489
27. Emamverdian A, Ding Y, Mokhberdoran F, Xie Y (2015) Heavy metal stress and some mechanisms of plant defense response. Sci World J
28. Fasani E, Manara A, Martini F, Furini A, DalCorso G (2018) The potential of genetic engineering of plants for the remediation of soils contaminated with heavy metals. Plant Cell Environ 41(5):1201–1232
29. Favas PJ, Pratas J, Varun M, D'Souza R, Paul MS (2014) Phytoremediation of soils contaminated with metals and metalloids at mining areas: potential of native flora. Environ Risk Assess Soil Contam 3:485–516
30. Feng NX, Yu J, Zhao HM, Cheng YT, Mo CH, Cai QY, Li YW, Li H, Wong MH (2017) Efficient phytoremediation of organic contaminants in soils using plant–endophyte partnerships. Sci Total Environ 583:352–368
31. Finkel OM, Castrillo G, Paredes SH, Gonz´ alez IS, Dangl JL (2017) Understanding and exploiting plant beneficial microbes. Curr Opin Plant Biol 38:155–163
32. Gautam N, Verma PK, Verma S, Tripathi RD, Trivedi PK, Adhikari B, Chakrabarty D (2012) Genome-wide identification of rice class I metallothionein gene: tissue expression patterns and induction in response to heavy metal stress. Funct Integr Genomics 12:635–647
33. Genchi G, Sinicropi MS, Lauria G et al (2020) The efects of cadmium toxicity. Int J Environ Res Public Health 17:3782
34. Goher ME, AM AEM, Abdel-Satar AM, Ali MH, Hussian AE, Napiorkowska- ´ Krzebietke A (2016) Biosorption of some toxic metals from aqueous solution using non-living algal cells of Chlorella vulgaris. J Elementology 21(3)
35. Gomes MA, da C, Hauser-Davis RA, de Souza AN, Vitoria AP (2016) Metal phytoremediation: general strategies, genetically modified plants and applications in metal nanoparticle contamination. Ecotoxicol. Environ Saf 134:133–147

36. Gong X, Huang D, Liu Y et al (2017) Stabilized nanoscale zerovalent iron mediated cadmium accumulation and oxidative damage of Boehmeria nivea (L.) gaudich cultivated in cadmium contaminated sediments. Environ Sci Technol 51:11308–11316
37. Guo J, Dai X, Xu W, Ma M (2008) Overexpressing gsh1 and AsPCS1 simultaneously increases the tolerance and accumulation of cadmium and arsenic in Arabidopsis thaliana. Chemosphere 72:1020–1026
38. Gupta P, Diwan B (2017) Bacterial exopolysaccharide mediated heavy metal removal: a review on biosynthesis, mechanism, and remediation strategies. Biotechnol Rep 13:58–71
39. Hasan M, Uddin M, Ara-Sharmeen I, Alharby FH, Alzahrani Y, Hakeem KR, Zhang L (2019) Assisting phytoremediation of heavy metals using chemical amendments. Plants 8(9):295
40. Hasegawa H, Rahman I, Rahman M (2016) Environmental remediation technologies for metal-contaminated soils. Springer
41. Hejna M, Moscatelli A, Stroppa N, Onelli E, Pilu S, Baldi A, Rossi L (2020) Bioaccumulation of heavy metals from wastewater through a Typha latifolia and Thelypteris palustris phytoremediation system. Chemosphere 241:125018
42. Hoang HG, Lin C, Tran HT, Chiang CF, Bui XT, Cheruiyot NK, Shern CC, Lee CW (2020) Heavy metal contamination trends in surface water and sediments of a river in a highly industrialized region. Environ Technol Innov 20:101043
43. Huang D, Qin X, Peng Z et al (2018) Nanoscale zero-valent iron assisted phytoremediation of Pb in sediment: Impacts on metal accumulation and antioxidative system of Lolium perenne. Ecotoxicol Environ Saf 153:229–237
44. Hussain I, Aleti G, Naidu R, Puschenreiter M, Mahmood Q, Rahman MM, Wang F, Shaheen S, Syed JH, Reichenauer TG (2018) Microbe and plant assisted remediation of organic xenobiotics and its enhancement by genetically modified organisms and recombinant technology: a review. Sci Total Environ 628:1582–1599
45. Ismail NI, Abdullah SRS, Idris M, Kurniawan SB, Halmi MIE, Sbani NHA, ..., Hasan HA (2020) Applying rhizobacteria consortium for the enhancement of Scirpus grossus growth and phytoaccumulation of Fe and Al in pilot constructed wetlands. J Environ Manage 267:110643
46. Jaskulak M, Grobelak A, Grosser A, Vandenbulcke F (2019) Gene expression, DNA damage and other stress markers in Sinapis alba L. exposed to heavy metals with special reference to sewage sludge application on contaminated sites. Ecotoxicol Environ Saf 181:508–517
47. Jeevanantham S, Saravanan A, Hemavathy RV, Kumar PS, Yaashikaa PR, Yuvaraj D (2019) Removal of toxic pollutants from water environment by phytoremediation: a survey on application and future prospects. Environ Technol Innov 13:264–276
48. Jegatheesan V, Ravishankar H, Shu L, Wang J (2016) Application of green and physico-chemical technologies in treating water polluted by heavy metals. In: Green technologies for sustainable water management. American Society of Civil Engineers, Reston, VA, pp 579–614
49. Jin Y, Luan Y, Ning Y, Wang L (2018) Effects and mechanisms of microbial remediation of heavy metals in soil: a critical review. Appl Sci 8(8):1336
50. Jing Y, Yan J, He H, Yang D, Xiao L (2014) Characterization of bacteria in the rhizosphere soils of Polygonum pubescens and their potential in promoting growth and Cd, Pb, Zn uptake by Brassica napus. Int J Phytoremediat 16(4):321–333
51. Karenlampi S, Schat H, Vangronsveld J, Verkleij JAC, van der Lelie D, Mergeay M, Tervahauta AI (2000) Genetic engineering in the improvement of plants for phytoremediation of metal polluted soils. Environ Pollut 107:225–231
52. Khalid HS, Akhtar MF, Ijaz M, Iqbal M, Bukhari SA, Mustafa G, Shaukat K (2021) Role of metal-binding proteins and peptides in bioremediation of toxic metals. In: Handbook of bioremediation. Academic Press, pp 437–444
53. Kumar A, Verma JP (2018) Does plant-microbe interaction confer stress tolerance in plants: a review? Microbiol Res 207:41–52
54. Kumari A, Lal B, Rai UN (2016) Assessment of native plant species for phytoremediation of heavy metals growing in the vicinity of NTPC sites, Kahalgaon India. Int J Phytoremediat 18:592–597

55. Leong YK, Chang JS (2020) Bioremediation of heavy metals using microalgae: recent advances and mechanisms. Bioresour Technol 303:122886
56. Li X, Yang Y, Gao B, Zhang M (2015) Stimulation of peanut seedling development and growth by zero-valent iron nanoparticles at low concentrations. PLoS ONE 10:0122884
57. Lim KT, Shukor MY, Wasoh H (2014) Physical, chemical, and biological methods for the removal of arsenic compounds. BioMed Res Int
58. Limmer M, Burken J (2016) Phytovolatilization of organic contaminants. Environ Sci Technol 50:6632–6643
59. Liu S, Yang B, Liang Y, Xiao Y, Fang J (2020) Prospect of phytoremediation combined with other approaches for remediation of heavy metal-polluted soils. Environ Sci Pollut Res 1–17
60. Liu X, Zhao J, Jiang P (2022) Easy removal of epiphytic bacteria on Ulva (Ulvophyceae, Chlorophyta) by Vortex with Silica Sands. Microorganisms 10(2):476. https://doi.org/10.3390/microorganisms10020476
61. Mahar A, Wang P, Ali A, Awasthi MK, Lahori AH, Wang Q, Li R, Zhang Z (2016) Challenges and opportunities in the phytoremediation of heavy metals contaminated soils: a review. Ecotoxicol Environ Saf 126:111–121
62. Man YB, Lei KM, Chow KL, Leung AOW, Mo WY, Wong MH (2020) Ecological risks of heavy metals/metalloid discharged from two sewage treatment works to Mai Po Ramsar site, South China. Environ Monit Assess 192:1–14
63. Manara A (2012) Plants and heavy metals. Signal Transduct 2754
64. Manoj SR, Karthik C, Kadirvelu K et al (2020) Understanding the molecular mechanisms for the enhanced phytoremediation of heavy metals through plant growth promoting rhizobacteria: a review. J Environ Manage 254:109779
65. Marques AP, Rangel AO, Castro PM (2009) Remediation of heavy metal contaminated soils: phytoremediation as a potentially promising clean-up technology. Crit Rev Environ Sci Technol 39(8):622–654
66. Mirza N, Mahmood Q, Maroof Shah M, Pervez A, Sultan S (2014) Plants as useful vectors to reduce environmental toxic arsenic content. Sci World J
67. Mokarram-Kashtiban S, Hosseini SM, Tabari Kouchaksaraei M, Younesi H (2019) The impact of nanoparticles zero-valent iron (nZVI) and rhizosphere microorganisms on the phytoremediation ability of white willow and its response. Environ Sci Pollut Res 26:10776–10789
68. Mukaro E, Nyakudya IW, Jimu L (2017) Edaphic conditions, aboveground carbon stocks and plant diversity on nickel mine tailings dump vegetated with Senegalia polyacantha (Willd.) Seigler & Ebinger. L Degrad Dev
69. Munir I, Faisal M (2016) Plant growth-promoting bacteria: a good source for phytoremediation of metal-contaminated soil. Phytoremediation. Springer International Publishing, Cham, pp 119–129
70. Mustafa A, Imran M, Ashraf M, Mahmood K (2018) Perspectives of using l-tryptophan for improving the productivity of agricultural crops: a review. Pedosphere 28:16–34
71. Naghipour D, Gharibi H, Taghavi K, Jaafari J (2016) Influence of EDTA and NTA on heavy metal extraction from sandy-loam contaminated soils. J Environ Chem Eng 4(3):3512–3518
72. Ngo HH, Bui XT, Nghiem LD, Guo W (2020) Green technologies for sustainable water. Bioresour Technol 317:123978
73. Nidhin S, Udayan A, Shantkriti S (2020) Algal bioremediation of heavy metals, removal of toxic pollutants through microbiological and tertiary treatment New perspectives, Chapter 11, 279–302. Elsevier
74. Nzediegwu C, Prasher S, Elsayed E, Dhiman J, Mawof A, Patel R (2020) Impact of soil biochar incorporation on the uptake of heavy metals present in wastewater by spinach plants. Water Air Soil Pollut 231:1–19
75. Oconnor D, Zheng X, Hou D, Shen Z, Li G, Miao G, Oconnell S, Guo M (2019) Phytoremediation: climate change resilience and sustainability assessment at a coastal brownfield redevelopment. Environ Int 130:104945
76. Pal R, Rai JPN (2010) Phytochelatins: peptides involved in heavy metal detoxification. Appl Biochem Biotechnol 160:945–963

77. Pandey VC, Bajpai O (2019) Phytoremediation: from theory toward practice. In: Phytomanagement of polluted sites. Elsevier, pp 1–49
78. Parnian A, Furze JN (2021) Vertical phytoremediation of wastewater using Vetiveria zizanioides L. Environ Sci Pollut Res 28(45):64150–64155
79. Parnian A, Furze JN, Chorom M, Jaafarzadeh N (2022) Competitive Bioaccumulation by Ceratophyllum demersum L. In: Earth systems protection and sustainability, vol 2. Springer International Publishing, Cham, pp 15–30
80. Parnian A, Mahbod M, Prasad MNV (2023) Microplastics remediation–possible perspectives for mitigating saline environments. Microplastics Ecosphere Air Water Soil Food 465–476
81. Parnian A, Parnian A, Pirasteh-Anosheh H, Furze JN, Prasad MNV, Race M, Hulisz P Ferraro A (2022) Full-scale bioremediation of petroleum-contaminated soils via integration of co-composting. J Soils Sediments 22(8):2209–2218
82. Polak N, Read DS, Jurkschat K, Matzke M, Kelly FJ, Spurgeon DJ, Stürzenbaum SR (2014) Metalloproteins and phytochelatin synthase may confer protection against zinc oxide nanoparticle induced toxicity in Caenorhabditis elegans. Comp Biochem Physiol C: Toxicol Pharmacol 160:75–85
83. Radziemska M, Koda E, Bilgin A, Vaverkov´ a MD (2018) Concept of aided phytostabilization of contaminated soils in postindustrial areas. Int J Environ Res Public Health 15(1):24
84. Rahman T, Seraj MF (2018) Available approaches of remediation and stabilisation of metal contamination in soil: a review. Am J Plant Sci 09:2033–2052
85. Rai PK, Kim KH, Lee SS, Lee JH (2020) Molecular mechanisms in phytoremediation of environmental contaminants and prospects of engineered transgenic plants/microbes. Sci Total Environ 705:135858
86. Rastogi A, Zivcak M, Sytar O et al (2017) Impact of metal and metal oxide nanoparticles on plant: a critical review. Front Chem 5
87. Rathankumar AK, Saikia K, Cabana H, Kumar VV (2022) Surfactant-aided mycoremediation of soil contaminated with polycyclic aromatic hydrocarbons. Environ Res 209:112926
88. Rehman F, Pervez A, Khattak BN, Ahmad R (2018) Plant growth promoting rhizobacteria impact on Typha latifolia and Phragmites australis growth and dissolved oxygen. CLEAN–Soil Air Water 46(8):1700353
89. Rostami S, Azhdarpoor A (2019) The application of plant growth regulators to improve phytoremediation of contaminated soils: a review. Chemosphere 220:818–827
90. Rue M, Paul AL, Echevarria G, van der Ent A, Simonnot MO, Morel JL (2020) Uptake, translocation and accumulation of nickel and cobalt in Berkheya coddii, a 'metal crop' from South Africa. Metallomics 12:1278–1289
91. Sarma H, Nava AR, Prasad MNV (2019) Mechanistic understanding and future prospect of microbe-enhanced phytoremediation of polycyclic aromatic hydrocarbons in soil. Environ Technol Innov 13:318–330
92. Selvi A, Rajasekar A, Theerthagiri J, Ananthaselvam A, Sathishkumar K, Madhavan J, Rahman PK (2019) Integrated remediation processes toward heavy metal removal/recovery from various environments-a review. Front Environ Sci 7:66
93. Senthil Kumar P, Saravanan A (2017) Sustainable wastewater treatments in textile sector. In: Sustainable Fibres and Textiles
94. Shadman SM, Daneshi M, Shafei F et al (2019) Aptamer-based electrochemical biosensors. In: Electrochemical biosensors. Elsevier, pp 213–251
95. Shah V, Daverey A (2020) Phytoremediation: a multidisciplinary approach to clean up heavy metal contaminated soil. Environ Technol Innov 18:100774
96. Sharma B, Shukla P (2020) A comparative analysis of heavy metal bioaccumulation and functional gene annotation towards multiple metal resistant potential by *Ochrobactrum intermedium* BPS-20 and *Ochrobactrum ciceri* BPS-26. Bioresour Technol 320:124330
97. Sharma P, Rath SK (2020) Potential applications of fungi in the remediation of toxic effluents from pulp and paper industries. In: Fungi bio-prospects in sustainable agriculture, environment and nanotechnology. Academic Press, pp 193–211

98. Sheoran V, Sheoran AS, Poonia P (2016) Factors affecting phytoextraction: a review. Pedosphere 26:148–166
99. Singh R, Gautam N, Mishra A, Gupta R (2011) Heavy metals and living systems: an overview. Indian J Pharmacol 43:2
100. Song B, Xu P, Chen M et al (2019) Using nanomaterials to facilitate the phytoremediation of contaminated soil. Crit Rev Environ Sci Technol 49:791–824
101. Souri Z, Karimi N, Sarmadi M, Rostami E (2017) Salicylic acid nanoparticles (SANPs) improve growth and phytoremediation efficiency of Isatis cappadocica Desv., under As stress. IET Nanobiotechnol 11:650–655
102. Spence C, Bais H (2015) Role of plant growth regulators as chemical signals in plantmicrobe interactions: a double-edged sword. Curr Opin Plant Biol 27:52–58
103. Sreekumar N, Haridas A, Godwin GS, Selvaraju N (2018) Lipid enhancement in microalgae by temporal phase separation: use of indigenous sources of nutrients. Chin J Chem Eng 26(1):75–182
104. Suman J, Uhlik O, Viktorova J, Macek T (2018) Phytoextraction of heavy metals: a promising tool for clean-up of polluted environment? Front Plant Sci 9:1476
105. Sun C, Zhang G, Zheng H, Liu N, Shi M, Luo X, Chen L, Li F, Hu S (2019) Fate of four phthalate esters with presence of Karenia brevis: uptake and biodegradation. Aquat Toxicol 206:81–90
106. Taj ZZ, Rajkumar M (2016) Perspectives of plant growth-promoting actinomycetes in heavy metal phytoremediation. In: Plant growth promoting Actinobacteria. Springer Singapore, Singapore, pp 213e231
107. Tarekegn MM, Salilih FZ, Ishetu AI (2020) Microbes used as a tool for bioremediation of heavy metal from the environment. Cogent Food Agric 6(1):1783174
108. Thakur S, Singh L, Ab Wahid Z, Siddiqui MF, Atnaw SM, Din MFM (2016) Plant-driven removal of heavy metals from soil: uptake, translocation, tolerance mechanism, challenges, and future perspectives. Environ Monit Assess 188(4):206
109. Udayan A, Arumugam M, Pandey A (2017) Nutraceuticals from algae and cyanobacteria. In: Algal green chemistry. Elsevier, pp 65–89
110. Ullah A, Heng S, Munis MFH, Fahad S, Yang X (2015) Phytoremediation of heavy metals assisted by plant growth promoting (PGP) bacteria: a review. Environ Exp Bot 117:28e40
111. ur Rehman Z, Rizwan M, Ali S, Ok YS, Ishaque W, Nawaz MF, Akmal F, Waqar M (2017) Remediation of heavy metal contaminated soils by using Solanum nigrum: a review. Ecotoxicol Environ Saf 143:236–248
112. Van Oosten MJ, Maggio A (2015) Functional biology of halophytes in the phytoremediation of heavy metal contaminated soils. Environ Exp Bot 111:135e146
113. Vandana UK, Gulzar ABM, Singha LP et al (2020) Hyperaccumulation of arsenic by Pteris vittata, a potential strategy for phytoremediation of arsenic-contaminated soil. Environ Sustain 3:169–178
114. Wang L, Hou D, Shen Z, Zhu J, Jia X, Ok YS, Tack FM, Rinklebe J (2020) Field trials of phytomining and phytoremediation: a critical review of influencing factors and effects of additives. Crit Rev Environ Sci Technol 50(24):2724–2774
115. Wang X, Shan T, Pang S (2022) Potential of Ulva prolifera in phytoremediation of seawater polluted by cesium and cobalt: an experimental study on the biosorption and kinetics. J Oceanology Limnol 40(4):1592–1599
116. Wani SH, Sanghera GS, Athokpam H et al (2012) Phytoremediation: curing soil problems with crops. Afr J Agric Research 7
117. Wani W, Masoodi KZ, Zaid A et al (2018) Engineering plants for heavy metal stress tolerance. Rend Lincei Sci Fis e Nat 29:709–723
118. Xiao R, Ali A, Wang P, Li R, Tian X, Zhang Z (2019) Comparison of the feasibility of different washing solutions for combined soil washing and phytoremediation for the detoxification of cadmium (cd)and zinc (zn)in contaminated soil. Chemosphere 230:510–518
119. Yadav M, Gupta R, Sharma RK (2019) Green and sustainable pathways for wastewater purification. In: Advances in water purification techniques. Elsevier, pp 355–383

120. Yan A, Wang Y, Tan SN, Yusof MLM, Ghosh S, Chen Z (2020) Phytoremediation: a promising approach for revegetation of heavy metal-polluted land. Frontiers Plant Sci 11
121. Yang Y, Liu Y, Li Z, Wang Z, Li C, Wei H (2020) Significance of soil microbe in microbial-assisted phytoremediation: an effective way to enhance phytoremediation of contaminated soil. Int J Environ Sci Technol 1–8
122. Zaid A, Ahmad B, Jaleel H et al (2020) A critical review on iron toxicity and tolerance in plants: role of exogenous Phytoprotectants. In: Plant micronutrients. Springer International Publishing, Cham, pp 83–99
123. Zaid A, Bhat JA, Wani SH (2020) Influence of metalloids and their toxicity impact on photosynthetic parameters of plants. In: Metalloids in plants. Wiley, pp 113–124
124. Zaid A, Wani SH (2019) Reactive oxygen species generation, scavenging and signaling in plant defense responses. In: Bioactive molecules in plant defense. Springer International Publishing, Cham, pp 111–132
125. Zango Usman U, Mukesh Y, Vandana S, Sharma JK, Sanjay P, Sidhartha D, Sharma Anil K (2020) Microbial bioremediation of heavy metals: emerging trends and recent advances. Res J Biotechnol 15:1
126. Zhang H, Zhao S, Li D, Xu X, Li C (2017) Genome-wide analysis of the ZRT, IRTLike Protein (ZIP) family and their responses to metal stress in Populus trichocarpa. Plant Mol Biol Reporter 35:534–549

Bioremediation of Heavy Metals—Its Pros and Cons

Seragadam Padma, Samaroha Ghosh, Badri Srinivas, Kartik Chandra Ghanta, and Susmita Dutta

Abstract Heavy metal pollution poses a substantial public health hazard, manifesting in diverse toxicological effects. Conventional remediation methods, while effective, are costly and generate environmentally harmful byproducts. In response, the adoption of green technologies utilizing biological agents, such as bacteria, algae, and fungi, has gained prominence for heavy metal removal due to their cost-effectiveness and efficiency. This paper explores the field of bioremediation, explicitly focusing on removing heavy metals like Lead(II), Nickel(II), and Chromium(VI) using those microorganisms. Bioremediation offers an eco-friendly and economically viable solution to combat the adverse impacts of heavy metal pollution. The paper elucidates the intricate mechanisms employed by these microorganisms, encompassing biosorption, bioaccumulation, biotransformation, and detoxification processes, all of which enable the conversion of toxic heavy metals into non-toxic forms or their sequestration. By comprehending the complex mechanisms harnessed by bacteria, fungi, and algae, this research seeks to contribute to developing innovative and effective bioremediation strategies for heavy metal-contaminated environments, with a strong emphasis on sustainability and environmental friendliness. This study underscores the advantages and drawbacks of bioremediation as a promising technology for mitigating heavy metal pollution.

Keywords Lead(II) · Nickel(II) · Chromium(VI) · Bioremediation

S. Padma · B. Srinivas
Department of Chemical Engineering, Gayatri Vidya Parishad College of Engineering, Vishakhapatnam 530048, India

S. Padma · S. Ghosh · K. C. Ghanta · S. Dutta (✉)
Department of Chemical Engineering, National Institute of Technology Durgapur, Durgapur 713 209, India
e-mail: susmita.dutta@che.nitdgp.ac.in

N. Kumar (ed.), *Heavy Metal Remediation*, Earth and Environmental Sciences Library,
https://doi.org/10.1007/978-3-031-53688-5_10

1 Introduction

Traditionally, heavy metals can be defined as metallic elements or metalloids with an atomic density greater than 4000 kg/m^3. Heavy metals are highly toxic and are known to cause severe health issues and illnesses in human beings, even at low concentrations in the surrounding environment. Although heavy metals are abundant in the Earth's crust, they pose the risk of contamination and exposure due to accelerated anthropogenic activities. Heavy metal water pollution is a significant and far-reaching global environmental issue that severely threatens ecosystems, human health, and aquatic life in water bodies worldwide [1]. Heavy metals, which are naturally occurring elements, become pollutants when they enter water bodies at elevated concentrations owing to various human activities [2, 3]. Common heavy metals of concern in water pollution encompass Lead (Pb(II)), Mercury (Hg(II)), Cadmium (Cd(II)), Arsenic (As(II)), Chromium (Cr(VI)), Copper (Cu(II)), and Nickel (Ni(II)), each with distinct sources, pollution mechanisms, and environmental and health impacts.

Industrial processes, such as mining, smelting, and manufacturing, play a prominent role in heavy metal pollution by releasing these elements into the environment [4, 5]. These activities frequently discharge pollutants directly into rivers, lakes, and oceans, contributing to widespread contamination. Besides these traditional industries, heavy metal pollution can also be accounted for by the discharges from dyes, paints, pigments, and leather industries [6]. Agricultural practices also contribute to the problem, with heavy metal-containing fertilizers and pesticides leading to contamination through runoff and leaching [7]. Volcanic and landfilling activities also contribute to the existing problem [8]. However, it is via seepage and leaching through heavy metal-rich rock sites that the heavy metals reach the aquatic systems. Most of them get dissolved and carried downwards into different segments of the hydrosphere, while others settle in sediments [9]. Improper disposal of electronic waste, batteries, and other heavy metal products poses a substantial risk of groundwater contamination [10]. Activities such as combustion of fossil fuels, like coal, by industrial and transport sectors, are also known to contribute to heavy metal contamination by large [11]. The discharged heavy metals are carried by flue gas streams and vehicle emissions to be released into the atmosphere and eventually recirculated throughout the aquatic and land environments by natural forces.

Additionally, several heavy metals naturally exist in the Earth's crust, and geological processes can transport them into water sources [12]. However, the specific type of metal contamination can be attributed to the specific operation and activities carried out at the site. The extent of the contamination would also depend on the scale of actions being commenced at the site and the disposal techniques and methods employed. Factors like groundwater-soil interaction chemistry, metal ion speciation, etc., also come into play while trying to understand the occurrence and fate of the said metals in the local environment.

Due to anthropogenic and natural processes, heavy metal pollution poses a severe environmental hazard with pervasive, long-term consequences. Most heavy metals such as lead, mercury, and arsenic are non-biodegradable and display accumulative behavior in the ecosystem. In aquatic systems, they can disrupt the natural balance inherent to the system, thereby leading to the death of marine organisms, a reduction in biodiversity, or even collapse of the existing food chains. In soil, heavy metals can inhibit plant growth and contaminate crops by uptake through the plant's internal transport system, threatening food safety. Humans risk exposure to heavy metals by consuming contaminated food and water, inhaling polluted air, or contacting contaminated soil. Heavy metals demonstrate profound and severe health impacts on human health. For example, lead exposure, primarily in children, has been known to hinder physical development, impairment of cognitive development, and behavioral issues. Mercury contamination, often from eating contaminated fish, can damage the nervous system and cause congenital disabilities in pregnant women. Exposure to cadmium has been linked to kidney damage and an increased risk of cancer, while exposure to arsenic can lead to skin damage, cardiovascular problems, and an increased risk of some cancers [13, 14].

On top of that, heavy metal pollutants are also persistent by nature, implying that they pose high resistance to natural degradation processes and can remain in the environment for decades, continuing to harm both ecosystems and human health. As a result, efficient remediation techniques and control measures must be implemented to contain the adverse impact of heavy metal pollution and protect public health. These measures may include strict guidelines on industrial emissions, better waste management practices, and the development of newer technologies for the remediation of contaminated sites.

2 Occurrence, Toxicity, and Effect of Pb(II), Cr(VI), Ni(II) Contamination

It is important to note that the toxicity of heavy metals varies, with Pb(II) ranking second, Cr(VI) seventeenth, and Ni(II) fifty-seventh in terms of their toxicity, according to the Agency for Toxic Substances and Disease Registry (ATSDR) [15].

Pb(II) is a naturally occurring metal, occurring in a combined state with other elements such as sulfur in the form of Lead Sulphates, Lead sulfides, and oxygen in the form of Lead carbonates. Lead can exist in several valence states- Pb(0)-metal, Pb(I), Pb(II), and Pb(IV). Speaking of potential environmental significance, Pb(I) is the only species among these that may not share the same level of importance. In urban environments, waste incineration increases the presence of Pb(II), which is utilized in various applications such as storage batteries, soldering materials, bearings, cable sheaths, plumbing, and ammunition [16]. In the United States, Pb(II) contamination is a raging issue, with an incident of the magnitude of the Flint-Michigan water crisis taking place as recently as 2014, plaguing nearly 140,000 individuals with Pb(II) and other contaminants in drinking water [17]. This incident occurred in April 2014 when Flint's water source transitioned from the Great Lakes' Lake Huron, supplied by the Detroit Water and Sewerage Department, to the Flint River. This change occurred without the essential corrosion control treatment to prevent the release of Pb(II) from pipes and plumbing. China is one of the world's largest producers and consumers of Pb(II), being a major manufacturer of Pb(II)-acid batteries, thus proving to be a significant contributor to Pb(II) pollution. Soil and water contamination due to Pb(II) smelting activities and battery manufacturing are a concern [18]. Nigeria did not fall behind in the list when an acute case of widespread Pb(II) poisoning took place in Zamfara State, Nigeria, in 2010. Over 735 children were reported dead, and thousands were sickened by the neurotoxin [19] . The source was traced to artisanal gold mining and processing in the villages. Bangladesh's burgeoning battery recycling industry is also a significant source of the problem. As per a research study titled 'Blood Lead Levels and Health Issues Among Lead Acid Battery Industry Workers in Bangladesh', published on Hindawi, individuals engaged in Pb(II) recycling were discovered to have blood Pb(II) levels (BLL) in the range of 66–78 μg per deciliter (μg/dL) [20]. In India, Pb(II) contamination in water primarily arises from human-made sources, including industries such as battery manufacturing, smelting, and mining. India's prominence in world Pb(II) production, ranking fourth in 2019, amplifies the issue. Pb(II) particles from various sources, including vehicle emissions and deteriorating paints, can wash into stormwater drains, polluting water during rainfall. Natural Pb(II) deposits in the Earth's crust also contaminate certain regions' groundwater. Pb(II)-based materials in pottery and cosmetics can introduce Pb(II) into water if not regulated or safely disposed of, proving to be of severe concern because of the adverse health effects even at low-level lead exposure, especially in children. Pb(II) may also affect the gastrointestinal tract, kidneys, and central nervous system. Pb(II) can accumulate in the body over time, leading to renal failure, cancer, and impairment of the reproductive system [21]. Exposure to Pb(II) in adults can result in various adverse effects, including reduced reaction time, memory impairment, nausea, insomnia, loss of appetite, joint weakness, reproductive issues, heme synthesis inhibition, irritation, and the development of tumors [22]. Encephalopathy is among the most severe acute clinical conditions associated with Pb(II) poisoning.

Hexavalent chromium, represented as Cr(VI), is a highly hazardous contaminant in water sources in India and globally. In humans, the primary source of Cr(VI) intake is consuming food containing Cr(III), a naturally occurring mineral in a wide range of vegetables and fruits. Cr(VI) is a significant threat to human health, particularly among workers in the steel and textile industries, since it is associated with various chronic health effects. Different Cr(VI) species, such as Cr(III) and Cr(VI), have distinct impacts on health. Cr(III) is a distinguished micronutrient benefiting human and animal systems and is significantly less toxic than Cr(VI). Furthermore, Cr(III) has limited solubility in water, making its transport in aquatic systems significantly less plausible.

Its sources include the leather industry's use of Cr(VI) compounds, industrial activities such as metal plating and alloy production, and the textile industry's processes [23]. Mining and processing of Cr(VI) ores can also contribute to contamination if not managed adequately [24]. South Africa, one of the largest producers of Cr(VI) globally, faces the risk of severe environmental contamination, particularly in areas like Rustenburg, due to Cr(VI) mining and processing operations [25]. China is the foremost producer of Cr(VI) slag globally, with approximately 25 enterprises engaged in the production of Cr(VI) salts. These enterprises collectively boast annual production capacities totaling 329,000 tons, resulting in yearly discharges of Cr(VI) slag amounting to 450,000 tons. Notably, the cumulative storage of untreated Cr(VI) slag from previous years has surpassed an impressive 400 million tons. A significant Cr(VI) slag contamination incident occurred on August 13, 2011, in Qujing, Yunnan province. This incident occurred due to the unlawful disposal of 5,000 tons of untreated Cr(VI) slag, which Lvliang Chemical Industry Co., Ltd produced. The slag was irresponsibly dumped in the mountains and Nanpan River, situated at the Zhujian River's source. As a result of this contamination, the concentration of Cr (VI) detected in the Nanpan River exceeded the standard limit by approximately 2000 times [26]. Significant Cr(VI) concentrations in the United States have also been reported in California and North Carolina [27]. In Dhaka city, the capital of Bangladesh, the Hazaribagh industrial zone is notorious for being a significant source of Cr(VI) (Cr) pollution. Over several decades, this pollution has arisen from the prolonged discharge of Cr-contaminated waste from around 150 tanneries. In 2000, heightened concentrations of Cr were detected in the underlying Dupi Tila Aquifer (DTA), which serves as a crucial source of the drinking water supply to Dhaka city's residents. Despite a temporary decrease in Cr levels within permissible limits, elevated concentrations were once more observed in 2010 [25]. The Aktobe region, a significant industrial area in Kazakhstan, has emerged as a focal point of Cr(VI) contamination within the country. This contamination can be attributed to Cr(VI) plants' substantial presence in this region [28]. In India, presently, mining operations for chromite are restricted in specific areas of India, such as the Sukinda ultramafic belt, Baula Nausahi chromite belt in Odisha, Hassan district of Karnataka, and minor quantities are produced in Maharashtra. Chromite, primarily used in the metallurgical industry for manufacturing ferroalloys like ferrochrome and silicochrome, plays a significant role. Companies like Vishnu Chemicals Ltd operate plants in Andhra Pradesh and Chhattisgarh, producing Cr(VI) products, including sodium dichromate, basic chromium sulfate, chromic acid, and potassium dichromate, which are used as additives in making stainless steel and special alloy steel [29]. Effective regulation and waste management are vital to combat Cr(VI) contamination, which poses substantial health risks to humans and ecosystems. Cr(VI), a toxic pollutant, can elevate cancer risk, cause respiratory issues, and lead to skin problems in humans [30]. In the environment, it harms aquatic life, disrupts ecosystems, and contaminates water sources, necessitating robust regulations and remediation efforts to mitigate these adverse effects.

Ni(II) is found in crude oil at differing concentrations, and it can also be released through the combustion of petroleum-based products. Ni(II) serves diverse purposes, such as contributing to the production of stainless steel, playing a catalytic role in the chemical industry, being utilized in electroplating materials, forming part of heat-resistant alloys, acting as an alloying element in non-ferrous metals, supporting applications in aerospace and defense, and being a key component in nickel–cadmium batteries [31]. Nickel sulfate, a crucial compound, is used commercially in India in nickel plating, dip baths for enameling, and the preparation of various nickel compounds, serving as a catalytic Ni(II) [32]. Ni(II) contaminations in the soil are metal plating industries, combustion of fossil fuels, nickel mining, and electroplating. It is released into the air by power plants and trash incinerators and settles to the ground after undergoing precipitation reactions. Humans may also be exposed to Ni(II) by inhalation, drinking water, and eating contaminated food. Russia, one of the most significant global Ni(II) producers, has demonstrated essential Ni(II) contamination in many areas, primarily the Norilsk region. High concentrations of heavy metals characterize the territory within 4 km of Norilsk, coupled with the absence of trees and the disturbance of organic matter mineralization. This zone corresponds to level 5 of the loss of environmental quality, owing to the operations of the mining and metallurgical organization Norilsk Nickel Company [33]. Indonesia is also infested with the problems of Ni(II) contamination, with the top soils in the province of South Sulawesi recording elevated levels [34]. Ni(II), occurring mainly as oxides, sulfides, and silicates, is found in India, with notable nickeliferous limonite deposits in Sukinda Valley, Odisha. It's also associated with uranium deposits in Jaduguda, Jharkhand. Odisha boasts the country's most considerable nickel ore resources, amounting to 175 million tonnes, followed by Jharkhand and Nagaland. Hindustan Copper Limited (HCL) produces nickel sulfate at Ghatsila Copper Smelter in Jharkhand. The copper ore extracted also contains trace amounts of Nickel, gold, and cobalt.

However, Ni(II) also significantly affects human health and the environment. Prolonged exposure can trigger skin allergies, respiratory problems, and an increased cancer risk. Ingesting Ni(II) through contaminated food or water can lead to gastrointestinal issues. Dermatitis (Ni itch) is the most frequent effect of exposure to Ni(II). Acute poisoning of Ni(II) causes headache, dizziness, nausea and vomiting, chest pain, tightness of the chest, dry cough and shortness of breath, rapid respiration, cyanosis, and extreme weakness. In the environment, Ni(II) pollution disrupts water quality, harms soil health, and adversely affects aquatic and terrestrial ecosystems. Due to its bioaccumulative nature, Ni(II) poses risks to various organisms in the food chain, emphasizing the importance of stringent regulations and responsible disposal practices to mitigate these adverse effects and safeguard human health and the environment.

3 Conventional Methods for the Removal of Heavy Metals

Conventional methods for removing heavy metals from water encompass a range of physical, chemical, and biological processes. Precipitation and coagulation involve adding chemicals to water to create solid metal particles that can be separated by settling or filtration. Chemical precipitation and sedimentation use chemical reactions to form insoluble compounds that settle out [35]. Filtration, using materials like sand or activated carbon, physically traps heavy metal particles as water passes through. Ion exchange involves replacing heavy metal ions with less harmful ions on exchange resins [36]. Adsorption uses activated carbon or zeolites to attract and capture heavy metals. Electrocoagulation employs an electric current to create metal hydroxide flocs that adsorb heavy metal ions [37]. Reverse osmosis uses a semipermeable membrane to reject heavy metal ions based on size and charge [38]. Distillation boils water to create vapor and leaves heavy metals behind. Chemical reduction converts heavy metals into less soluble forms for easier removal [39]. Solvent extraction employs organic solvents to selectively bind with heavy metal ions [40]. Neutralization adjusts the pH of acidic waters to precipitate heavy metals.

These methods offer various levels of effectiveness and are chosen based on factors like the specific heavy metals present and water quality conditions. First, using physical and chemical processes in water treatment usually requires much energy for filtering, settling, and adding chemicals. Some physio-chemical methods, such as electrokinetic remediation and chemical precipitation, consume considerable energy. This can contribute to higher operational costs and have environmental implications, mainly if the energy source is not sustainable. Second, these methods can create extra pollutants like sludge or leftover chemicals that must be dealt with, which adds to the overall environmental impact. For example, using strong acids or bases in leaching processes can cause damage to the ecosystem [41].

4 Biological Methods for the Removal of Heavy Metals

Bioremediation of heavy metals in wastewater treatment is an emerging technology of great significance. It offers an environmentally friendly approach by harnessing the natural abilities of microorganisms to remove and detoxify heavy metals from contaminated water. This method is cost-effective, highly versatile, and specific, allowing for targeted removal of a wide range of heavy metals. On the other hand, biological methods, like using microorganisms in processes like activated sludge or constructed wetlands, are generally cheaper and use less energy. They use nature's techniques to break down waste and remove pollutants. Also, sometimes chemical methods don't work well for certain types of contaminants, like complex organic compounds or heavy metals. For these, biological processes are needed to get rid of them completely. It is a promising solution to mitigate the environmental impact of heavy metal pollution, offering a sustainable and efficient alternative to conventional, often more expensive, and less eco-friendly treatment methods [42]. As the need for sustainable wastewater management grows, bioremediation continues to gain prominence as a vital tool in safeguarding our ecosystems and water resources. The present review paper aims to delve into the bioremediation of heavy metals found in industrial wastewater, explicitly focusing on Pb(II), Cr(VI), and Ni(II). This investigation encompasses the utilization of various microorganisms, including bacteria, fungi, and algae, as tools for removing these contaminants. It will comprehensively examine bioremediation principles, shedding light on the intricate mechanisms involved. Furthermore, the review will give practical examples, offering real-world insights into applying these bioremediation techniques. The paper will also evaluate the advantages and disadvantages of each approach, contributing to a nuanced understanding of their efficacy and feasibility in addressing heavy metal pollution in industrial wastewater settings.

5 Mechanism of Microorganisms in Removing Heavy Metals

Heavy metals, typically poisonous to organisms, have evolved specific resistance mechanisms and complex intracellular pathways that enable organisms to utilize and detoxify these heavy metals for cellular reproduction [43]. Among these mechanisms, extracellular polysaccharides (EPS) play a vital role, as certain bacteria and cyanobacteria have developed the ability to synthesize EPS, which readily binds to heavy metals, preventing their entry into the cells. In contrast, other bacterial, fungal, and algal cells allow heavy metals to cross the extracellular membrane, leading to their accumulation in the cytoplasm. Within the cytoplasm, these heavy metals undergo detoxification through intrinsic metabolic reactions or are expelled from cellular structures using efflux pumps [5]. These resistance mechanisms vary depending on the organism's nature and essential proficiencies. These capabilities are integral to the field of bioremediation, which encompasses various mechanisms such as biosorption, bioaccumulation, bioreduction, bioprecipitation, biovolatilization, bioleaching, bioreactors and biosparging [44]. Two significant mechanisms, biosorption, and bioaccumulation, are widely observed across various organisms, including plants are shown in Fig. 1. Biosorption is a complex process where biosorbents interact with and adsorb metal ions in their surrounding environment. These biosorbents can include biomass from bacteria, fungi, algae, agricultural waste, plants, and industrial byproducts.

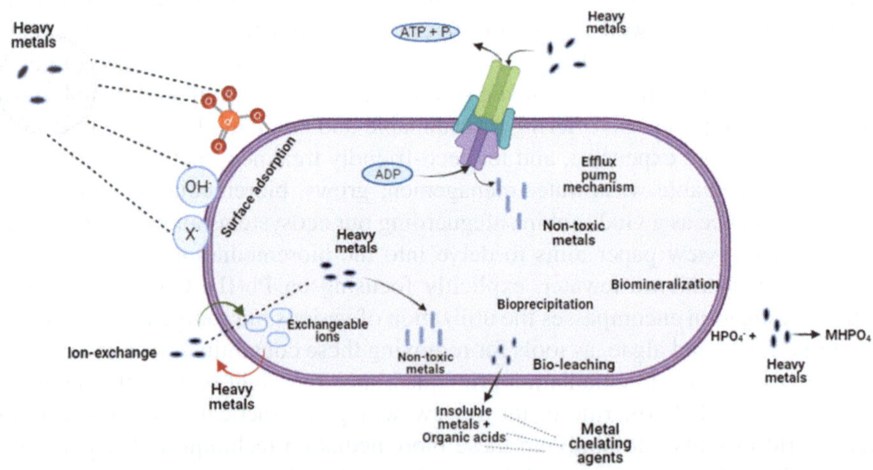

Fig. 1 Different bioremediation mechanisms for heavy metals (Created with BioRender.com)

In some cases, live cellular biomass removes heavy metals through biosorption [42]. Whether the biomass is living or dead, biosorption is a surface-based phenomenon involving physical and chemical interactions, such as electrostatic or Van der Waals interactions, ion exchange, complexation, diffusion, or precipitation. The cellular surface typically carries a negative charge, attracting heavy metals, and, in some cases, a mucus or polysaccharide layer enhances physical interactions. Heavy metals are adsorbed on the surface passively until equilibrium is reached. Functional groups on the cell surface, like phosphates, sulfates, amides, hydroxyl, or negatively charged proteins, exchange ions with metal ions, resulting in strong ionic interactions. In live cells, metal ions may reach the periplasmic space through surface porins and establish interactions with cellular membranes. This ionic interaction is crucial in reducing heavy metals and detoxifying them. Complexation, the aggregation of metal species and functional groups on the cell surface, is a common outcome of biosorption.

Bioaccumulation, conversely, refers to the absorption of contaminants by living biomass or cells. It involves metabolism-dependent active uptake of heavy metals. It employs passive diffusion, ion pumps, protein channels, and carrier-mediated transport to remove heavy metals from the environment [41]. Heavy metal transport across the phospholipid bilayers of living cells occurs in two phases: adherence to the cell surface in a metabolism-independent fashion, followed by the internalization of metal ions via the cell membrane. This resembles the cellular uptake of essential ions like Na^{2+}, K^{+}, and Ca^{2+}. Active transporters have been identified for various heavy metals in different microorganisms like cadmium, chromium, silver, lead, mercury, strontium, uranium, and thorium in *Pseudomonas putida, Bacillus subtilis, Thiobacillus ferrooxidans, Rhizopus arrhizus, Micrococcus luteus, Saccharomyces cerevisiae*, and *Aspergillus niger*, respectively [45] Bioprecipitation is another essential microbial activity that transforms soluble metal species and metalloids into insoluble forms like hydroxides, carbonates, phosphates, and sulfides. These metal precipitates remain attached to microbial cells, and slight pH changes can facilitate the recovery of heavy metal ions. Microbes often release organic acids and electron donors into the extracellular medium, precipitating heavy metal ions. In some cases, enzymes also play a role in this process, such as alkaline phosphatase derived from the calf intestine, which can precipitate ions like cadmium, Ni(II), cobalt, and Cr(VI) from industrial sources [46].

Mineral components often contain substantial levels of insoluble metals that are not biodegradable. In such cases, microbial mobilization can help regulate the release and transformation of these insoluble metal species. For soluble metal species, microbial changes are essential for modifying the speciation, affecting factors like oxidation/reduction states, solubility, mobility, bioavailability, and toxicity. An example of this is the bioreduction of Cr(VI), which involves direct and indirect reduction to Cr(III) and is influenced by biological processes, including the role of bio-produced Fe(II) [47].

6 Bioremediation of Nickel(II) from the Wastewater

6.1 Bioremediation of Ni (II) Using Bacterial Cells

In bioremediation, various bacterial strains have demonstrated significant potential for removing Ni(II) from contaminated solutions, reflecting the diverse strategies employed by different microorganisms in this context [48] identified Ni(II)-resistant strains from industrial effluent, among which *Klebsiella oxytoca* ATHA6 emerged as a notably efficient performer. This strain exhibited the capacity to reduce 12% of Ni(II) from a solution with a concentration of 704 mg/l in just three days [21] isolated two metal-resistant strains, *Sphingomonas melonis* E8 and *Enterobacter hormaechei* WW28, from soil samples near coal mines. In a medium containing 50 mg/l of Ni(II) at pH 6, *Sphingomonas melonis* E8 effectively removed 78% of the Ni(II) after 48 h, while *Enterobacter hormaechei* WW28 achieved a removal rate of 63%. [49] discovered an indigenous bacterium, *Bacillus sp.* KL1, within a soil sample obtained from municipal solid waste leachate. This Ni(II)-resistant strain exhibited the highest removal efficiency at 55.06% when the initial Ni(II) concentration was 100 mg/l after 24 h of incubation at 30 °C and a pH of 7. In a separate investigation, with an initial solution characterized by a pH of 7 and a Ni(II) concentration of 50 mg/l, *Micrococcus sp.* bacteria demonstrated the capability to remove 55% of the Ni(II) within a 20 h incubation period [50]. Furthermore, studies have highlighted the remarkable potential of *Bacillus megaterium* EMCC, which, with an inoculum size of 0.1×10^{29} CFU, exhibited the highest Ni(II) removal capacity from a solution containing 600 mg/l of Ni(II). After 24 h of incubation at 30 °C and a pH of 7, this bacterium achieved a substantial removal rate of 26.67% [51]. Likewise, bioremediation with 102.5 cells/mL of *Bacillus subtilis* resulted in a remarkable reduction of the initial Ni(II) concentration, with an 85.61% removal rate when the initial concentration was 2.1425 mg/l [52]. These findings collectively illustrate the diverse array of microbial strategies and capabilities in the context of Ni(II) removal, providing valuable insights for potential bioremediation applications. In an experimental investigation, the removal of Ni(II) ions utilizing *Bacillus pumilus*, with an initial concentration of 100 mg/l at pH 7.5 and a 6% inoculum, yielded an 83.02% removal rate over 48 h [53]. *Stenotrophomonas sp.* MB339 accumulated 48.78% of Ni(II) at 37 °C for 24 h, at neutral pH, and with the maximum inoculum size [54].

6.2 Bioremediation of Ni (II) Using Algal Cells

In recent years, several studies have made significant contributions to water and wastewater treatment, particularly in the context of heavy metal and pollutant removal. In 2021, [55] conducted research emphasizing the effectiveness of the "growing-algae-while-removing-Ni" approach compared to "biomass-physical-adsorption" for Ni(II) removal. Their findings revealed that this innovative method yielded superior results, and they showcased the potential of the Revolving Algal Bioreactor system, which achieved over 90% efficiency in Ni(II) removal, with an impressive rate of 534 mg/L-day Ni(II) removal. Building on this, [56] contributed to understanding Ni(II) removal processes. Their study, conducted over 12 days at 22.2 °C, found that *Chlorella vulgaris* removed 32% of Ni(II), while *Desmodesmus sp.* removed 39% of this heavy metal. Furthermore, research by [57] explored the capabilities of *Oedogonium westi* algae in Ni(II) removal. Their investigations under specific conditions, including a temperature of 18 ± 1 °C and a pH of 5, demonstrated that this algal species could effectively remove Ni(II), with removal rates ranging from 59 to 89% in 7 days. Notably, the study by [58] expanded the scope to encompass the removal of malathion, a pesticide, and heavy metals, particularly Ni(II). They utilized a consortium of microorganisms, including *Chlorella vulgaris, Scenedesmus quadricuda*, and cyanobacteria *Spirulina platensis,* to address this challenge. Their work demonstrated remarkable efficiency, removing 95% (± 0.7%) of Ni(II) from wastewater collected from various sources. Despite the initial high concentrations of 25 mg/l of malathion and 10 mg/l of Ni(II), this purification process was completed within 28 days. These collective findings underline the significant advancements in water treatment strategies and the promising potential of various organisms and innovative systems in addressing pollution and enhancing water quality. Such research is critical in pursuing sustainable environmental practices and cleaner water resources.

6.3 *Bioremediation of Ni (II) Using Fungal Cells*

A comprehensive study systematically evaluated the effectiveness of various fungal species under specific environmental conditions within heavy metal removal. The parameters investigated encompassed consistent pH levels (5 and 6), a standard incubation period of 144 h, a temperature set at 30 °C, and the utilization of an inoculum size of 6%. This investigation unveiled the remarkable heavy metal removal capabilities of *Neurospora crassa* and *Aspergillus flavus*, achieving maximum removal rates of 83.0% and 85.18%, respectively, when exposed to a Ni(II) concentration of 100 mg/l. Moreover, growing cells of *Rhizopus delemar* were harnessed for removing Ni(II) and Cu(II) in single metal systems, underscoring the adaptability of fungal species in addressing heavy metal contamination [59]. In a different study, *Aspergillus niger* demonstrated a maximum uptake of 0.55 mg/g in the presence of 50 mg/l of Ni under specific conditions, including an incubation period of 96 h, a temperature of 28 °C, and a suspension volume of 1 ml [60]. Furthermore, research led by [50] delved into Ni(II) removal by *Aspergillus sp.* fungi at a pH of 5, revealing an impressive removal rate of 90% within a 20-h timeframe under the specified conditions.

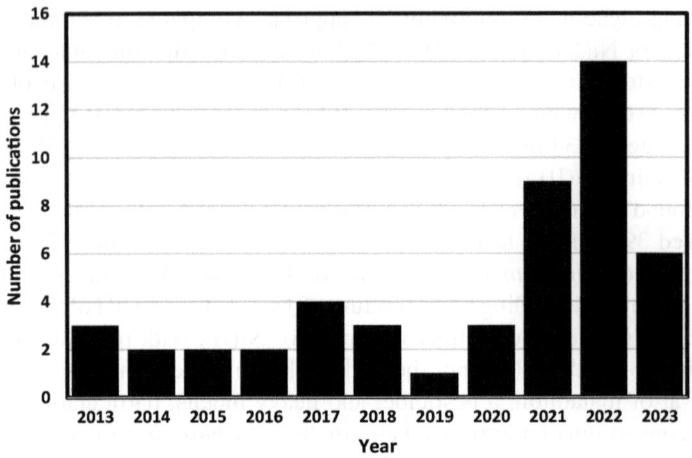

Fig. 2 Publications per year for bioremediation of Ni(II) during the period January 2013 to October 2023 (Data collected from dimensions database)

Various filamentous fungi from genera such as Aspergillus, Penicillium, Geotrichum, Alternaria, Fusarium, Rhizopus, Monilia, and Trichoderma were identified in soil ecosystems. These fungal strains underwent testing for the presence of heavy metals, including Co(II), Cr(VI), Cu(II), Cd(II), and Ni(II). This extensive investigation, conducted by [5], illuminates the potential roles of these fungi in biogeochemical cycling and their capacity for metal sequestration. In aggregate, these findings offer valuable insights into the potential applications of fungal species in removing and sequestrating heavy metals, enriching the comprehension of their contributions to bioremediation and environmental protection. Figure 2 draws a comparison between number of publications every year dedicated to Bioremediation of Ni(II) between the period of January 2013 to October 2023.

7 Bioremediation of Hexavalent Chromium

7.1 Bioremediation of Cr(VI) Using Bacterial Cells

Within the context of metal removal by bacteria, a fascinating array of research findings highlight the diverse strategies employed by different microbial species to effectively address the challenge of removing metal ions from aqueous solutions. These investigations shed light on the factors influencing metal removal efficiency, including pH, cell viability, and the specific mechanisms at play. One noteworthy example is *Pseudochrobactrum saccharolyticum* LY10, which demonstrated remarkable efficiency by removing 95% of a substance when tested at a pH of 8.3, with an initial concentration of 130 mg/l, utilizing a cell concentration of 1.47 × 10^9 cells/

ml over 72 h. This underscores the crucial role of environmental conditions in determining the effectiveness of metal removal [61]. Distinguishing between live and dead cells, the research unveiled that live *Ochrobactrum intermedium* LBr exhibits superior efficiency in removing Cr(VI) metal. This observation suggests the involvement of biological activities, potentially metabolic processes, in successfully removing metal ions [62]. Cr(VI)-tolerant *Bacillus amyloliquefaciens*, isolated from soil in a chromate mine, exhibited rapid chromate reduction capabilities during its growth phase, facilitated by chromate reductase enzymes [63]. *Bacillus methylotrophicus*, isolated from tannery sludge, harnessed extracellular chromate reductase enzymes for Cr(VI) reduction [64]. These examples underscore the pivotal role of enzymes in transforming and removing metal ions [63, 64]. *Bacillus firmus* TE7, as another intriguing case, displayed the remarkable ability to simultaneously reduce Cr(VI) and oxidize As(III), illustrating the multifunctionality of specific bacterial strains in managing multiple metal species [65]. In bacterial metal removal mechanisms, active uptake often plays a crucial role. Bacteria reduce Cr(VI) to the less toxic Cr(III), with enzymatic reduction being instrumental in this process [62]. As exemplified by *Ochrobactrum intermedium* LBr and *Cupriavidus metallidurans* CH34, passive uptake mechanisms have also been observed. These bacteria employ negatively charged functional groups on their cell walls to bind positively charged metal ions. The efficiency of this passive uptake is influenced by pH, with lower pH values leading to reduced metal removal due to competition with hydronium ions (H_3O^+) for binding sites on the bacterial surface. An increase in pH enhances the binding of metals to cell wall groups, effectively increasing removal efficiency [62].

Furthermore, [66] reported the reduction of Cr(VI) by the extracellular chromate reductase enzyme produced by *Bacillus amyloliquefaciens* (CSB 9), suggesting either enzymatic or intracellular uptake mechanisms for metal removal. This highlights the role of specific enzymes in the metal removal process, potentially within the membrane or cytosolic fractions of chromate-reducing bacteria [66]. Exploring the isolation of metal-resistant strains from contaminated domestic industrial effluent, Bestawy et al. [67] identified several indigenous strains, including *Enterobacter sp.*, *Stenotrophomonas sp.*, and *Ochrobactrum sp.*, each exhibiting unique resistance profiles. *Ochrobactrum sp.*, for instance, demonstrated resistance to 29 mg/l of Cr(VI). This indicates the potential for naturally occurring microbial communities to adapt to challenging environmental conditions and metal stress [67]. Additionally, *Micrococcus sp.* bacteria, in an initial solution with a pH of 7 and a Cr(VI) concentration of 100 mg/l, achieved an impressive 90% removal within 18 h, showcasing the capabilities of different microbial species in addressing metal pollution challenges [50]. These diverse findings offer insights into the potential applications of various bacteria's remarkable adaptability and strategies in addressing the removal of metal ions from aqueous solutions, contributing to bioremediation and environmental protection. At 37 °C for 24 h, with a neutral pH and maximum inoculum size, *Klebsiella pneumonia* MB361 exhibited a removal of 83.51% with an initial concentration of 200 μg/ml Cr(VI), as shown by [54].

7.2 Bioremediation of Cr(VI) Using Algal Cells

Several notable findings have shed light on the efficiency and mechanisms underlying this bioremediation process in the context of utilizing microalgae to remove Cr(VI) (Cr). *Chlamydomonas reinhardtii*, for instance, exhibited a remarkable Cr uptake capacity, reaching 18.2 mg/g under specific conditions, including a pH of 2, an algae biomass concentration of 0.6 g/l, a temperature of 25 °C, and a duration of 2 h [68]. In the case of *Dunaliella* sp.1, the Cr(VI) uptake capacity was even more impressive, standing at 58.3 mg/g. These findings were obtained under conditions that included a pH of 2, an initial Cr(VI) concentration of 100 mg/l, an algae biomass concentration of 1 g/l, a temperature of 25 °C, and a duration of 72 h. Furthermore, the wastewater sample collected from a tannery industry, diluted with tap water at a 1:1 ratio, contained an initial Cr(VI) concentration of 3.22 mg/l. Notably, microalgae *C. vulgaris* exhibited exceptional efficacy by completely removing Cr(VI) from the diluted water sample at 28 °C within 12 days [68]. Raw *P. canaliculata* biomass, when packed in a column reactor, displayed a noteworthy Cr(VI) uptake of 2.1 mmol/g [69], emphasizing the potential of this biomass in metal removal applications. In a 2017 study, Sen et al. [70] examined the efficacy of a cyanobacterial consortium (*Limnococcus limneticus and Leptolyngbya subtilis*) from East Kolkata Wetland in removing Cr(VI) from water. They found that starting with 15 mg/L Cr(VI), a pH of 9, and a 10% inoculum size achieved 50% removal within 12 days. Also examined the removal of Cr(VI) from synthetic wastewater using *Limnococcus sp.* cyanobacteria. They achieved a maximum Cr(VI) removal of 63.23% with an initial concentration of 20 mg/l, pH 9.0, a 10% inoculum size, and a 16-day incubation period [71]. One of the intriguing aspects of microalgae's role in removing Cr is their capacity to release electrons through photosynthesis. This unique metabolic process allows them to compensate for the electrons required to reduce Cr(VI) [72]. These findings collectively underscore the potential of microalgae as effective agents for removing Cr(VI) and provide valuable insights into the various factors and mechanisms that influence their performance in this bioremediation process.

7.3 Bioremediation of Cr(VI) Using Fungal Cells

Research has shed light on the remarkable resilience and mechanisms underlying these processes in the context of fungal adaptation and bioaccumulation of heavy metals. Fungi's capacity to survive in Cr(VI) presence is contingent upon their structural attributes, biochemical properties, genetic adaptations, and physiological characteristics [73]. Moreover, these fungi demonstrate an exceptional ability to flourish in environments characterized by extreme variations in pH, temperature, nutrient availability, and elevated concentrations of metals. The pivotal role of metal-fungus interactions in facilitating the adaptation of fungi to diverse environments is underscored by the correlation between the presence of metals and the components of the microbial cell, which include the cell wall, periplasm, plasmatic membrane, and cytoplasm [73]. One species, *Galerina vittiformis*, discovered in a municipal landfill, has been documented as a proficient bioaccumulator of several heavy metals, including Cr(VI), Cu(II), Zn(II), Cd(II), and Pb(II) [74].

Further exploration into the subject reveals specific instances of fungal bioaccumulation. For example, *Trichoderma longibrachiatum* exhibited a notable maximum Cr uptake of 0.55 mg/g under particular conditions, with an initial concentration of 50 mg/l, a duration of 96 h, a temperature of 28 °C, and a suspension volume of 1 mL [60]. Another *Aspergillus sp.* effectively removed 92% of Cr(VI) from a solution with an initial concentration of 100 mg/l, a pH of 5, over 18 h [50]. In the context of wastewater treatment, a fungal consortium comprising *Cladosporium perangustum, Penicillium commune, Paecilomyces lilacinus*, and *Fusarium equiseti* immobilized on a nylon mesh in a stirred tank bioreactor achieved significant reductions in various pollution parameters. This innovative approach led to complete Cr(VI) removal (100%), a nearly total reduction in Cr(VI) (99.92%), and a substantial decrease in Pb(II) (95.91%) following 120 h of treatment [75, 76] conducted experiments by introducing varying concentrations of Cr(VI) (ranging from 25 to 500 mg/ml) to Sabouraud dextrose agar and allowing a 5-day incubation period at a temperature of 25 °C. Their findings revealed that *A. niger* exhibited remarkable tolerance, with the ability to withstand concentrations ranging from 350 to 400 μg per milliliter of Cr(VI). In a similar approach, M. [77] investigated the tolerance of *Penicillium sp.* and *A. niger* to hexavalent Cr(VI), exploring different concentrations of Cr(VI) ranging from 100 to 1500 mg/ml. Their research reported that Penicillium sp. exhibited a notable tolerance level of up to 800 mg/ml, while A. niger displayed tolerance up to 512 mg/ml. These observations collectively emphasize fungi's remarkable adaptability in heavy metal tolerance and bioaccumulation, making them valuable candidates for bioremediation and environmental applications. Figure 3 draws a comparison between number of publications every year dedicated to Bioremediation of Cr(VI) between the period of January 2013 to October 2023.

8 Bioremediation of Lead(II)

8.1 Bioremediation of Pb(II) Using Bacterial Cells

In the context of heavy metal removal, several studies have explored the effectiveness of various bacterial strains under specific environmental conditions. These investigations have shed light on the potential applications of these bacteria in bioremediation and environmental protection. *Enterobacter hormaechei* WW28, in a medium with a pH of 6 and containing 100 mg/L of Pb(II), exhibited the capacity to remove 48% of Pb(II) within 48 h. In contrast, *Sphingomonas melonis* E8 only removed 23% of Pb(II) under similar conditions [21]. A study used *Bacillus cereus* to eliminate Pb(II) ions from a solution with an initial concentration of 100 mg/l and a pH of 7, using

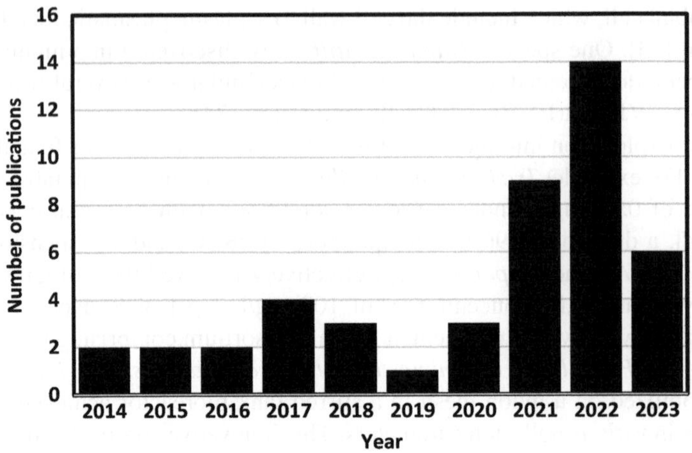

Fig. 3 Publications per year for bioremediation of Cr(VI) during the period January 2013 to October 2023 (Data collected from dimensions database)

a 4% inoculum. This method achieved a removal efficiency of 90.13% within 24 h [53]. The study by [78] demonstrated the noteworthy Pb(II) uptake capacities of *Bacillus xiamenensis* bacteria. Under the best conditions with an initial concentration of 100 mg/l, Pb(II) removal was 99.19%. Importantly, this study highlighted that live bacteria outperformed their non-living counterparts due to a mechanism that accumulated the metal inside the cells. *Acinetobacter calcoaceticus* NCCB 22,016 strain exhibited remarkable efficiency in Pb(II) removal. Under stress conditions of 100 mg/l Pb(II), the removal efficiency reached 75.1%, with the highest efficiency observed at a pH of 5 and a temperature of 35 °C [79] in the study of [54], *Stenotrophomonas sp.* MB339 accumulated 85.30% of Pb under neutral pH and a temperature of 37 °C, with an initial concentration of 200 μg/ml and a 500 μl/l inoculum size over 24 h. *Staphylococcus sp.* MB371 exhibited strong performance by removing 88.33% of Pb(II) at a pH of 6, with an initial concentration of 200 μg/ml. These findings collectively underscore the potential of various bacterial strains in removing Pb(II) and offer valuable insights into their applications in bioremediation and environmental protection.

8.2 Bioremediation of Pb(II) Using Algal Cells

A comprehensive review of studies focusing on removing heavy metal Pb(II) from wastewater reveals various methods and organisms employed. Experiments were also conducted where live Oscillatorialaete-virens were exposed to Pb(II) under conditions of pH 5, resulting in an impressive Pb(II) uptake of 21.6 mg/g and live Pseudochlorococcum typicum at pH 7, observed a Pb(II) uptake of 4.49, showcasing the influence of pH on uptake efficiency. Another remarkable finding for Pb(II) uptake rate of 155.63 mg/g using live Synechocystis sp. was recorded. Furthermore, [80] studied the incubation of *Chlorella sp.* and *Chaetoceros sp.* over 14 days, with an initial Pb concentration of 0.5 mg/l. Their findings revealed that Chlorella sp. achieved an impressive 90% removal rate, while Chaetoceros sp. attained an 81% removal rate, indicating the efficiency of these microorganisms in Pb removal. In a separate study by [57], using *Oedogonium westi* algae for Pb removal proved highly effective. Under the experimental conditions of a temperature of 18 ± 1 °C and a pH of 5, the removal rates ranged from 61% to an impressive 96% within just 7 days, underlining the potential of this algal species for rapid and efficient Pb removal. [58] conducted an intriguing study in which a consortium of microorganisms, including *Chlorella vulgaris, Scenedesmus quadricuda,* and cyanobacteria *Spirulina platensis,* was employed to cleanse water contaminated with malathion and Pb. Over 28 days, these microorganisms demonstrated their remarkable ability to remove 87% ($\pm$0.4%) of Pb from water initially containing 25 mg/l of malathion and 10 mg/l of Pb. These findings underscore the diverse strategies and organisms available for effective heavy metal removal from wastewater, offering promising prospects for environmentally friendly water treatment and pollution control.

8.3 Bioremediation of Pb(II) Using Fungal Cells

Within the bioremediation domain, many studies have uncovered intriguing revelations regarding the potential of various fungal species in removing Pb(II) [60] notably achieved a remarkable breakthrough by reporting 59.67 mg/g Pb(II) uptake in *Aspergillus terreus.* This notable accomplishment was realized under specific conditions, which included an initial Pb concentration of 50 mg/l, a duration of 96 h, and a temperature of 28 °C, using a 1 ml suspension. Moreover, another compelling study conducted by [5] unveiled the resilience of fungal strains when subjected to elevated concentrations of heavy metals. Among the four fungi scrutinized—*A. niger, A. fumigatus, T. asperellum,* and *Penicillium simplicissimum*—*T. asperellum* exhibited exceptional resistance to levels as high as Cu(II) 800 mg/l and Pb(II) 1000 mg/l in a specialized medium. This remarkable resistance and efficient Pb(II) absorption suggest the potential of *T. asperellum* as a promising biosorption agent for environmental Pb(II) removal. In addition to *A. terreus* and *T. asperellum,* other fungal species have also demonstrated their proficiency in Pb(II) removal. For instance, [81] investigated the removal of Pb(II) using live and dead biomass of *Saccharomyces cerevisiae* AUMC 3875, establishing a minimum inhibitory concentration (MIC) value of 600 mg/l for Pb(II) resistance. This underscores the potential application of *S. cerevisiae* AUMC 3875 for Pb(II) removal from aqueous solutions. Furthermore, the mycelia of *P. chrysosporium* have shown an impressive capacity to accumulate Pb ions under varying culture conditions, which could be harnessed for the removal of Pb(II) from heavy metal-polluted wastewater, as highlighted by [82]. Lastly, the

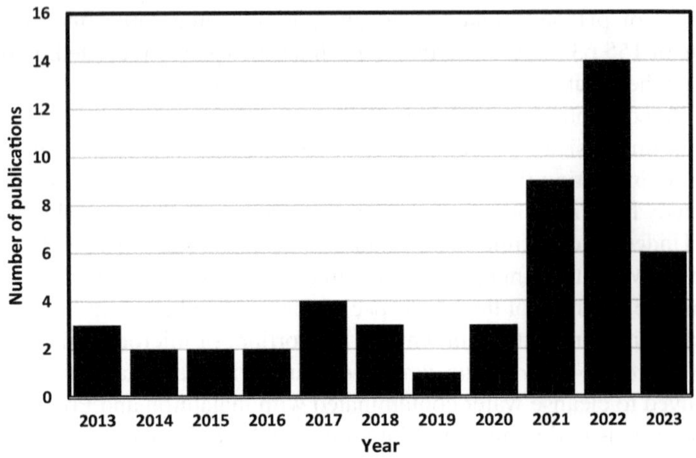

Fig. 4 Publications per year for bioremediation of Pb(II) during the period January 2013 to October 2023 (Data collected from dimensions database)

speciation transformation of Pb in P. oxalicum SL2 mycelium was found to predominantly involve Pb-hydrogen phosphate (40.5%) and Pb citrate (40.2%) as the main Pb species, as observed by [83]. These findings collectively illuminate the diverse and promising strategies within fungal-based bioremediation for Pb(II) removal, addressing crucial environmental concerns related to heavy metal pollution in water. Figure 4 draws a comparison between number of publications every year dedicated to Bioremediation of Pb(II) between the period of January 2013 to October 2023.

9 Conclusion

Bioremediation remains a valuable, natural, and eco-friendly process for the biological degradation of polluted environments. Microorganisms play a pivotal role in eliminating heavy metal pollutants that pose a toxic threat to living cells. Anaerobic bacteria address issues such as biphenyls, dechlorination, and chloroform. Fungal microorganisms are effective in breaking down various toxic pollutants. Phytoremediation is an emerging technology that employs plants to remediate soil, water, and other environmental pollution. Bioremediation is a cost-effective, sustainable, and eco-friendly approach, requiring minimal labor and effort, making it relatively straightforward to implement.

However, it is important to note that bioremediation has some drawbacks, primarily its time-consuming nature. In some cases, the byproducts of biodegradation can become more toxic than the original compounds. Completeness and performance evaluation can be challenging due to unclear endpoints. Further research is essential to advance bioremediation technologies and discover additional biological solutions for mitigating heavy metal contamination in various environmental systems.

References

1. Tchounwou PB, Yedjou CG, Patlolla AK, Sutton DJ (2012) Heavy metal toxicity and the environment. In: Luch A (ed) Molecular, clinical and environmental toxicology, vol 3. Environmental Toxicology. Springer Basel, Basel, pp 133–164
2. Deepa CN, Suresha S (2014) Biosorption of lead (II) from aqueous solution and industrial effluent by using leaves of Araucaria cookii: application of response surface methodology. IOSR J Environ Sci Toxicol Food Technol 8:67–79. https://doi.org/10.9790/2402-08716779
3. Okolo V, Olowolafe E, Okoduwa SIR (2016) Effects of industrial effluents on soil resources in Challawa Industrial Area, Kano, Nigeria. Global J Environ Sci Manage 5:1–10
4. Igiri BE, Okoduwa SIR, Idoko GO, Akabuogu EP, Adeyi AO, Ejiogu IK (2018) Toxicity and bioremediation of heavy metals contaminated ecosystem from tannery wastewater: a review. J Toxicol 2018:2568038. https://doi.org/10.1155/2018/2568038
5. Siddiquee S, Kobun R, Azad S, Saallah S (2015) Heavy metal contaminants removal from wastewater using the potential filamentous fungi biomass: a review. J Microb Biochem Technol 07.https://doi.org/10.4172/1948-5948.1000243
6. Naidu R, Bolan N (2008) Chapter 2 contaminant chemistry in soils: key concepts and bioavailability. Chem Bioavailab Terrestr Environ 32. https://doi.org/10.1016/S0166-2481(07)32002-3
7. Rai PK, Lee SS, Zhang M, Tsang YF, Kim K-H (2019) Heavy metals in food crops: Health risks, fate, mechanisms, and management. Environ Int 125:365–385. https://doi.org/10.1016/j.envint.2019.01.067
8. Yang J, Sun Y, Wang Z, Gong J, Gao J, Tang S, Ma S, Duan Z (2022) Heavy metal pollution in agricultural soils of a typical volcanic area: risk assessment and source appointment. Chemosphere 304:135340. https://doi.org/10.1016/j.chemosphere.2022.135340
9. Álvarez EA, Mochón MC, Sánchez JCJ, Rodríguez MT (2002) Heavy metal extractable forms in sludge from wastewater treatment plants. Chemosphere 47:765–775. https://doi.org/10.1016/S0045-6535(02)00021-8
10. Chakraborty SC, Qamruzzaman M, Zaman MWU, Alam MM, Hossain MD, Pramanik BK, Nguyen LN, Nghiem LD, Ahmed MF, Zhou JL, Mondal MdIbrahimH, Hossain MA, Johir MAH, Ahmed MB, Sithi JA, Zargar M, Moni MA (2022) Metals in e-waste: Occurrence, fate, impacts and remediation technologies. Process Safety Environ Protect 162:230–252. https://doi.org/10.1016/j.psep.2022.04.011
11. Anastasiadou K, Christopoulos K, Mousios E, Gidarakos E (2012) Solidification/stabilization of fly and bottom ash from medical waste incineration facility. J Hazard Mater 207–208:165–170. https://doi.org/10.1016/j.jhazmat.2011.05.027
12. Malik B, Kaur Sandhu K (2023) Occurrence and impact of heavy metals on environment. Mater Today Proc. https://doi.org/10.1016/j.matpr.2023.01.317
13. Molalign Medfu Tarekegn FZS, Ishetu AI (2020) Microbes used as a tool for bioremediation of heavy metal from the environment. Cogent Food Agric 6:1783174. https://doi.org/10.1080/23311932.2020.1783174
14. Su C, Jiang L, Zhang W (2014) A review on heavy metal contamination in the soil worldwide: Situation, impact and remediation techniques. Environ Skept Crit 3:24–38
15. ATSDR (Agency for Toxic Substances and Disease Registry) (2017) ATSDR's substance priority list. Accessed June 12, 2023
16. Seaman JC, Arey JS, Bertsch P (2001) Immobilization of nickel and other metals in contaminated sediments by hydroxyapatite addition. J Environ Qual 30:460–469. https://doi.org/10.2134/jeq2001.302460x

17. Ruckart PZ, Ettinger AS, Hanna-Attisha M, Jones N, Davis SI, Breysse PN (2019) The flint water crisis: a coordinated public health emergency response and recovery initiative. J Public Health Manage Pract 25
18. Ji A, Wang F, Luo W, Yang R, Chen J, Cai T (2011) Lead poisoning in China: a nightmare from industrialisation. Lancet 377:1474–1476. https://doi.org/10.1016/S0140-6736(10)60623-X
19. Udiba UU, Udofia UU, Akpan ER, Antai EE (2019) Assessment of lead (Pb) uptake and hazard potentials of cassava plant (Manihot esculentus cranz), Dareta Village, Zamfara, Nigeria. Int Res J Public Environ Health 6:115–126. https://doi.org/10.15739/irjpeh.19.014
20. SkA A, Khan MH, Khandker S, Sarwar AFM, Yasmin N, Faruquee MH, Yasmin R (2014) Blood lead levels and health problems of lead acid battery workers in Bangladesh. Scientific World J 2014:974104. https://doi.org/10.1155/2014/974104
21. Heidari P, Sanaeizade S, Mazloomi F (2020) Removal of Nickel, Copper, Lead and Cadmium by New Strains of *Sphingomonas melonis* E8 and *Enterobacter hormaechei* WW28. J Appl Biotechnol Rep 7:208–214. https://doi.org/10.30491/jabr.2020.120185
22. Drexler JW, Brattin WJ (2007) An in vitro procedure for estimation of lead relative bioavailability: with validation. Hum Ecol Risk Assess Int J 13:383–401. https://doi.org/10.1080/10807030701226350
23. Marzan LW, Hossain M, Mina SA, Akter Y, Chowdhury AMMA (2017) Isolation and biochemical characterization of heavy-metal resistant bacteria from tannery effluent in Chittagong city, Bangladesh: Bioremediation viewpoint. The Egyptian J Aquatic Res 43:65–74. https://doi.org/10.1016/j.ejar.2016.11.002
24. Prabhakaran DC, Bolaños-Benitez V, Sivry Y, Gelabert A, Riotte J, Subramanian S (2019) Mechanistic studies on the bioremediation of Cr(VI) using Sphingopyxis macrogoltabida SUK2c, a Cr(VI) tolerant bacterial isolate. Biochem Eng J 150:107292. https://doi.org/10.1016/j.bej.2019.107292
25. Gzik A, Kuehling M, Schneider I, Tschochner B (2003) Heavy metal contamination of soils in a mining area in South Africa and its impact on some biotic systems. J Soils Sediments 3:29–34. https://doi.org/10.1007/BF02989466
26. Gao Y, Xia J (2011) Chromium contamination accident in China: viewing environment policy of China. Environ Sci Technol 45:8605–8606. https://doi.org/10.1021/es203101f
27. Coyte RM, McKinley KL, Jiang S, Karr J, Dwyer GS, Keyworth AJ, Davis CC, Kondash AJ, Vengosh A (2020) Occurrence and distribution of hexavalent chromium in groundwater from North Carolina, USA. Science of the Total Environment 711.https://doi.org/10.1016/j.scitotenv.2019.135135
28. Adotey EK, Burkutova L, Tastanova L, Bekeshev A, Balanay MP, Sabanov S, Rule AM, Hopke PK, Amouei Torkmahalleh M (2022) Quantification and the sources identification of total and insoluble hexavalent chromium in ambient PM: a case study of Aktobe, Kazakhstan. Chemosphere 307:136057. https://doi.org/10.1016/j.chemosphere.2022.136057
29. Indian Minerals Yearbook (2019) Part-III: Mineral Reviews, 58th Edition, Chromite, Government of India, Ministry of Mines, Indian Bureau of Mines, August, 2021
30. Fernández PM, Viñarta SC, Bernal AR, Cruz EL, Figueroa LIC (2018) Bioremediation strategies for chromium removal: current research, scale-up approach and future perspectives. Chemosphere 208:139–148. https://doi.org/10.1016/j.chemosphere.2018.05.166
31. Gautam PK, Gautam R, Banerjee S, Chattopadhyaya M, Pandey J (2016) Heavy metals in the environment: Fate, transport, toxicity and remediation technologies. In: Heavy metals: sources, toxicity and remediation techniques, pp 101–130
32. Indian Minerals Yearbook (2020) Part- II: Metals & Alloys, 59th Edition, Nickel, Government Of India, Ministry of Mines, Indian Bureau of Mines, August 2021
33. Yakovlev AS, Plekhanova IO, Kudryashov SV, Aimaletdinov RA (2008) Assessment and regulation of the ecological state of soils in the impact zone of mining and metallurgical enterprises of Norilsk Nickel Company. Eurasian Soil Sci 41:648–659. https://doi.org/10.1134/S1064229308060100
34. Artiningsih A, Zubair H, Imran AM, Widodo S (2019) Contamination and characteristic of Ni and Cr metal on top soil from Antang landfill, Makassar City, South Sulawesi Province,

Indonesia. In: IOP Conference Series: Earth and Environmental Science. Institute of Physics Publishing
35. Wuana RA, Okieimen FE, Imborvungu JA (2010) Removal of heavy metals from a contaminated soil using organic chelating acids. Int J Environ Sci Technol 7:485–496. https://doi.org/10.1007/BF03326158
36. Agelidis T, Fytianos K, Vasilikiotis G, Jannakoudakis D (1988) Lead removal from wastewater by cementation utilising a fixed bed of iron spheres. Environ Pollut 50:243–251. https://doi.org/10.1016/0269-7491(88)90226-6
37. Varma AK, Chouhan A, Shankar R, Mondal P, Rathore AK, Thakur LS (2021) Simultaneous removal of lead and copper from synthetic water by electrocoagulation and techno-economic evaluation: optimization through response surface methodology. Int J Eng Sci Technol 13:61–68. https://doi.org/10.4314/ijest.v13i1.9s
38. Pirsaheb M, Naderi S, Lorestani B, Khosrawi T, Sharafi K (2014) Efficiency of reverse osmosis system in the removal of lead, cadmium, chromium and zinc in feed water of dialysis instruments in kermanshah hospitals. J Mazand Univ Med Sci 24:151–157
39. Kousi P, Remoundaki E, Hatzikioseyian A, Tsezos M (2007) A study of the operating parameters of a sulphate-reducing fixed-bed reactor for the treatment of metal-bearing wastewater
40. Chang Y-K, Chang J-E, Lin T-T, Hsu Y-M (2002) Integrated copper-containing wastewater treatment using xanthate process. J Hazard Mater 94:89–99. https://doi.org/10.1016/S0304-3894(02)00060-2
41. Parasakthi J, Chandrani D, Vijayaraghavan R, Muthusivaramapandian M (2023) Trends in bioremediation of heavy metal contaminations. Environ Eng Res 28:220631. https://doi.org/10.4491/eer.2021.631
42. Priyadarshanee M, Das S (2021) Biosorption and removal of toxic heavy metals by metal tolerating bacteria for bioremediation of metal contamination: A comprehensive review. J Environ Chem Eng 9:104686. https://doi.org/10.1016/j.jece.2020.104686
43. de Alencar FLS, Navoni JA, do Amaral VS (2017) The use of bacterial bioremediation of metals in aquatic environments in the twenty-first century: a systematic review. Environ Sci Pollut Res 24:16545–16559.https://doi.org/10.1007/s11356-017-9129-8
44. Jasrotia S, Kansal A, Mehra A (2017) Performance of aquatic plant species for phytoremediation of arsenic-contaminated water. Appl Water Sci 7:889–896. https://doi.org/10.1007/s13201-015-0300-4
45. Mir-Tutusaus JA, Baccar R, Caminal G, Sarrà M (2018) Can white-rot fungi be a real wastewater treatment alternative for organic micropollutants removal? a review. Water Res 138:137–151. https://doi.org/10.1016/j.watres.2018.02.056
46. Kapahi M, Sachdeva S (2019) Bioremediation Options for Heavy Metal Pollution. J Health Pollut 24
47. Rahman Z, Thomas L (2021) Chemical-Assisted Microbially Mediated Chromium (Cr) (VI) Reduction Under the Influence of Various Electron Donors, Redox Mediators, and Other Additives: An Outlook on Enhanced Cr(VI) Removal. Front Microbiol 11. https://doi.org/10.3389/fmicb.2020.619766
48. Alboghobeish H, Tahmourespour A, Doudi M (2014) The study of Nickel Resistant Bacteria (NiRB) isolated from wastewaters polluted with different industrial sources. J Environ Health Sci Eng 12:44. https://doi.org/10.1186/2052-336X-12-44
49. Taran M, Sisakhtnezhad S, Azin T (2015) Biological removal of Nickel (II) by sp. KL1 in different conditions: optimization by Taguchi statistical approach. Polish Journal of Chemical Technology 17:29–32. https://doi.org/10.1515/pjct-2015-0046
50. Congeevaram S, Dhanarani S, Park J, Dexilin M, Thamaraiselvi K (2007) Biosorption of chromium and Nickel by heavy metal resistant fungal and bacterial isolates. J Hazard Mater 146:270–277. https://doi.org/10.1016/j.jhazmat.2006.12.017
51. El-Barbary T, Hafez M (2018) Bio Removal Potential of Nickel(II) by Different Bacterial Species

52. Mardiyono S, Masykuri M, Setyono P (2019) Bioremediation of nickel heavy metals in electroplating industrial liquid waste with Bacillus subtilis. AIP Conf Proc 2202:020084. https://doi.org/10.1063/1.5141697
53. Sharma R, Jasrotia T, Umar A, Sharma M, Sharma S, Kumar R, Alkhanjaf AAM, Vats R, Beniwal V, Kumar R, Singh J (2022) Effective removal of Pb(II) and Ni(II) ions by Bacillus cereus and Bacillus pumilus: an experimental and mechanistic approach. Environ Res 212:113337. https://doi.org/10.1016/j.envres.2022.113337
54. Aslam F, Yasmin A, Sohail S (2020) Bioaccumulation of lead, chromium, and Nickel by bacteria from three different genera isolated from industrial effluent. Int Microbiol 23:253–261. https://doi.org/10.1007/s10123-019-00098-w
55. Zhou H, Zhao X, Kumar K, Kunetz T, Zhang Y, Gross M, Wen Z (2021) Removing high concentration of Nickel (II) ions from synthetic wastewater by an indigenous microalgae consortium with a Revolving Algal Biofilm (RAB) system. Algal Res 59:102464. https://doi.org/10.1016/j.algal.2021.102464
56. Goswami RK, Agrawal K, Shah MP, Verma P (2022) Bioremediation of heavy metals from wastewater: a current perspective on microalgae-based future. Lett Appl Microbiol 75:701–717. https://doi.org/10.1111/lam.13564
57. Shamshad I, Khan S, Waqas M, Asma M, Nawab J, Gul N, Raiz A, Li G (2016) Heavy metal uptake capacity of fresh water algae (Oedogonium westti) from aqueous solution: A mesocosm research. Int J Phytoremediation 18:393–398. https://doi.org/10.1080/15226514.2015.1109594
58. Abdelrazek M, Abozeid A, Eltholth M, Abouelenien F, El-Midany S, Moustafa N, Mohamed R (2019) Bioremediation of a pesticide and selected heavy metals in wastewater from various sources using a consortium of microalgae and cyanobacteria. Slov Vet Res 56:61–74. https://doi.org/10.26873/SVR-744-2019
59. Açikel U, Alp T (2009) A study on the inhibition kinetics of bioaccumulation of Cu(II) and Ni(II) ions using Rhizopus delemar. J Hazard Mater 168:1449–1458
60. Joshi PK, Swarup A, Maheshwari S, Kumar R, Singh N (2011) Bioremediation of heavy metals in liquid media through fungi isolated from contaminated sources. Indian J Microbiol 51:482–487. https://doi.org/10.1007/s12088-011-0110-9
61. Long D, Tang X, Cai K, Chen G, Chen L, Duan D, Zhu J, Chen Y (2013) Cr(VI) reduction by a potent novel alkaliphilic halotolerant strain Pseudochrobactrum saccharolyticum LY10. J Hazard Mater 256–257:24–32. https://doi.org/10.1016/j.jhazmat.2013.04.020
62. Fan J, Onal Okyay T, Frigi Rodrigues D (2014) The synergism of temperature, pH and growth phases on heavy metal biosorption by two environmental isolates. J Hazard Mater 279:236–243. https://doi.org/10.1016/j.jhazmat.2014.07.016
63. Das S, Mishra J, Das SK, Pandey S, Rao DS, Chakraborty A, Sudarshan M, Das N, Thatoi H (2014) Investigation on mechanism of Cr(VI) reduction and removal by Bacillus amyloliquefaciens, a novel chromate tolerant bacterium isolated from chromite mine soil. Chemosphere 96:112–121. https://doi.org/10.1016/j.chemosphere.2013.08.080
64. Sandana Mala JG, Sujatha D, Rose C (2015) Inducible chromate reductase exhibiting extracellular activity in Bacillus methylotrophicus for chromium bioremediation. Microbiol Res 170:235–241. https://doi.org/10.1016/j.micres.2014.06.001
65. Ghosh A, Dastidar MG, Sreekrishnan TR (2016) Recent advances in bioremediation of heavy metals and metal complex dyes: review. J Environ Eng 142:C4015003. https://doi.org/10.1061/(ASCE)EE.1943-7870.0000965
66. Rath BP, Das S, Mohapatra PK Das, Thatoi H (2014) Optimization of extracellular chromate reductase production by Bacillus amyloliquefaciens (CSB 9) isolated from chromite mine environment. Biocatal Agric Biotechnol 3:35–41. https://doi.org/10.1016/j.bcab.2014.01.004
67. Bestawy EEl, Helmy S, Hussien H, Fahmy M, Amer R (2013) Bioremediation of heavy metal-contaminated effluent using optimized activated sludge bacteria. Appl Water Sci 3:181–192. https://doi.org/10.1007/s13201-012-0071-0
68. Das C, Naseera K, Ram A, Meena RM, Ramaiah N (2017) Bioremediation of tannery wastewater by a salt-tolerant strain of Chlorella vulgaris. J Appl Phycol 29:235–243. https://doi.org/10.1007/s10811-016-0910-8

69. Hackbarth F V, Maass D, de Souza AAU, Vilar VJP, de Souza SMAGU (2016) Removal of hexavalent chromium from electroplating wastewaters using marine macroalga Pelvetia canaliculata as natural electron donor. Chemical Engineering Journal 290:477–489. https://doi.org/10.1016/j.cej.2016.01.070
70. Sen S, Dutta S, Guhathakurata S, Chakrabarty J, Nandi S, Dutta A (2017) Removal of Cr(VI) using a cyanobacterial consortium and assessment of biofuel production. Int Biodeterior Biodegradation 119:211–224. https://doi.org/10.1016/j.ibiod.2016.10.050
71. Sen G, Sen S, Thakurta SG, Chakrabarty J, Dutta S (2018) Bioremediation of Cr(VI) using live cyanobacteria: experimentation and kinetic modeling. J Environ Eng 144:4018089. https://doi.org/10.1061/(ASCE)EE.1943-7870.0001425
72. Majhi P, Nayak S, Samantaray SM (2021) Microalgal Bioremediation of Toxic Hexavalent Chromium: A Review. In: Environmental and Agricultural Microbiology. John Wiley & Sons, Ltd, pp 25–37
73. García-Hernández MA, Villarreal-Chiu JF, Garza-González MT (2017) Metallophilic fungi research: an alternative for its use in the bioremediation of hexavalent chromium. Int J Environ Sci Technol 14:2023–2038. https://doi.org/10.1007/s13762-017-1348-5
74. Damodaran D, Vidya Shetty K, Raj Mohan B (2013) Effect of chelaters on bioaccumulation of Cd (II), Cu (II), Cr (VI), Pb (II) and Zn (II) in Galerina vittiformis from soil. Int Biodeterior Biodegradation 85:182–188. https://doi.org/10.1016/j.ibiod.2013.05.031
75. Anwer M, Bushra A (2015) Aspergillus niger-a novel heavy metal bio-absorbent and pesticide tolerant fungus. Res J Chem Environ 19
76. Sharma S, Malaviya P (2016) Bioremediation of tannery wastewater by chromium resistant novel fungal consortium. Ecol Eng 91:419–425. https://doi.org/10.1016/j.ecoleng.2016.03.005
77. M. Jayanthi, D. Kanchana, P. Saranraj, D. Sujitha (2014) Bioadsorption of Chromium by Penicillium chrysogenum and Aspergillus niger Isolated from Tannery Effluent. International Journal of Microbiological Research
78. Mohapatra RK, Parhi PK, Pandey S, Bindhani BK, Thatoi H, Panda CR (2019) Active and passive biosorption of Pb(II)using live and dead biomass of marine bacterium Bacillus xiamenensis PbRPSD202: kinetics and isotherm studies. J Environ Manage 247:121–134. https://doi.org/10.1016/j.jenvman.2019.06.073
79. Ma Y, Wang P, Wang C, Zhang S, Cheng S (2015) Isolation and characterization of Pb-resistant strains and the removal of Pb(II). Fresenius Environ Bull 24:1150–1157
80. Soeprobowati T, Hariyati R (2012) The Potential Used of Microalgae for Heavy Metals Remediation
81. El-Sayed MT (2013) Removal of lead(II) by Saccharomyces cerevisiae AUMC 3875. Ann Microbiol 63:1459–1470. https://doi.org/10.1007/s13213-013-0609-x
82. Zeng G, Li N, Huang D, Lai C, Zhao M, Huang C, Wei Z, Xu P, Zhang C, Cheng M (2015) The stability of Pb species during the Pb removal process by growing cells of Phanerochaete chrysosporium. Appl Microbiol Biotechnol 99:3685–3693. https://doi.org/10.1007/s00253-014-6275-5
83. Ye B, Luo Y, He J, Sun L, Long B, Liu Q, Yuan X, Dai P, Shi J (2018) Investigation of lead bioimmobilization and transformation by Penicillium oxalicum SL2. Bioresour Technol 264:206–210. https://doi.org/10.1016/j.biortech.2018.05.066

Adsorptive Removal of Heavy Metals from Wastewater Using Low-Cost Adsorbents Derived from Agro-based Materials

Jonas Bayuo, Mwemezi J. Rwiza, Kelvin Mark Mtei, and Joon Weon Choi

Abstract Recently, anthropogenic activities have evolved into sources of pollution, particularly when it comes to the discharge of harmful heavy metals into the natural environment. As a result, the concentration of diverse heavy metal ions in surface and ground waters significantly increases, compromising aquatic life. Given that toxic heavy metals have undesirable consequences on the health of all living things, their presence in the aquatic environment is a major worry. However, the drawbacks of conventional wastewater treatment technologies, such as their high consumption of energy, production of hazardous secondary sludge, and high operating costs, made them uneconomical and non-sustainable for developing nations. This book chapter presents and discusses the most recent developments and advances in the adsorptive removal of various heavy metals from aquatic systems through the application of low-cost adsorbents derived from agricultural waste materials. The influence of independent adsorption parameters as well as the mechanism of heavy metals removal from aqueous media have been explained using adsorption isotherm and kinetic models. This book chapter has demonstrated that the adsorptive removal of heavy metals using low-cost adsorbents derived from agricultural waste materials has several advantages. Almost all the studies on the adsorptive decontamination of various heavy metals from aqueous solutions revealed that adsorbents synthesized from agro-based materials are promising, eco-friendly, and cost-effective. However, several gaps exist, which need to be addressed to increase the application of the adsorption technology in treating industrial wastewater at a large scale. Hence, at the

J. Bayuo (✉) · M. J. Rwiza · K. M. Mtei
School of Materials, Energy, Water, and Environmental Sciences (MEWES), The Nelson Mandela African Institution of Science and Technology (NM-AIST), P.O. Box 447, Arusha, Tanzania
e-mail: bayuoj@nm-aist.ac.tz; jbayuo@cktutas.edu.gh

J. Bayuo
Department of Science Education, School of Science, Mathematics, and Technology Education (SoSMTE), C. K. Tedam University of Technology and Applied Sciences (CKT-UTAS), Postal Box 24, Navrongo, Upper East Region, Ghana

J. Bayuo · J. W. Choi
Institute of Green-Bio Science and Technology, Seoul National University, Pyeongchang-daero 1447, Gangwon-do, Korea

N. Kumar (ed.), *Heavy Metal Remediation*, Earth and Environmental Sciences Library,
https://doi.org/10.1007/978-3-031-53688-5_11

end of this book chapter, some future perspectives providing knowledge gaps that require consideration and further research have been enumerated.

Keywords Adsorbent · Adsorption · Agricultural waste · Heavy metal · Wastewater

1 Introduction

The presence of hazardous heavy metals in aquatic systems is caused by human-induced processes including urbanization, industrialization, technological development, and farming. Industrial, municipal, and domestic wastewater that emanates from a variety of anthropogenic activities contains heavy metals such as cadmium (Cd), arsenic (As), zinc (Zn), lead (Pb), copper (Cu), manganese (Mn), mercury (Hg), and chromium (Cr), which are toxic to animals, plants, and mankind [34, 45, 57, 84, 98]. Heavy metal contamination in aquatic environments is mainly caused by industries engaging in mining, metal plating, smelting, and alloying as well as manufacturing of batteries, plastics, textile, and agricultural fertilizers, pesticides, insecticides, herbicides, and fungicidal sprays [42, 58].

Heavy metals are not subjected to chemical or biological decomposition to become less harmful, in contrast to organic contaminants. These lethal heavy metals, which cannot decay, remain in the natural environment and pose major health risks to living organisms. Within the food chain, these poisonous heavy metal species continuously accrue in the cells, tissues, organs, and systems of all life forms through ingestion [51]. The accumulation of excessive amounts of heavy metal ions in biological cells and tissues is capable of instigating cancer and damage to the reproductive, renal, and nervous systems, which could be fatal [16]. Heavy metals need to be decontaminated to a tolerable range when found in aqueous systems because of their negative effects on the environment as stated by the Agency for Toxic Substances and Disease Registry [161]. Even at considerably smaller concentrations, prolonged exposure to toxic metal species can cause both short and long-term ailments like cancer, cardiac issues, and even mortality. It is well known that high concentrations of toxic metal ions in aquatic systems pose deadly threats to all living organisms [132]. This is because they can accumulate via food webs after being ingested by living things and are extremely deadly [130, 153].

Following their toxic effects, heavy metals have been removed from wastewater using an array of purification techniques, such as ion exchange, nano-filtration, chemical precipitation, and coagulation as well as membrane processes and adsorption [145]. Among all the purification technologies, the adsorption technique is more efficient because the majority of other remediation techniques have numerous serious drawbacks that prevent their widespread usage, including high operation and maintenance costs, the necessity for skilled labor, the creation of secondary pollutants and sludges, high energy and chemical consumption [115, 142, 146]. Hence, it is of the utmost need to minimize the high concentration of heavy metals in aqueous media using natural and low-cost sorbent materials, which have been proven to have multiple functional groups that are capable of forming complexes with heavy metal ions [14]. The effectiveness of many potential sources of low-cost agro-based adsorbents in detoxifying heavy metal ions from aquatic systems is therefore currently being investigated [6, 57, 58, 75, 155, 163].

In this book chapter, we present the most recent developments and advances in the adsorptive removal of various heavy metals from aquatic environments by utilizing low-cost adsorbents derived from agro-based materials. In addition, the impacts of independent influencing adsorption parameters and the mechanism of heavy metals removal from water systems have been explained using various adsorption equilibrium isotherm and kinetic models.

2 Heavy Metal Pollution in the Environment

Environmental protection is becoming more of a public concern in recent years, both locally as well as worldwide. The release of heavy metals into the natural environment from diverse anthropogenic activities has unexpectedly raised concerns about aquatic pollution globally [47, 94, 125, 144, 159, 172, 173]. These heavy metals are persistent in the ecosystem because they cannot decompose. Many environmental factors, including various human activities and natural occurrences, play a part in the distribution of noxious heavy metal ions in the natural environment [139].

The ecosystem, which includes the water, land, and living organisms, is badly impacted by heavy metals, and their easy solubility in aqueous media affects the physiological functions in humans and results in many kinds of health issues [53]. Due to their non-degradable nature, heavy metal ions build up in the cells and tissues of all organisms in significant proportions causing serious threats to their well-being. Some of its harmful effects include cancer, stunted growth, nervous system, and organ impairment, and eventually death [129].

The World Health Organization (WHO) has established regulations for the maximum allowable limits of heavy metal ions intake by humans due to their hazardous nature [113]. Additionally, as shown in Table 1, the United States Environmental Protection Agency (USEPA) outlined the considerable acceptable contamination levels of toxic heavy metals released into the natural environment [25, 62, 86]. According to Abidli et al. [5], the release of heavy metals into the ecosystem happens through a variety of processes and pathways, including natural phenomena (1) and Human-made activities (2). This has negative impacts on surface and groundwater (3), vegetation (4), and humans (5) through the food chain, air, and skin contact (6) as shown in Fig. 1.

Humans and other organisms have health issues because of toxic heavy metal pollution in sediments, soils, air, surface, and groundwater [81, 141, 149, 170]. After entering the food chain, there may be a strong propensity for heavy metal species to bioaccumulate in the tissues of living organisms and, when ingested at levels beyond the safe limit, cause significant health problems [88, 122, 150]. Furthermore, it is

Table 1 Maximum allowable contamination levels for various heavy metals

Heavy metal	MAL (mg/L)	Sources of heavy metal pollution
As	0.05	Mining, steel factories, pesticides, pigments, smelting, and paints industries
Hg	0.001	Mining, ceramics, battery, pigments, and paints industries
Cr	0.05	Metallurgical, steel, chemical, cement, printing, paints, pesticide, pigments, and graphics industries
Cd	0.01	Metal smelting, fertilizer, pesticide, pigments, printing, paints, graphics, and battery industries
Ni	0.2	Metallurgical, electroplating, steel, battery, mining, and paper mills industries
Zn	0.8	Galvanizing, brass, printing, ceramic, electrical, steel, pesticide, battery, pigments, paints, graphics, and metal plating industries
Pb	0.06	Pesticide, battery, electrical, printing, mining, ceramics, steel, graphics, paints, pigments, plumbing, and lead smelting industries
Cu	0.25	Mining, metal smelting, ceramics, machinery, pesticide, fly ash, steel, fertilizer, electrical, and iron industries

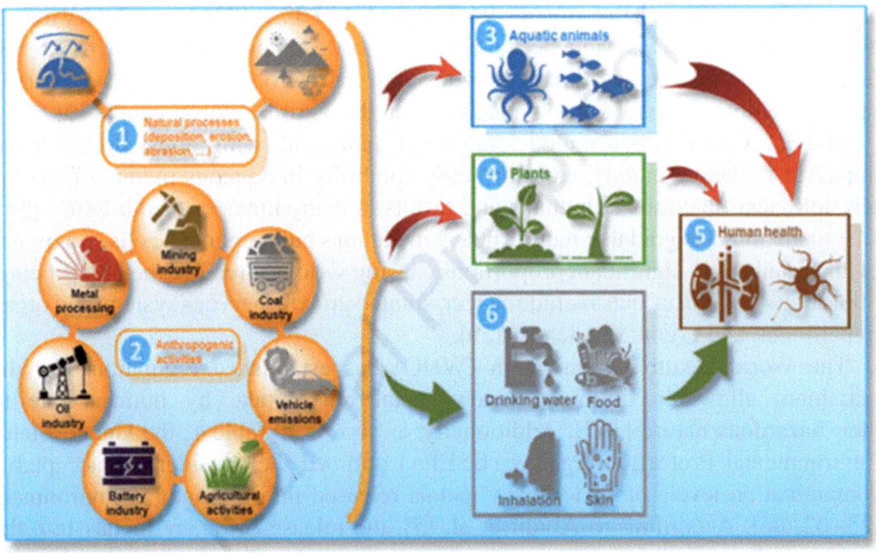

Fig. 1 A schematic diagram demonstrating the various pathways of heavy metals and metalloids in the ecosystem

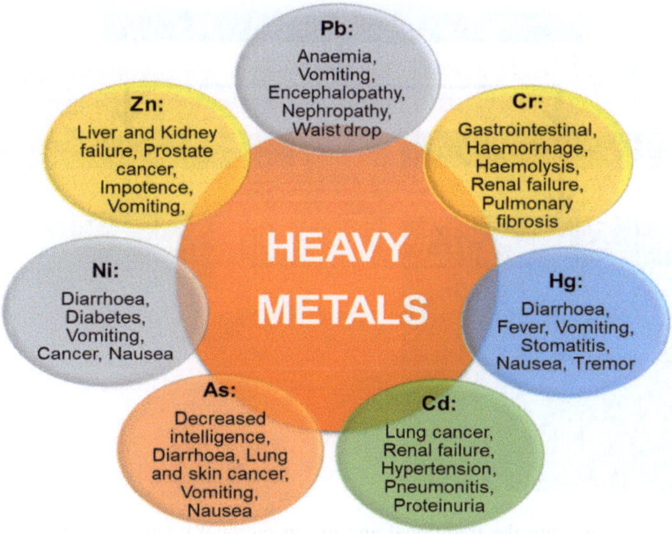

Fig. 2 The effects of different heavy metal contamination on the well-being of humans

established that extremely toxic heavy metal ions may destroy red blood cells and induce hypertension, anemia, and neurological dysfunction in people despite small amounts [140, 165].

Many heavy metals are recognized to be harmful to humans and other living things when taken in amounts greater than those needed for normal growth. Figure 2 illustrates several of these toxic heavy metals, including chromium (Cr), arsenic (As), lead (Pb), mercury (Hg), nickel (Ni), cadmium (Cd), and zinc (Zn), along with the negative effects these metal ions have on living things when they bioaccumulate in concentrations that are over their permitted levels [102]. Recently, environmentalists and scientists have become much more concerned about heavy metals due to their noxious nature when found in aquatic environments. Therefore, efficient wastewater treatment technology is required to decontaminate heavy metal ions from our natural environment [62].

3 Technologies for Heavy Metals Removal

Figure 3 illustrates some of the main traditional and unconventional treatment techniques employed to eliminate heavy metals from water and wastewater such as chemical precipitation, reverse osmosis, electrodialysis, ion exchange, adsorption, and biosorption [34, 102]. However, the treatment method utilized for the decontamination of heavy metals depends on a number of variables, including environmental impact, cost of operation, and the treatment technique effectiveness. Some of the

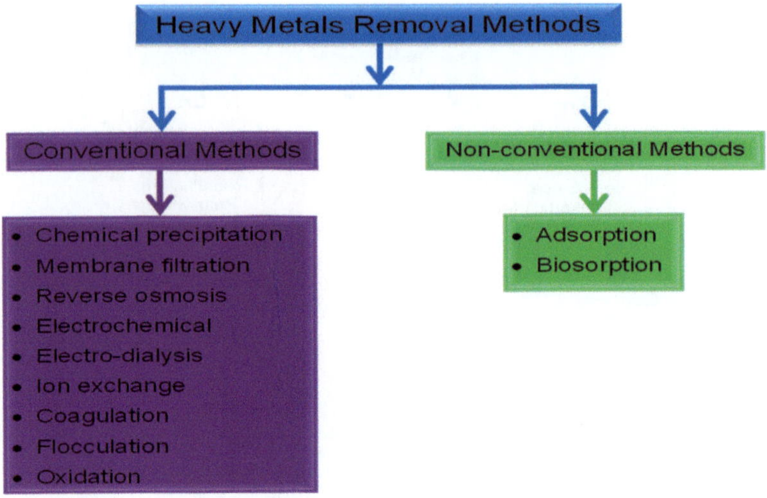

Fig. 3 A diagram showing the traditional and unconventional treatment methods for removing heavy metals from water systems

traditional and unconventional treatment methods for removing heavy metal ions from aqueous systems are presented in the subsequent sections of this book chapter.

3.1 Electro-Dialysis Method

By combining ion-exchange membranes with electrical potential differences, the electro-dialysis treatment method eliminates metal ions from aqueous media [157]. There are arrays of anion and cation exchange membranes found between the anode and cathode. In contrast to cations, which move toward the cathode, the anions migrate to the anode. The ions pass through the membrane before being taken up by the electrodes with charges that are opposite [49]. The disadvantages of this treatment method in the sequestration of heavy metal ions in water systems are its high cost and poor performance [171].

3.2 Ion Exchange

A solid material that exchanges either anions or cations from an aquatic system is known as an ion exchanger. Ion exchange is the process of replacing an undesirable ion with a desirable ion present in a liquid phase. The ion exchange treatment process of heavy metals goes on till the resin's capability for exchanging ions is exhausted. It is necessary to recycle and renew the exhausted resin using additional chemicals to restore the resin to its original form and facilitate reusability in the subsequent cycle [166].

Polluted water typically traverses an ion-exchange resin, which consists of a substance that is insoluble in water but has the ability to exchange some of its ions in aqueous systems for other ions that have the same charge [85]. The electrical charges of the hazardous metal ions as well as the ions inside the resin must be identical for the exchange of ions to take place [99]. According to (Kumar and Jain 2013), Ion-exchange characteristics can either be naturally present in various kinds of organic materials or chemically introduced to them. Ion exchangers are made from organic materials including fibers, wood, peat, and coal by using strong sulphuric or nitric acid as an oxidant.

Ion exchange may efficiently and selectively eliminate heavy metal ions to parts per million levels, despite the high cost of the resins [118]. The ion-exchange technology for wastewater treatment is also friendly to the ecosystem. However, ion exchange has many limitations when it comes to eliminating heavy metals found in aqueous systems, including insignificant surface area, poor selectivity and wettability, low uptake rates, and renewal problems. The existence of particles in suspension may reduce the efficacy of the ion-exchange treatment system. In reality, prefiltration is necessary to get rid of suspended particles that can mechanically block the resin bed because ion exchange is unable to achieve the treatment target. Additionally, it causes water to get fouled by suspended particles and is ineffective in getting rid of small amounts of pollutants [171].

3.3 Chemical Precipitation

The chemical precipitation technique comprises transforming dissolved components into insoluble substances before filtering or sedimentation is applied to remove them [28, 66]. The chemical precipitation approach is the most commonly and widely utilized technique for the decontamination of dissolved heavy metal ions from water systems. Using precipitating chemicals like sodium hydroxide, calcium hydroxide, zinc chloride, or ferrous sulphate, heavy metal ions are transformed into suspended particles. The particles that have been formed in the process are removed from the aqueous solution through sedimentation and/or filtration [93].

The chemical precipitation procedure is very simple and economical but this technology cannot be employed in wastewater treatment since it is not flexible. Chemical precipitation necessitates the use of numerous chemicals to abate the amount of heavy metals to a tolerable level for safe disposal. The primary drawback of chemical precipitation is that it creates toxic sludge that needs to be discarded in a landfill, which is the target of the European Union's waste management policy. Furthermore, a safe disposal method for the produced sludge is required. The fact that the concentration of heavy metal ions in the wastewater that has been treated cannot be lowered below the soluble state of the precipitate is another issue with the chemical precipitation treatment method. Additionally, it alters the water pH, requiring pH adjustment.

Adsorption is one of the best alternate treatment strategies that should be applied for the elimination of heavy metals from aquatic environments.

3.4 Reverse Osmosis

The reverse osmosis treatment technique employs semi-permeable membranes to eliminate heavy metals present in aqueous systems [2]. This treatment method involves applying higher pressure to a semi-permeable membrane to push molecules of water through it [97]. This pressure ought to be adequate to balance osmotic pressure. The total amount of contaminants present in the water system to be treated determines the required pressure [131]. While other contaminants and salts are forbidden from passing the membranes and are poured into the concentrated stream, the membranes allow the passage of water molecules [56]. Based on their sizes and charges, toxins are rejected by the reverse osmosis membrane. The water contaminants may be in the form of ions, colloids, particles, or microorganisms [162]. This method is incredibly effective, it generates no waste; and it requires no pH adjustment or regeneration. However, it has a high start-up and recurring operating cost, poses a disposal problem for effluent with heavy metal ions, and causes water loss [109].

3.5 Membrane Separation Techniques

Water and wastewater treatment techniques include electrodialysis, membrane processes driven by electricity, and membrane processes driven by temperature. Some other treatment techniques employed in water and wastewater treatment include reverse osmosis, ultrafiltration, nanofiltration, high pressure, and low-pressure filtration. Membrane filtration can be used to filter out contaminants from wastewater, including color, dissolved solids, and heavy metal ions. One distinctive quality of this treatment method is its capacity to generate solid-free waste products and secondary sludge [160]. Membrane separation techniques also have many benefits, such as simplicity, minimal chemical usage, and ease of operation and maintenance [35]. Despite its advantages, the membrane separation method has the drawback of insufficiently removing small molecular weight molecules and uses considerable amounts of energy. Another limitation of the membrane separation technique is fouling, which could lead to the blockage of both inorganic and organic substances in the pores of the membrane. Membrane fouling decreases the volume of produced purified water and decreases the life of the membrane, diminishing the economic effectiveness of membrane processes [91].

3.6 Coagulation and Flocculation Technologies

When colloids are destabilized through coagulation, the forces holding them apart are broken. Through the provision of positive charges, cationic coagulants decrease the negative charges of colloids. Due to particle collisions, enormous particles called flocs are created. To get rid of particles that filtration or sedimentation alone is incapable of getting rid of, coagulation is always employed in conjunction with flocculation [102]. Colloids are microscopic spheres that are less than one meter and due to their poor sedimentation, they result in coloring and increasing the turbidity of water. Among them are clays, bacteria, proteins, metal oxides, and some other organic compounds. The interactions among the colloidal particles and the molecules of water, along with the fact that they all have a negative charge, prevent the colloidal particles from aggregating and settling in water systems. Chemical coagulants, such as colloids and trivalent ions like Fe(III) and Al(III), which are found in aluminum and ferric salts, can be utilized to aggregate the particles [56].

Municipal wastewater treatment facilities use coagulation and flocculation to deal with both treated sewage effluent and industrial wastewater. However, the sector of potable water filtration has the most widespread coagulation and flocculation technologies [46]. However, it has been discovered that coagulation and flocculation techniques for water treatment are extremely expensive when chemicals and sludge disposal are taken into consideration. It is crucial to develop and utilize low-cost techniques for wastewater treatment [60].

3.7 Electrochemical Method

Electrochemical processes like electro-deposition, electro-oxidation, electro-flotation, and electrocoagulation have been employed to treat wastewater [133]. To clean domestic, industrial, and municipal sewages, a mixture of sewage and salt-water is usually electrolyzed. The electrochemical treatment method is frequently applied to clean industrial wastewater and effluent. Electrolytic techniques in water and wastewater purification are expensive with very high maintenance costs. Therefore, it is crucial to develop alternative useful water and wastewater purification approaches.

Heavy metal ions could also be removed through various treatment methods including solvent extraction, complexation, foam flotation, electrodeposition, and cementation. However, these treatment methods have limitations such as slow kinetics and limited selectivity. Adsorption technology has distinguished itself among other techniques for the efficient elimination of heavy metals that are resistant to conventional treatment technologies [102, 145].

3.8 Adsorption Technique

The most secure and suitable separation, purification, waste treatment, and chemical ions recovery technique is adsorption, especially at industrial levels. Due to its incredible efficiency in the sequestration of various contaminants from aqueous systems, the adsorption technology has largely replaced practically all other purification and separation techniques [1]. The disadvantages of the traditional treatment approaches for water and wastewater are their high cost, time, and energy consumption as well as the production of secondary wastes [29, 37]. Adsorption is the mass transfer of the heavy metal ions to be eliminated from the gaseous or liquid phases onto the sorbent surface where it accumulates due to chemical and physical interactions, as shown in Fig. 4. Adsorption can therefore be triggered by neither physical nor chemical interactions, or it can involve both processes. As seen in Fig. 4, chemical adsorption leads to the formation of a monolayer by the heavy metal ions, as opposed to physical adsorption, where the heavy metal ions form a multilayer on the adsorbent surface [148].

The primary obstacle to industrial applications of adsorption technology in water purification is the high cost of commercial adsorbents [38]. Commercial adsorbents ought not to be employed in the elimination of heavy metal ions from water and wastewater, due to their expensive nature and industrial applications. The use of natural materials as well as agricultural by-products, which are readily available can be synthesized into low-cost sorbents for the adsorptive sequestration of heavy metals in water systems. Most of the time, agricultural biomass or wastes are inexpensive, readily accessible, and of little economic worth [71]. The main advantages

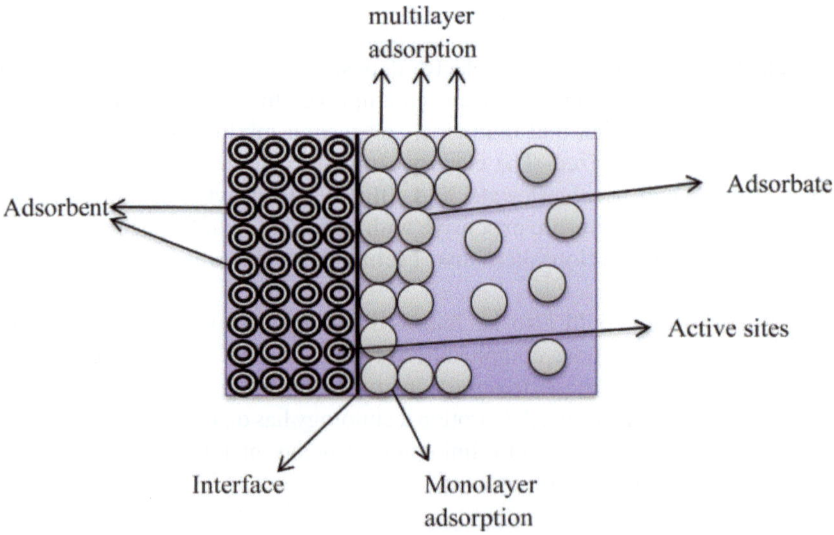

Fig. 4 A diagram showing the adsorptive removal of heavy metal ions on the surface of the adsorbent

of the adsorption technique over traditional water and wastewater treatment methods are cost-effectiveness, high efficacy, less chemical consumption, and adsorbent regeneration and heavy metal ions recovery abilities.

Adsorption using natural and agro-based materials as adsorbents is therefore strongly encouraged [101, 119, 154]. Agricultural waste materials, including rice husk, sugarcane bagasse, coconut husk, sawdust, neem bark, oil palm shell, papaya seeds, jack fruit as well as orange peels, and spent tea, have been found to be effective in the decontamination of heavy metal ions [126, 128]. The removal and uptake rates of heavy metal ions from aquatic environments by different sorbent materials can be affected by a number of independent affecting parameters as a function of two dependent parameters. Time of contact/shaking time, pH of the solution, adsorbent particle size, adsorbent dose, initial concentration, and reaction temperature are some of these independent affecting parameters.

4 Independent Adsorption Influencing Parameters

The adsorption capacities and removal efficiencies of potential adsorbents are influenced by a variety of independent parameters throughout the sorption process. Earlier studies have confirmed that parameters, including solution pH, agitation speed, initial heavy metal ion concentration, contact time, adsorbent particle size, adsorbent dosage, interfering ions, and temperature, have a substantial influence on the performance of any sorbent material [32, 33, 43]. These adsorption process parameters influence the concentration of various heavy metals to be reduced as well as the adsorbent capacity and uptake rate. Numerous adsorption studies have been carried out to explore the interactive effect of the various independent parameters on the percentage removal and uptake capacity of heavy metal ions from aqueous systems.

4.1 Contact/Shaking Time

The period of time that the adsorbent is in contact with the adsorbate molecules in aqueous systems affects the sorption process. The likelihood that equilibrium will be reached and the maximum uptake rate of the adsorbate molecules achieved, takes place with increasing shaking time [21]. Insufficient time for contact between the sorbent material and the adsorbate molecules in the aqueous systems may result in an incomplete adsorption process [138]. Studies showed that at the beginning of the sorption process, sorbent materials are capable of decontaminating more adsorbate molecules from the aqueous media. However, at longer contact times, the uptake capacity of the adsorbate molecules drops due to the adsorbent binding sites becoming saturated [126]. The ideal time of contact shows resilience on the amount of adsorbent applied, the amount of adsorbate molecules present as well as the solution temperature, and the adsorbent surface functional groups.

4.2 Agitation/Shaking Speed

The elimination of heavy metal ions from aqueous media has been discovered to be at maximum at a particular agitation speed that needs to be established through preliminary experimentations. The agitation/shaking speed can either accelerate or retard the heavy metal's adsorption process. On the contrary, a higher agitation speed may not always be associated with a higher adsorption rate [79]. However, Dhoble et al. [55] found that the removal efficiency of adsorbate molecules improves with increasing agitation speed. This might be due to greater collisions occurring between the adsorbent surface and the adsorbate molecules, which opens up extra active pores on the adsorbent surface for binding heavy metal ions present in aqueous media [40].

4.3 pH of the Solution

Due to the release of H^+ and OH^- ions into the aqueous media, pH affects the adsorption process. At a low pH level, more H^+ ions are found in the aqueous media, which could interact with both the adsorbate molecules and adsorbent thereby affecting the adsorption rate. Likewise, an alkaline pH leads to the production of OH^- in aqueous media, which may also influence the removal and uptake rate of the adsorbate molecules by the sorbent material. Therefore, the pH has a substantial influence on the sequestration of heavy metals from aquatic environments. It has an impact on the speciation of the adsorbate molecules, degree of ionization, and adsorbent surface charge [92]. Consequently, almost all of the sorption processes of adsorbate molecules enhance with an upsurge in pH till a point in which an increase results in a decline in the uptake rate of the adsorbate molecules. This means that the solution pH of any aqueous medium, being examined for its capability to remove heavy metals has an undeniable effect on the elimination rate of the metal ions. Also, the adsorbent surface functional groups responsible for forming complexes with the heavy metal species show dependence on the solution pH [87].

It is crucial to understand how pH affects the decontamination of harmful heavy metals from aquatic environments because, at low pH levels, heavy metal species interact with hydrogen ions to form hydrogen-heavy metal complexes. A surplus of hydroxide ions is also present at elevated pH levels, which hinders the diffusion of heavy metal ions [29, 37]. Hence, an aqueous solution pH exhibits a greater influence on the removal of heavy metal ions from wastewater because it controls the surface charges of sorbent material.

4.4 Adsorbent Particle Size

One of the crucial parameters that significantly affect a sorbent material's ability to absorb adsorbate molecules is its particle size. The uptake and removal rates of an adsorbent change with particle size. The uptake capacity and removal rate of adsorbate molecules declines as particle size upsurges and improves as particle size decreases. An adsorbent has a greater specific surface area as its particle size decreases, and a higher specific surface area results in a higher uptake rate of the adsorbate molecules [100]. To develop a full-scale adsorption system for commercial and industrial applications, it is important to determine how different particle sizes of the adsorbent affect the adsorption capability of various adsorbate species.

Apart from the adsorption of the adsorbate molecules at the sorbent surface, intra-particle diffusion of these molecules through the sorbent pores and accumulates on its surface is a possibility. Therefore, the larger the particle size of sorbent material, the higher the resistance to mass transfer [10]. The internal adsorbent surface area might not be completely utilized, causing low uptake of the adsorbate molecules, due to a number of inhibiting parameters, including mass transfer resistivity, agitation time, and obstruction of the diffusional pathway. Notwithstanding, the adsorptive elimination efficiency of adsorbate molecules depends on the accessible specific surface area [59].

4.5 Adsorbent Dosage

The ability of any adsorbent to decontaminate adsorbate molecules present in aqueous media at a particular concentration is determined by the adsorbent load. As the concentration of the adsorbent is increased, extra active binding sites are created, which results in a greater removal of the adsorbate molecules from the aqueous system. This might be because higher dosages allow for better surface and pore volume availability as well as the accessibility of a higher surface area [124]. Reducing the quantity of adsorbent dosage needed to remove the adsorbate molecules can be accomplished in part by boosting the adsorbent surface area. By using adsorbents with incredibly small particle sizes, higher uptake and removal rates of the adsorbate molecules could be attained. Given that larger surface areas are more efficient at the adsorptive removal of adsorbate molecules, the same adsorbent requires lesser doses than its counterparts with smaller surface areas [24]. The saturation of the accessible binding sites caused by high adsorbent doses and adsorbate molecules interaction, however, causes the uptake capacity to typically decrease [99].

4.6 Heavy Metal Ion Concentration

Due to several variables, including the accessibility of particular adsorbent surface functional groups and their potential to bind heavy metal ions from aquatic environments, the initial amount of heavy metal ions might alter the decontamination and uptake capacity. To overcome the resistance of mass transfer between the sorbent and adsorbate in the aqueous media, the initial heavy metal ion concentration can be a powerful driving force [78]. The adsorbent capacity rapidly decreases by raising the initial heavy metal ion concentration in the aqueous system. This might be because there are only a limited number of adsorption sites that can be accessed for any given amount of the adsorbent, and at larger concentrations, these binding sites become fully saturated [175]. The lower the heavy metal ion concentration, the greater the uptake capacity since there are fewer adsorbate molecules for any given quantity of the sorbent material to remove [9]. Besides, some sorbent materials perform very well at high initial heavy metal ion concentrations, leading to an increased uptake capacity [79].

Furthermore, the initial concentration is one means by which the transport of heavy metal ions to the sorbent surface may be facilitated [137, 151]. Numerous experiments have been carried out to investigate the consequence of initial heavy metal ion amount on their abatement from aqueous systems using agro-based sorbents. A general pattern has been found regarding the optimum uptake capacity and heavy metal ion initial concentration. The patterns showed that increasing heavy metal ion concentration leads to rising uptake capability of the metal ions [82, 95, 164].

4.7 Temperature

Adsorption has been found to show dependence on solution temperature and can change the process to be either endothermic or exothermic [74]. Numerous adsorption investigations on the impact of solution temperature on the decontamination efficiency of heavy metals have been carried out. The temperature has an effect on the diffusion rate of heavy metal species as well as their solubility in aqueous systems [104]. It has been established that temperature is a crucial component in the reduction of heavy metal ions in water systems since it can influence the variety of functional groups on the surface of the sorbent material [59, 148].

Upsurging the solution temperature enhances the removal efficiency of heavy metal ions as well as boosts the adsorption kinetic energy [92]. Nevertheless, the rise in temperature has the potential to physically distort the adsorbent properties [120]. As a consequence, room temperature is often employed in carrying out most adsorption tests.

4.8 *Interfering Ions*

Aqueous solutions may contain anions like chlorides, carbonates, phosphates, nitrates, and bicarbonates in addition to heavy metal ions. Geographical disparities can affect the concentration of various contaminants coexisting in the aqueous media and how well these contaminants are being adsorbed by the sorbent material [64]. For the different adsorbents, the sequence of inhibition of anions in the decontamination of heavy metal species is indicated as $HCO_3^- > SO_4^{2-} > Cl^-$. In most adsorption tests, bicarbonate drastically reduced the uptake capacity of heavy metals while chloride, sulfate, and nitrate ions have little influence. This is explained by the bicarbonate's tendency to outcompete heavy metal ions for available active binding sites on the sorbent surface [80].

The ionic strength of the aqueous system is often inversely proportional to the removal rate of the heavy metal ions attained by the adsorbent. The capacity of the adsorbents to eliminate heavy metal species decreases as the ionic strength increases. This might be caused by an upsurge in the competitive cationic amount in the aqueous media, which influences the removal efficiency of the heavy metal species [69]. In addition to heavy metals, a variety of other ions are present in actual wastewater. For instance, Na^+, K^+, Mg^{2+}, and Ca^{2+} are frequent cations present in actual wastewater, while chemical substances like KCl, $CaCl_2$, $MgCl_2$, and $NaNO_3$ are typically added to water to spike it to examine the effect of ionic strength on the competitive adsorptive removal of heavy metals.

5 Agro-based Materials as Low-Cost Adsorbents in Heavy Metals Removal

Even though they are frequently employed in the purification of both wastewater and water, commercially available activated carbons are costly sorbent materials. To decontaminate heavy metals from contaminated water systems, there is an increasing focus on the development of simple, safe, affordable, and cost-effective sorbent materials that are more efficient than commercial activated carbon in the market [36, 96]. The main sources of natural and agro-based sorbent materials displayed in Fig. 5 are commonly modified and transformed into activated carbons, which are capable of the adsorptive reduction of heavy metal ions in water and wastewater [40, 174].

For example, Fig. 6 illustrates the use of activated carbons as a sorbent material obtained from agricultural wastes for the decontamination of toxic heavy metals from water and wastewater [70]. Considering the fact that adsorption capacity is directly proportional to the total amount of accessible surface area, activated carbon should have a high adsorption rate, be simple to renew or retrieve from the liquid phase, possess little pore diameter, and have high porosity, [4]. The adsorption characteristics of sorbent material are obtained from complex constituents such as simple

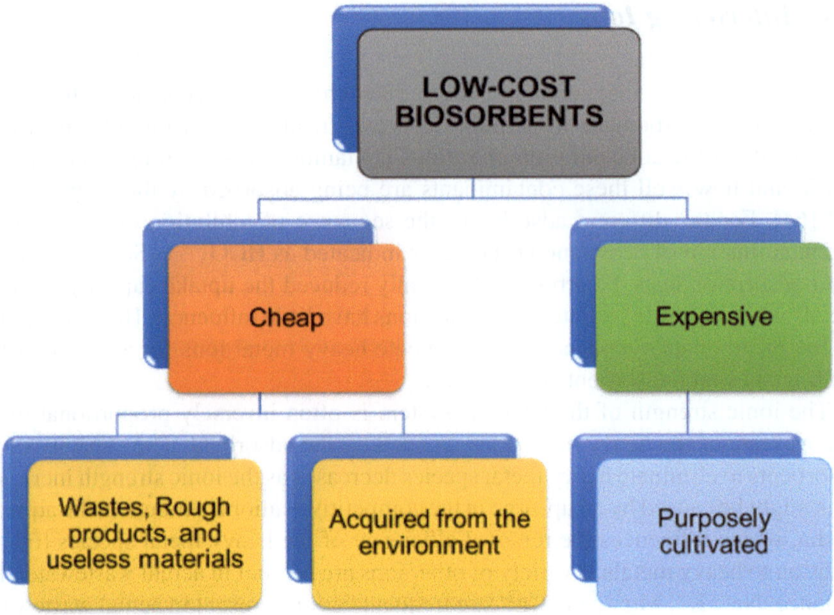

Fig. 5 Different sources of agro-based sorbent materials for adsorptive removal of heavy metal species from aquatic systems

sugars, lignin, hemicellulose, lipids, cellulose, proteins, and starch with diverse surface functional groups [14, 30]. Due to their capacity to form complexes and chelates with heavy metal species, these constituents are essential for the adsorptive elimination of heavy metal ions in water systems. Functional groups on sorbent material surfaces bind to dangerous heavy metal ions or provide electrons to replace hydrogen ions in aqueous media [18].

Notably among the surface functional groups found in agricultural waste materials are alcohols, phenolics, esters, carbonyl, sulphydryl, carboxyl, amido, acetamido, and amido [23]. These are the surface functional groups that have a preference for complexing with diverse heavy metal species. Several adsorption studies have established the presence of various surface functional groups on agro-based sorbent materials as well as the capability of these groups to bind to heavy metals through spectroscopic analysis [26, 29, 37, 121].

Using inexpensive adsorbents to eliminate toxic heavy metal ions is thought to be more economically feasible and promising because there are so many locally accessible and abundant natural and agricultural residues that could be utilized in the synthesis of these cheap sorbent materials [41, 54]. It is widely recognized as a potential and cost-effective method for treating wastewater in which agricultural-based products can be converted into less expensive adsorbents. The growing interest in the search for more simple, innovative, and less costly sorbent materials that are effective in the sequestration of various heavy metal ions in aqueous systems is suggested by

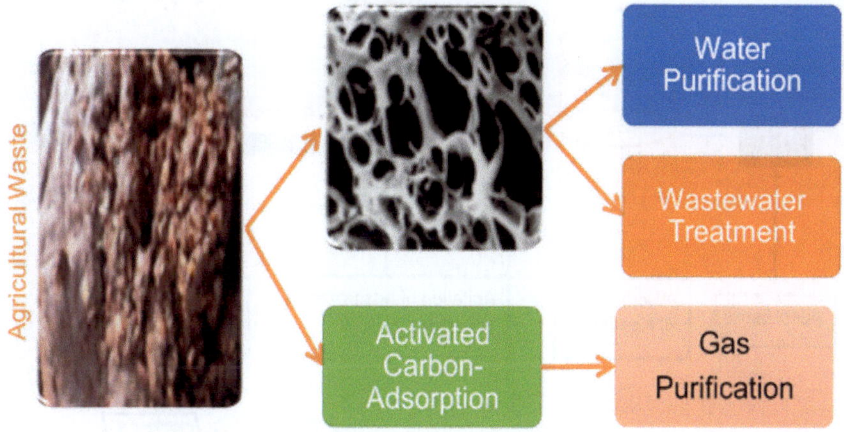

Fig. 6 A diagram illustrating the application of activated carbon derived from biomass feedstocks for the adsorptive removal of various environmental contaminants

the increasing number of research publications on the adsorption–desorption process depicted in Fig. 7 utilizing low-cost sorbent materials. The decontamination of heavy metal species from water, wastewater, or effluents has been reported in numerous studies using agro-based adsorbents. Table 2 provides a summary of the removal efficiency of several heavy metal ions from aqueous media through the application of low-cost adsorbents synthesized from agro-based and other natural materials.

6 Biosorption Isotherm and Kinetic Modeling

Adsorption isotherm describes how much adsorbate is accumulated per unit mass of the sorbent material with respect to the equilibrium adsorbate concentration in the bulk aqueous phase at a constant temperature [83]. Therefore, the adsorption isotherm elucidates how the adsorbate molecules are distributed in the aqueous and solid phases when the sorption system reaches equilibrium [61]. Thus, adsorption isotherms are crucial in determining the interactive behavior between the adsorbate molecules and the adsorbent material and help maximize the efficiency of sorbent material.

In addition to providing comprehensive insights on the uptake capacity of the investigated sorbent material, adsorption isotherm has a considerable role in predicting the adsorption mechanisms [108]. Also, the surface characteristics of the sorbent material, selectivity and affinity of adsorbate molecules for adsorption sites, and description of equilibrium sorption data are all frequently explained using isotherm models [22]. When the adsorption and desorption rates are the same, the dynamic equilibrium sorption is said to be established. To reach the sorption equilibrium, the quantity of adsorbate molecules at the sorbent interface must be in

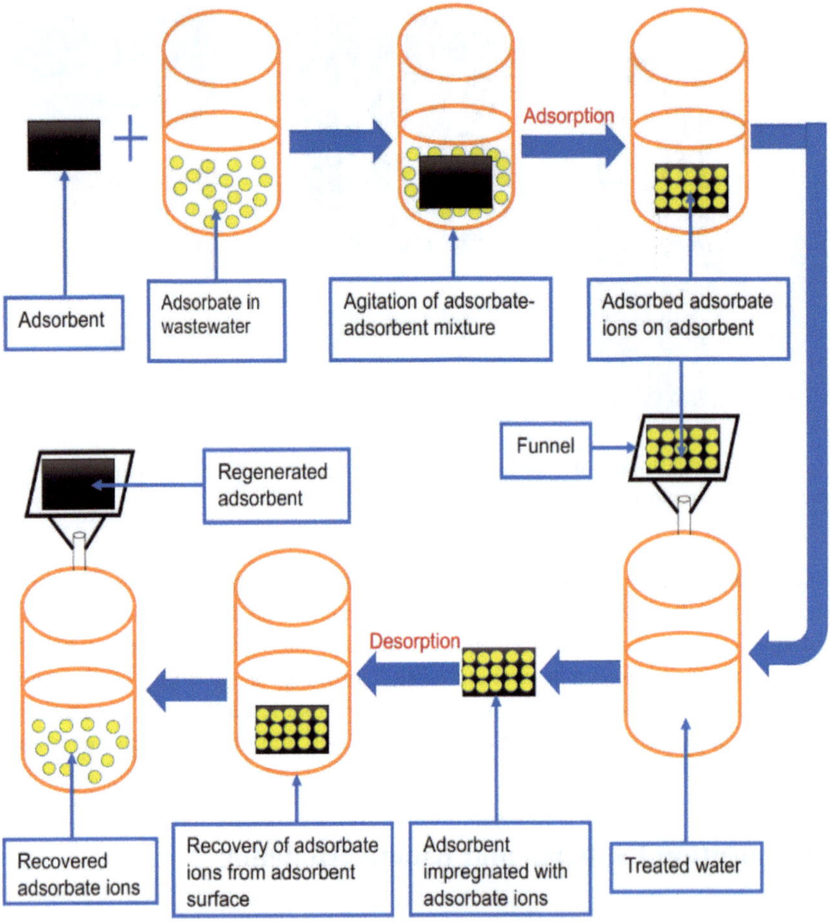

Fig. 7 A schematic diagram showing the adsorption–desorption process of heavy metal ions using low-cost adsorbents produced from agricultural and natural waste materials

dynamic equilibrium with the adsorbate concentration in the solute phase, and every active site on the sorbent material has the potential to interact with the molecules of the adsorbate. The most common equilibrium models utilized to comprehend the mechanism of heavy metals adsorptive removal include the Langmuir, Temkin, Sips, Dubinin-Radushkevich (D-R), Freundlich, and Redlich-Peterson isotherm models. The assumptions of the various adsorption isotherm models including their non-linearized equations are summarized in Table 3.

Table 2 Application of agro-based adsorbents for the adsorptive sequestration of heavy metal ions in water and wastewater

Adsorbent	Heavy metal	Removal efficiency (%)	References
Papaya seeds	Pb(II) and Cu(II)	99.96 and 97.55	[65]
Banana peel	Cd(II) and Pb(II)	93.2 and 83.78	[107]
Mustard waste	Pb(II), Zn(II), and Cd(II)	94.56, 96.15 and 76.48	[112]
Persian *Eucalyptus* leaves	As(III) and Hg(II)	> 94.00	[17]
Ficus benghalensis leaf	Co(II)	98.73	[73]
Rice husk ash	Pb(II)	75.00	[48]
Mung bean husk	As(V)	98.75	[106]
Neem bark	As(III)	89.96	[134]
Typha angustifolia and *Salix matsudana* branches	Cd(II) and Pb(II)	90.00	[152]
Lobeira fruit	Pb(II)	93.00	[20]
Spent tea leaves	Cr(III)	95.42	[114]
Bagasse biochar	Pb(II)	75.37	[123]
Corn and rice husks	Pb(II)	> 90.00	[135]
Egg shell	Cd(II)	98.76	[136]
Pomelo peels	Pb(II)	99.90	[167]
Leucaena leucocephala seed shell	Cr(VI)	95.62	[169]
Groundnut shell	Pb(II)	96.61	[39]
Cow bone	Cd(II) and Pb(II)	88.00 and 84.00	[3]
Cotton shells	Pb(II)	90.00	[13]
Tamarind fruit shell	Pb(II)	83.50	[27]
Turkish coffee	Cd(II)	96.00	[52]
Rice husk	Pb(II)	90.98	[63]
Saccharomyces cerevisiae	As(III) and As(V)	66.20 and 15.80	[68]
Sugarcane bagasse	Hg(II)	61.00	[77]
Magnolia leaf	Cr(VI)	98.80	[105]
Rice straw	As(V)	69.60	[110]
Fish scale and bean husk	Pb(II)	90.00	[117]
Teff husk	Cr(VI)	95.60	[7]
Olive Stones	Cd(II), Cu(II), Pb(II) and Cr(VI)	77.40, 80.50, and 94.50	[18]
Wood biomass	Pb(II)	100.00	[44]
Sesame hull	Cr(VI)	100.00	[67]
Bagasse	Pb(II), Cu(II), Zn(II) and Cd(II)	94.53, 92.74, 70.46 and 83.30	[72]
Banana peels	Cu(II) and Pb(II)	99.79 and 88.94	[8]

(continued)

Table 2 (continued)

Adsorbent	Heavy metal	Removal efficiency (%)	References
Shea fruit biomass	Cd(II)	76.86	[35, 36]
Spent tea leaves	As(V)	87.00	[50]
Sunflower waste carbon	Cd(II)	99.80	[76]
Coconut shell	Cr(VI)	99.00	[89]
Orange zest	Cu(II)	99.61	[147]
Onion skin	Pb(II) and Cd(II)	92.05 and 94.89	[168]
Peanut shells	Pb(II)	87. 27	[40]
Maize residues	Hg(II) and As(III)	85.68–98.13 and 93.14	[30, 31]

Table 3 Heavy metals adsorption isotherm models

Isotherm model	Isotherm model assumptions	Equation	Notation	References
Langmuir	Following the Langmuir model, the metal ions accumulation occurred at limited discrete localized active sites on the adsorbent surfaces, which is homogeneous in nature leading to the occurrence of monolayer adsorption	$q_e = \frac{q_{\max} K_L C_e}{(1+K_L C_e)}$	$\mathbf{K_L}$: Net adsorption enthalpy $\mathbf{q_e}$: Equilibrium uptake capacity $\mathbf{q_{max}}$: Maximum uptake capacity $\boldsymbol{C_e}$: Concentration at equilibrium	[158]
Freundlich	The Freundlich model presumes that the sequestration of heavy metals took place on the heterogeneous surface of the sorbent material leading to multi-layer adsorption. As the initial heavy metal ion concentration upsurges, more ions are adsorbed by the sorbent. The surface heterogeneity indicator is $\frac{1}{n}$ and the adsorbent surface becomes increasingly heterogeneous as $\frac{1}{n} \approx 0$. While favorable multilayer adsorption is achieved when n exceeds 1 ($n > 1$)	$q_e = K_F C_e^{\frac{1}{n}}$	$\mathbf{K_F}$: Sorption capacity $\boldsymbol{n}$: Sorption intensity	[35, 36]

(continued)

Table 3 (continued)

Isotherm model	Isotherm model assumptions	Equation	Notation	References
Dubinin-Radushkevich (D-R)	The D-R model is analogous to the Langmuir model. However, this model assumes that the adsorptive elimination of heavy metal ions could occur on both the homogeneous and heterogeneous sorbent material surfaces. The Dubinin-Radushkevich model is usually employed to classify the type of heavy metal adsorption as being either physisorption or chemisorption	$q_e = q_s \exp(-K_{DR}\varepsilon^2)$	$\mathbf{q_s}$: Uptake capacity $\mathbf{K_{DR}}$: D-R constant $\mathcal{E}$: Polanyi Potential	[15]
Temkin	The Temkin model takes into account the interactive behavior between the sorbent material surface and metal species in the aqueous system and the non-uniform distribution of the adsorption heat	$q_e = \frac{RT}{b} \ln AC_e$	***b*** and ***A***: Temkin's constants T:Absolute temperature ***R*** : Gas constant	[103]

The efficiency and industrial applications of various sorbent materials in heavy metals as well as other contaminants removal from water systems required careful consideration of the process design, operational control, and sorption kinetics. This is because sorption kinetics is essential in providing knowledge about the mechanisms of adsorption of various heavy metals, which is essential for the viability of the process [111]. The reaction time experimental data are employed to identify the rate-limiting step during the adsorption process [127]. Numerous adsorption kinetic equations might be applied to analyze the sorption data obtained by a particular sorbent material during the adsorptive decontamination of heavy metal species from aqueous media to describe the sorption mechanisms and the rate-controlling step.

Generally, sorption in porous adsorbents is normally modeled using adsorption kinetic models. In kinetic modeling, it is presumed that the uptake rate of the adsorbate species on the sorbent material active site is the rate-controlling step and this rate could be explained by the three most common models including the pseudo-first-order, pseudo-second-order, and Elovich's kinetic models. As summarized in Table 4, the kinetic models are normally applied to investigate the adsorption of heavy metal ions from aqueous systems and quantify the uptake capacity of the sorbent material [156]. However, neither the pseudo-first-order reaction nor the pseudo-second-order reaction elucidates the diffusion of the adsorbate molecules into the sorbent material. Thus, it is impossible to draw any conclusions about adsorption mechanisms without taking into account the intraparticle diffusion model [12].

The determination of an appropriate model among the various models presented in Tables 3 and 4 requires analyzing isothermal and kinetic data by fitting them to the different types of models. Generally, in selecting the well-fit adsorption isotherm and kinetic models, the coefficient of determination (R^2) and error functions such as sum square error, average relative error, hybrid fractional error, the sum of absolute error, standard deviation of relative error, nonlinear chi-square value, and the sum of normalized errors are the most important parameters to take into consideration. When the coefficient of determination is close to unity and the minor, the error function values, the attained sorption data would be more compatible with the investigated equilibrium sorption isotherm and kinetic models [19, 143].

Table 4 Heavy metals adsorption kinetic models

Kinetic model	Model description	Equation	Notation	References
Pseudo-first-order	This kinetic model presumes that the sequestration of adsorbates from sorption systems onto the adsorbent surface is reversible at equilibrium. Hence, this model describes the physisorption of various adsorbates onto the surface of the adsorbent for which diffusion is the rate-controlling step	$\frac{dq_t}{dt} = k_{p_1}(q_e - q_t)$	k_{p1}: Rate of Pseudo-first-order constant t: Reaction time q_t: Uptake capacity at a particular time	[11]
Pseudo-second-order	This model also presumes that during the adsorptive decontamination of adsorbates from aqueous media, electrons are shared between the adsorbates and sorbent material functional groups. The sharing of these electrons between the adsorbate molecules and the sorbent surface functional groups leads to the formation of a chemical bond, in which chemical adsorption is the rate-limiting step	$\frac{dq_t}{dt} = k_{p_2}(q_e - q_t)^2$	k_{p2}: Pseudo-second-order rate constant	[100]
Elovich	The Elovich model assumes a chemical bonding between the adsorbate species in the water systems and the sorbent surface functional groups	$\frac{dq_t}{dt} = \alpha \exp(-\beta q_t)$	α: Initial adsorption rate β: Desorption constant	[90]
Intra-particle diffusion	The adsorption mechanism of adsorbate species on the sorbent material surface is explained by the intra-particle diffusion model. The rate-controlling step is intra-particle diffusion if the graph between q_t and $t^{1/2}$ shows a straight line. Also, for the intra-particle diffusion to be the only rate-limiting step, the plot must pass transverse to the origin	$q_t = k_{id}t^{1/2} + C$	k_{id}: Intra-particle diffusion constant C: Intercept	[116]

7 Conclusions and Future Perspectives

This book chapter has demonstrated that the adsorptive decontamination of heavy metal ions by low-cost sorbents derived from agricultural waste materials has several advantages. Almost all of the investigations on the adsorption of various heavy metal species from aqueous solutions revealed that adsorbents synthesized from agro-based materials are simple, promising, eco-friendly, and cost-effective. However, several gaps exist, which need to be addressed to increase the application of the adsorption technology in treating industrial wastewater at a large scale. The following future perspectives present knowledge gaps that call for thoughtful consideration and further studies.

1. The capability of inexpensive adsorbents to effectively decontaminate heavy metal ions from aqueous systems has been shown in numerous laboratory-scale studies. However, they are not employed on an industrial scale level due to insufficient assurance in the engineering processes and viability of these sorbent materials. To abate the amount of hazardous heavy metal ions in aquatic settings, it is imperative to employ multifunctional and interdisciplinary research designs and methodologies in the preparation of sorbent materials for both industrial and commercial applications.
2. A further consideration that should be taken into account is the fact that industrial wastewater or effluents typically contain many different heterogeneous pollutants rather than just one, which could have a greater impact on adsorption technique and selectivity. Therefore, considerable efforts should be taken to synthesize and utilize new, inexpensive adsorbents made from agricultural wastes or by-products for the sequestration of diverse harmful heavy metal species from aqueous media. Hence, more research into the mechanics of the adsorption technology is needed, with a focus on the concurrent decontamination of several heavy metals by utilizing simple and low-cost adsorbents.
3. Additionally, the majority of studies focused on heavy metal adsorption utilizing just one component of an agricultural by-product. To synthesize a hybrid adsorbent that could successfully eliminate heavy metal species from water systems, further studies ought to explore various combinations of agro-based waste materials and by-products.
4. For the retrieval of various heavy metal ions, renewal of depleted sorbent materials, and secure disposal of used adsorbents, other practical solutions, and reliable approaches should be employed to prevent further pollution of the natural environment.
5. Adsorption parameters and adsorbent production processes should be optimized utilizing efficient optimization strategies, such as the design of experiments, to improve the yield, characteristics, recovery, and recycling of the spent adsorbent to ensure economic and environmental sustainability.

Acknowledgements The authors express their gratitude to the Partnership for Applied Sciences, Engineering, and Technology (PASET), who funded this study.

References

1. Abd El-Ghany NA, Elella MHA, Abdallah HM, Mostafa MS, Samy M (2023) Recent advances in various starch formulation for wastewater purification via adsorption technique: a review. J Polym Environ 31(7):2792–2825. https://doi.org/10.1007/s10924-023-02798-x
2. Abdi O, Kazemi M (2015) A review study of biosorption of heavy metals and comparison between different biosorbents. J Mater Environ Sci 6(5):1386–1399
3. Abdulrahman OA, Abd Latiff AA, Daud Z, Saphira Radin Mohamed RM, Ismail N, Ab Aziz A, Rafatullah M, Hossain K, Ahmad A, Kamoldeen Abiodun A (2019) Adsorption of cadmium and lead from palm oil mill effluent using bone-composite: optimisation and isotherm studies. Int J Environ Anal Chem 99(8):707–725https://doi.org/10.1080/03067319.2019.1607318
4. Abegunde SM, Idowu KS, Adejuwon OM, Adeyemi-Adejolu T (2020) A review on the influence of chemical modification on the performance of adsorbents. Resour Environ Sustain 1(100001). https://doi.org/10.1016/j.resenv.2020.-100001
5. Abidli A, Huang Y, Ben Rejeb Z, Zaoui A, Park CB (2021) Sustainable and efficient technologies for removal and recovery of toxic and valuable metals from wastewater: recent progress, challenges, and future perspectives. Chemosphere 292:133102. https://doi.org/10.1016/j.chemosphere.2021.133102
6. Abou Taleb MF, Albalwi H, Abou El Fadl FI (2021) Removal of mercury (II) from aqueous solution using silver nanocomposite: synthesis and adsorption mechanism. J Inorg Organomet Polym Mater 31(4):1825–1835. https://doi.org/10.1007/s10904-020-01839-5
7. Adane T, Haile D, Dessie A, Abebe Y, Dagne H (2020) Response surface methodology as a statistical tool for optimization of removal of chromium (VI) from aqueous solution by Teff (Eragrostis teff) husk activated carbon. Appl Water Sci 10(1):1–13. https://doi.org/10.1007/s13201-019-1120-8
8. Afolabi FO, Musonge P, Bakare BF (2021) Bio-sorption of a bi-solute system of copper and lead ions onto banana peels: characterization and optimization. J Environ Health Sci Eng 19(1):613–624. https://doi.org/10.1007/s4020102100632x
9. Afroze S, Sen TK (2018) A review on heavy metal ions and dye adsorption from water by agricultural solid waste adsorbents. Water Air Soil Pollut 229(7):1–50. https://doi.org/10.1007/s11270-018-3869-z
10. Ahmaruzzaman M (2011) Industrial wastes as low-cost potential adsorbents for the treatment of wastewater laden with heavy metals. Adv Coll Interface Sci 166(1–2):36–59. https://doi.org/10.1016/j.cis.2011.04.005
11. Ahmed S, Unar IN, Khan HA, Maitlo G, Mahar RB, Jatoi AS, Memon AQ, Shah AK (2020) Experimental study and dynamic simulation of melanoidin adsorption from distillery effluent. Environ Sci Pollut Res 27(9):9619–9636. https://doi.org/10.1007/s11356-019-07441-8
12. Ajiboye TO, Oyewo OA, Onwudiwe DC (2021) Simultaneous removal of organics and heavy metals from industrial wastewater: a review. Chemosphere 262:1–20. https://doi.org/10.1016/j.chemosphere.2020.128379
13. Akram M, Khan B, Imran M, Ahmad I, Ajaz H, Tahir M, Rabbani F, Kaleem I, Nadeem Akhtar M, Ahmad N, Samad Shah N (2019) Biosorption of lead by cotton shells powder: characterization and equilibrium modeling study. Int J Phytorem 21(2):138–144. https://doi.org/10.1080/15226514.2018.1488810
14. Alalwan HA, Kadhom MA, Alminshid AH (2020) Removal of heavy metals from wastewater using agricultural byproducts. J Water Supply: Res Technol AQUA 69(2):99–112. https://doi.org/10.2166/aqua.2020.133
15. Alam MA, Shaikh WA, Alam MO, Bhattacharya T, Chakraborty S, Show B, Saha I (2018) Adsorption of As (III) and As (V) from aqueous solution by modified Cassia fistula (golden shower) biochar. Appl Water Sci 8(7):1–14. https://doi.org/10.1007/s13201-018-0839-y
16. Ali H, Khan E, Ilahi I (2019) Environmental chemistry and ecotoxicology of hazardous heavy metals: environmental persistence, toxicity, and bioaccumulation. J Chem 2019:1–14. https://doi.org/10.1155/2019/6730305

17. Alimohammadi M, Saeedi Z, Akbarpour B, Rasoulzadeh H, Yetilmezsoy K, Al-Ghouti MA, Khraisheh M, McKay G (2017) Adsorptive removal of arsenic and mercury from aqueous solutions by eucalyptus leaves. Water Air Soil Pollut 228(11):1–27. https://doi.org/10.1007/s11270-017-3607-y
18. Amar MB, Walha K, Salvadó V (2020) Evaluation of olive stones for Cd(II), Cu(II), Pb(II) and Cr(VI) biosorption from aqueous solution: equilibrium and kinetics. Int J Environ Res 14(2):193–204. https://doi.org/10.1007/-s41742-020-00246-5
19. Ameri A, Tamjidi S, Dehghankhalili F, Farhadi A, Saati MA (2020) Application of algae as low cost and effective bio-adsorbent for removal of heavy metals from wastewater: a review study. Environ Technol Rev 9(1):85–110. https://doi.org/10.1080/-21622515.2020.1831619
20. Araújo CST, Almeida ILS, Rezende HC, Marcionilio SMLO, Léon JJL, de Matos TN (2018) Elucidation of mechanism involved in adsorption of Pb(II) onto lobeira fruit (Solanum lycocarpum) using Langmuir Freundlich and Temkin isotherms. Microchem J 137(2):348–354. https://doi.org/10.1016/j.microc.2017.11.009
21. Azadegan F, Bidhendi ME, Badiei A (2019) Removal of Hg(II) ions from aqueous environment with the use of modified LUS-1 as new nanostructured adsorbent. Int J Environ Res 13(3):557–569. https://doi.org/10.1007/-s41742-019-00195-8
22. Babapoor A, Rafiei O, Mousavi Y, Azizi MM, Paar M, Nuri A (2022) Comparison and optimization of operational parameters in removal of heavy metal ions from aqueous solutions by low-cost adsorbents. Int J Chem Eng 2022:1–21. https://doi.org/10.1155/2022/3282448
23. Baby R, Saifullah B, Hussein MZ (2019) Palm Kernel Shell as an effective adsorbent for the treatment of heavy metal contaminated water. Sci Rep 9(1):1–11. https://doi.org/10.1038/s41598-019-55099-6
24. Baby SR, Saifullah B, Rehman FU (2018) Greener method for the removal of toxic metal ions from the wastewater by application of agricultural waste as an adsorbent. Water 10(10):1316. https://doi.org/10.3390/w10101316
25. Bakshe P, Jugade R (2023) Phytostabilization and rhizofiltration of toxic heavy metals by heavy metal accumulator plants for sustainable management of contaminated industrial sites: a comprehensive review. J Hazard Mater Adv 10:1–16. https://doi.org/10.1016/j.hazadv.2023.100293
26. Banchhor A, Pandey M, Pandey PK (2021) Optimization of adsorption parameters for effective removal of hexavalent chromium using simarouba glauca from aqueous solution. Water Conserv Sci Eng 6(3):127–144. https://doi.org/10.1007/s41101-021-00106-z
27. Bangaraiah P, Sarathbabu B (2019) Optimization of process parameters in removal of lead from aqueous solution through response surface methodology. Chem Eng Commun 206(8):986–993. https://doi.org/10.1080/00986445.2018.1541800
28. Barakat MA (2011) New trends in removing heavy metals from industrial wastewater. Arab J Chem 4(4):361–377. https://doi.org/10.1016/j.arabjc.-2010.07.019
29. Bayuo J, Pelig-Ba KB, Abukari MA (2019) Adsorptive removal of chromium(VI) from aqueous solution unto groundnut shell. Appl Water Sci 9(4):1–11. https://doi.org/10.1007/s13201-019-0987-8
30. Bayuo J, Rwiza MJ, Mtei KM (2023a) Adsorption and desorption ability of divalent mercury from an interactive bicomponent sorption system using hybrid granular activated carbon. Environ Monit Assess 195(935):1–17. https://doi.org/10.1007/s10661-023-11540-y
31. Bayuo J, Rwiza MJ, Mtei KM (2023) Modeling and optimization of trivalent arsenic removal from wastewater using activated carbon produced from maize plant biomass: a multivariate experimental design approach. Biomass Convers Biorefinery 0123456789:1–24. https://doi.org/10.1007/s13399-023-04494-1
32. Bayuo, J, Rwiza, MJ, and Kelvin, MM (2023) Optimization of divalent mercury removal from synthetic wastewater using desirability function in central composite design of response surface methodology. J Environ Health Sci Eng 1–19. https://doi.org/10.1007/s40201-023-00888-5
33. Bayuo J, Rwiza MJ, Mtei KM (2023c) Non-competitive and competitive detoxification of As(III) ions from single and binary biosorption systems and biosorbent regeneration. Biomass Convers Biorefinery 1–28. https://doi.org/10.1007/-s13399-022-03734-0

34. Bayuo J, Rwiza MJ, Sillanpää M, Mtei KM (2023) Removal of heavy metals from binary and multicomponent adsorption systems using various adsorbents - a systematic review. RSC Adv 13(19):13052–13093. https://doi.org/10.1039/d3ra01660a
35. Bayuo J (2021) An extensive review on chromium (vi) removal using natural and agricultural wastes materials as alternative biosorbents. J Environ Health Sci Eng 19(1):1193–1207. https://doi.org/10.1007/s40201-021-00641-w
36. Bayuo J (2021) Decontamination of cadmium(II) from synthetic wastewater onto shea fruit shell biomass. Appl Water Sci 11(5):1–8. https://doi.org/10.1007/s13201-021-01416-2
37. Bayuo J, Abukari MA, Pelig-Ba KB (2019) Equilibrium isotherm studies for the sorption of hexavalent chromium (VI) onto groundnut shell. IOSR J Appl Chem (IOSR-JAC). 11(12):40–46. https://doi.org/10.9790/5736-1112014046
38. Bayuo J, Abukari MA, Pelig-Ba KB (2020) Optimization using central composite design (CCD) of response surface methodology (RSM) for biosorption of hexavalent chromium from aqueous media. Appl Water Sci 10(6):1–12. https://doi.org/10.1007/s13201-020-01213-3
39. Bayuo J, Pelig-Ba KB, Abukari MA (2018) Isotherm modeling of lead (II) adsorption from aqueous solution using groundnut shell as a low-cost adsorbent. IOSR J Appl Chem (IOSR-JAC). 11(11):18–23. https://doi.org/10.9790/5736-1111011823
40. Bayuo J, Rwiza M, Abukari MA, Pelig-Ba KB, Mtei K (2022) Modeling and optimization of independent factors influencing lead(II) biosorption from aqueous systems: a statistical approach. Sci Afr 16:1–16. https://doi.org/10.1016/j.-sciaf.2022.-e01270
41. Bayuo J, Rwiza MJ, Mtei KM (2023) Applicability of bio-adsorbents synthesized from maize/corn plant residues for heavy metals removal from aquatic environments: an insight review. EQA Int J Environ Qual 56:15–35. https://doi.org/10.6092/issn.2281-4485/16951
42. Bayuo J, Rwiza M, Mtei K (2022) A comprehensive review on the decontamination of lead(II) from water and wastewater by low-cost biosorbents. RSC Adv 12(18):11233–11254. https://doi.org/10.1039/d2ra00796g
43. Bayuo J, Rwiza M, Mtei K (2022) Response surface optimization and modeling in heavy metal removal from wastewater—a critical review. Environ Monit Assess 194(5):1–34. https://doi.org/10.1007/s10661-022-09994-7
44. Boni MR, Chiavola A, Marzeddu S (2020) Remediation of lead-contaminated water by virgin coniferous wood biochar adsorbent: batch and column application. Water Air Soil Pollut 231(4):9–16. https://doi.org/10.1007/s11270-020-04496-z
45. Boulaiche W, Belhamdi B, Hamdi B, Trari M (2019) Kinetic and equilibrium studies of biosorption of M(II) (M = Cu, Pb, Ni, Zn and Cd) onto seaweed Posidonia oceanica fibers. Appl Water Sci 9(8):1–11. https://doi.org/10.1007/s13201-019-1062-1
46. Bratby J (2006) Coagulation and flocculation in water and wastewater treatment. In: Water. IWA Publishing, p 21. https://doi.org/10.2166/9781780407500
47. Calderón ROA, Abdeldayem OM, Pugazhendhi A, Rene ER (2020) Current updates and perspectives of biosorption technology: an alternative for the removal of heavy metals from wastewater. Curr Poll Rep 6(1):8–27. https://doi.org/10.1007/s40726-020-00135-7
48. Chidozie CN, Chinwe JE, Emmanuel IU (2017) Physicochemical conditions for adsorption of lead from water by rice husk ash. BioResources 12:799–818
49. Choi SH (2016) On the brine re-utilization of a multi-stage flashing (MSF) desalination plant. Desalination 398:64–76. https://doi.org/10.1016/j.desal.2016.07.020
50. Dakhil IH, Naser GF, Ali AH (2021) Response surface modeling of arsenic adsorption by modified spent tea leaves. IOP Conf Ser Mater Sci Eng 1090(1):1–11 https://doi.org/10.1088/1757-899x/1090/1/012129
51. Dawodu FA, Akpan BM, Akpomie KG (2020) Sequestered capture and desorption of hexavalent chromium from solution and textile wastewater onto low cost Heinsia crinita seed coat biomass. Appl Water Sci 10(1):1–15. https://doi.org/10.1007/s1320101911146
52. Demir DA, Gülçiçek O, Gören N (2019) Optimization of adsorption for the removal of cadmium from aqueous solution using Turkish coffee grounds. Int J Environ Res 13(5):861–878. https://doi.org/10.1007/s41742-019-00224-6

53. Dewi AK, Sharma RK, Das K, Sukul U, Lin PY, Huang YH, Lu CM, Lu CK, Chen TH, Chen CY (2023) Biologically-induced synthetic manganese carbonate precipitate (BISMCP) for potential applications in heavy metal removal. Heliyon 9(5):1–12. https://doi.org/10.1016/j.heliyon.2023.e15919
54. Dhiman N (2021) Binary adsorption of [Pb(II) + Co(II)] from aqueous solution using thiolated saw dust. Water Sci Technol 84(9):2591–2600. https://doi.org/10.2166/wst.-2021.454
55. Dhoble RM, Maddigapu PR, Bhole AG, Rayalu S (2018) Development of bark-based magnetic iron oxide particle (BMIOP), a bio-adsorbent for removal of arsenic (III) from water. Environ Sci Poll Res 25(20):19657–19674. https://doi.org/10.1007/s11356-018-1792-x
56. Dula T, Duke TN (2019) Removal methods of heavy metals from laboratory wastewater. J Nat Sci Res 9(2):36–42. https://doi.org/10.7176/jnsr/9-2-04
57. Egirani D, Latif MT, Wessey N, Poyi NR, Shehata N (2021) Preparation and characterization of powdered and granular activated carbon from Palmae biomass for mercury removal. Appl Water Sci 11(1):1–11. https://doi.org/10.1007/s13201-020-01343-8
58. El-Bouhy ZM, Reda RM, Mahboub HH, Gomaa FN (2021) Bioremediation effect of pomegranate peel on subchronic mercury immunotoxicity on African catfish (Clarias gariepinus). Environ Sci Poll Res 28(2):2219–2235. https://doi.org/10.1007/s11356-020-10599-1
59. Emenike PC, Omole DO, Ngene BU, Tenebe IT (2016) Potentiality of agricultural adsorbent for the sequestering of metal ions from wastewater. Global J Environ Sci Manage 2(4):411–442. https://doi.org/10.22034/gjesm.2016.02.04.010
60. Fabre E, Lopes CB, Vale C, Pereira E, Silva CM (2020) Valuation of banana peels as an effective biosorbent for mercury removal under low environmental concentrations. Sci Total Environ 709:135883. https://doi.org/10.1016/j.scitotenv.-2019.135883
61. Farnane M, Machrouhi A, Elhalil A, Abdennouri M, Qourzal S, Tounsadi H, Barka N (2018) New Sustainable biosorbent based on recycled deoiled carob seeds : optimization of heavy metals remediation. J Chem 2018:1–17. https://doi.org/10.1155/2018/5748493
62. Fiyadh SS, Alardhi SM, Al Omar M, Aljumaily MM, Al Saadi MA, Fayaed SS, Ahmed SN, Salman AD, Abdalsalm AH, Jabbar NM, El-Shafi A (2023) A comprehensive review on modelling the adsorption process for heavy metal removal from waste water using artificial neural network technique. Heliyon 9(4):1–11. https://doi.org/10.1016/j.heliyon.2023.e15455
63. Fooladgar S, Teimouri A, Ghanavati Nasab S (2019) Highly efficient removal of lead ions from aqueous solutions using chitosan/rice husk ash/nano alumina with a focus on optimization by response surface methodology: Isotherm, kinetic, and thermodynamic studies. J Polym Environ 27(5):1025–1042. https://doi.org/10.1007/s10924-019-01385-3
64. Foroutan R, Mohammadi R, Adeleye AS, Farjadfard S, Esvandi Z, Arfaeinia H, Sorial GA, Ramavandi B, Sahebi S (2019) Efficient arsenic(V) removal from contaminated water using natural clay and clay composite adsorbents. Environ Sci Pollut Res 26(29):29748–29762. https://doi.org/10.1007/s11356-019-06070-5
65. Garba ZN, Bello I, Galadima A, Lawal AY (2016) Optimization of adsorption conditions using central composite design for the removal of copper (II) and lead (II) by defatted papaya seed. Karbala Int J Mod Sci 2(1):20–28. https://doi.org/10.1016/j.kijoms.2015.12.002
66. Gautam PK, Gautam RK, Banerjee S, Chattopadhyaya MC, Pandey JD (2016) Heavy metals in the environment: fate, transport, toxicity and remediation technologies. In: Heavy metals: sources, toxicity and remediation techniques, pp 101–130
67. Ghasemi SM, Ghaderpoori M, Moradi B, Taghavi M, Karimyan K, Mehdipour F (2020) Optimization of Cr(VI) adsorption by modified sesame hull from aqueous solutions using response surface methodology. Int J Environ Anal Chem 1–15. https://doi.org/10.1080/03067319.2020.1763972
68. Hadiani MR, Khosravi-Darani K, Rahimifard N (2019) Optimization of As (III) and As (V) removal by Saccharomyces cerevisiae biomass for biosorption of critical levels in the food and water resources. J Environ Chem Eng 7(2):1–9. https://doi.org/10.1016/j.jece.2019.102949
69. He S, Li Y, Weng L, Wang J, He J, Liu Y, Zhang K, Wu Q, Zhang Y, Zhang Z (2018) Competitive adsorption of Cd(II), Pb(II) and Ni(II) onto Fe(III)-modified argillaceous limestone: Influence

of pH, ionic strength and natural organic matters. Sci Total Environ 637–638:69–78. https://doi.org/10.1016/j.scitotenv.-2018.04.300
70. Heidarinejad Z, Dehghani MH, Heidari M, Javedan G, Ali I, Sillanpää M (2020) Methods for preparation and activation of activated carbon: a review. Environ Chem Lett 18(2):393–415. https://doi.org/10.1007/s10311-019-00955-0
71. Hu C, Hu H, Tang Y, Dai Y, Wang Z, Yan R (2020) Comparative study on adsorption and immobilization of Cd(II) by rape component biomass. Environ Sci Pollut Res 27(8):8028–8033. https://doi.org/10.1007/s11356-019-07535-3
72. Huang S, Jin S, Wang Y, Liu J, Yu J, Liu D, Chi R (2020) Selective adsorption of heavy metal ions from aqueous solution by modified bagasse. Chem Ecol 839–854. https://doi.org/10.1080/02757540.2020.1787998
73. Hymavathi D, Prabhakar G (2017) Studies on the removal of Cobalt(II) from aqueous solutions by adsorption with Ficus benghalensis leaf powder through response surface methodology. Chem Eng Commun 204(12):1401–1411. https://doi.org/10.1080/00986445.2017.1365063
74. Iftekhar S, Ramasamy DL, Srivastava V, Asif MB, Sillanpää M (2018) Understanding the factors affecting the adsorption of Lanthanum using different adsorbents: a critical review. Chemosphere 204:413–430. https://doi.org/10.1016/j.-chemosphere.2018.04.053
75. Jagirani MS, Balouch A, Mahesar SA, Kumar A, Abdullah, Mustafai FA, Bhanger MI (2020) Preparation of novel arsenic-imprinted polymer for the selective extraction and enhanced adsorption of toxic As(III) ions from the aqueous environment. Polym Bull 77(10):5261–5279.https://doi.org/10.1007/s00289-019-03008-2
76. Jain M, Garg VK, Paliwal R, Kadirvelu K, Chaudhry S (2021) Optimization of cadmium(II) removal from water using sunflower waste carbon–a statistical approach. Toxin Rev 40(4):1373–1382. https://doi.org/10.1080/15569543.2020.1718163
77. Javidi Alsadi K, Esfandiari N (2019) Synthesis of activated carbon from sugarcane bagasse and application for mercury adsorption. Pollution 5(3):585–596. https://doi.org/10.22059/poll.2019.269364.540
78. Jjagwe J, Olupot PW, Menya E, Kalibbala HM (2021) Synthesis and application of granular activated carbon from biomass waste materials for water treatment: a review. J Bioresources Bioprod 6(4): 292–322. https://doi.org/10.1016/j.jobab.-2021.03.003
79. Kaakani M (2012) Heavy metal removal from wastewater using novel adsorbent. American University of Sharjah, College of Engineering, Thesis
80. Kanaujia S, Singh B, Singh SK (2015) Removal of fluoride from groundwater by carbonised Punica granatum carbon bio-adsorbent. J Geosci Environ Prot 03(04):1–9. https://doi.org/10.4236/gep.2015.34001
81. Khan FSA, Mubarak NM, Khalid M, Walvekar R, Abdullah E, Mazari SA, Nizamuddin S, Karri RR (2020) Magnetic nanoadsorbents' potential route for heavy metals removal—a review. Environ Sci Pollut Res 27(19):24342–24356. https://doi.org/10.1007/s11356-020-08711-6
82. Khan N, Kumar D, Kumar P (2019) Microwave assisted synthesis of polyvinylbutyral-silica composites for mercury removal application. ChemistrySelect 4(6):1979–1984. https://doi.org/10.1002/slct.201803426
83. Khan SA, Imteaz MA (2021) Experimental studies on arsenic removal efficiencies through adsorption using different natural adsorbents. Water Air Soil Pollut 232(1):1–10. https://doi.org/10.1007/s11270-020-04977-1
84. Khorshidi P, Shirazi RHSM, Miralinaghi M, Moniri E, Saadi S (2020) Adsorptive removal of mercury (II), copper (II), and lead (II) ions from aqueous solutions using glutathione-functionalized $NiFe_2O_4$/graphene oxide composite. Res Chem Intermed 46(7):3607–3627. https://doi.org/10.1007/s11164-020-04164-1
85. Kumar S, Jain S (2013) History, introduction, and kinetics of ion exchange materials. J Chem 1–13. https://doi.org/10.1155/2013/957647
86. Kumar A, Sidharth S, Kandasubramanian B (2023) A review on algal biosorbents for heavy metal remediation with different adsorption isotherm models. Environ Sci Poll Res 39474–39493. https://doi.org/10.1007/s11356-023-25710-5

87. Kumar D, Pandey LK, Gaur JP (2016) Metal sorption by algal biomass: from batch to continuous system. Algal Res 18:95–109. https://doi.org/10.1016/j.algal.2016.05.026
88. Kumar S, Prasad S, Yadav KK, Shrivastava M, Gupta N, Nagar S, Bach QV, Kamyab H, Khan SA, Yadav S, Malav LC (2019) Hazardous heavy metals contamination of vegetables and food chain: role of sustainable remediation approaches—a review. Environ Res 179:108792. https://doi.org/10.1016/j.-envres.2019.108792
89. Kumari B, Tiwary RK, Yadav M, Singh KMP (2021) Nonlinear regression analysis and response surface modeling of Cr (VI) removal from synthetic wastewater by an agro-waste Cocos Nucifera: Box-Behnken design (BBD). Int J Phytorem 23(8):791–808. https://doi.org/10.1080/15226514.2020.1858399
90. Labied R, Benturki O, Eddine Hamitouche AY, Donnot A (2018) Adsorption of hexavalent chromium by activated carbon obtained from a waste lignocellulosic material (Ziziphus jujuba cores): Kinetic, equilibrium, and thermodynamic study. Adsorpt Sci Technol 36(3–4):1066–1099. https://doi.org/10.1177/0263-617417750739
91. László Z, Hodúr C (2007) Purification of thermal wastewater by membrane separation and ozonation. Desalination 206(1–3):333–340. https://doi.org/10.1016/j.desal.-2006.03.572
92. Lenka SP, Shaikh WA, Owens G, Padhye LP, Chakraborty S, Bhattacharya T (2021) Removal of copper from water and wastewater using dolochar. Water Air Soil Pollut 232(5):1–15. https://doi.org/10.1007/s11270-021-05135-x
93. Li C, Zhou K, Qin W, Tian C, Qi M, Yan X, Han W (2019) A review on heavy metals contamination in soil: effects, sources, and remediation techniques. Soil Sediment Contam 28(4):380–394. https://doi.org/10.1080/15320383.2019.-1592108
94. Li Y, Yu H, Liu L, Yu H (2021) Application of co-pyrolysis biochar for the adsorption and immobilization of heavy metals in contaminated environmental substrates. J Hazard Mater 420:1–12. https://doi.org/10.1016/j.jhazmat.2021.126655
95. Liu G, Li Z, Xu L, Xu X, Huang Q, Zeng Y, Wen M (2018) The dynamics and adsorption of Cd (II) onto hydroxyapatite attapulgite composites from aqueous solution. J Sol-Gel Sci Technol 87(2):269–284. https://doi.org/10.1007/s10971-018-4717-8
96. Liu X, Xu X, Dong X, Park J (2020) Competitive adsorption of heavy metal ions from aqueous solutions onto activated carbon and agricultural waste materials. Pol J Environ Stud 29(1):749–761. https://doi.org/10.15244/pjoes/104455
97. Malik AH, Nasreen S, Mahmood Q, Khan ZM, Sarwar R, Jilani G, Khan A (2010) Strategies for low-cost water defluoridation of drinking water—a review of progress. J Chem Soc Pak 32(4):550–558
98. Manjuladevi M, Anitha R, Manonmani S (2018) Kinetic study on adsorption of Cr(VI), Ni(II), Cd(II) and Pb(II) ions from aqueous solutions using activated carbon prepared from Cucumis melo peel. Appl Water Sci 8(1):1–8. https://doi.org/10.1007/s13201-018-0674-1
99. Mbugua M, Mbuvi H, Muthengia J (2014) Rice husk ash derived zeolite blended with water hyacinth ash for enhanced adsorption of cadmium Ions. Curr World Environ 9(2):280–286. https://doi.org/10.12944/cwe.9.2.08
100. Memon AQ, Ahmed S, Bhatti ZA, Maitlo G, Shah AK, Mazari SA, Muhammad A, Jatoi AS, Kandhro GA (2021) Experimental investigations of arsenic adsorption from contaminated water using chemically activated hematite (Fe_2O_3) iron ore. Environ Sci Pollut Res 28(10):12898–12908. https://doi.org/10.1007/s11356-020-11208-x
101. Milani PA, Consonni JL, Labuto G, Carrilho ENVM (2018) Agricultural solid waste for sorption of metal ions, part II: competitive assessment in multielemental solution and lake water. Environ Sci Pollut Res 25(36):35906–35914. https://doi.org/10.1007/s11356-018-1726-7
102. Mohd I, Ahamed I, Lichtfouse E (2021) Water pollution and remediation: heavy metals. Environ Chem Sustain World 53. http://link.springer.com/10.1007/9783030547233%0Ahttp://link.springer.com/10.1007/9783030524210
103. Mohdee V, Maneeintr K, Wannachod T, Phatanasri S, Pancharoen U (2018) Optimization of process parameters using response surface methodology for Pd(II) extraction with quaternary ammonium salt from chloride medium: kinetic and thermodynamics study. Chem Pap 72(12):3129–3139. https://doi.org/10.1007/s11696-018-0542-3

104. Mohubedu RP, Diagboya PN, Abasi CY, Dikio ED, Mtunzi F (2019) Magnetic valorization of biomass and biochar of a typical plant nuisance for toxic metals contaminated water treatment. J Clean Prod 209:1016–1024. https://doi.org/10.1016/j.-jclepro.2018.-10.215
105. Mondal NK, Samanta A, Roy P, Das B (2019) Optimization study of adsorption parameters for removal of Cr(VI) using Magnolia leaf biomass by response surface methodology. Sustain Water Resour Manag 5(4):1627–1639. https://doi.org/10.1007/s40899-019-00322-5
106. Mondal S, Aikat K, Halder G (2017) Biosorptive uptake of arsenic(V) by steam activated carbon from mung bean husk: equilibrium, kinetics, thermodynamics and modeling. Appl Water Sci 7(8):4479–4495. https://doi.org/10.1007/s13201-017-0596-3
107. Motaghi M, Ziarati P (2016) Adsorptive removal of cadmium and lead from Oryza sativa rice by banana peel as bio-sorbent. Biomed Pharmacol J 9(2):739–749. https://doi.org/10.13005/bpj/998
108. Mukherjee S, Thakur AK, Goswami R, Mazumder P, Taki K, Vithanage M, Kumar M (2021) Efficacy of agricultural waste derived biochar for arsenic removal: tackling water quality in the Indo-Gangetic plain. J Environ Manage 281:1–12. https://doi.org/10.1016/j.jenvman.2020.111814
109. Mulopo J (2015) Continuous pilot scale assessment of the alkaline barium calcium desalination process for acid mine drainage treatment. J Environ Chem Eng 3(2):1295–1302. https://doi.org/10.1016/j.jece.2014.12.001
110. Mustapha S, Shuaib DT, Ndamitso MM, Etsuyankpa MB, Sumaila A, Mohammed UM, Nasirudeen MB (2019) Adsorption isotherm, kinetic and thermodynamic studies for the removal of Pb(II), Cd(II), Zn(II) and Cu(II) ions from aqueous solutions using Albizia lebbeck pods. Appl Water Sci 9(6):1–11. https://doi.org/10.1007/s132010191021x
111. Nazerdeylami S, Zare-Dorabei R (2019) Simultaneous adsorption of Hg(II), Cd(II) and Cu(II) ions from aqueous solution with mesoporous silica/DZ and conditions optimise with experimental design: Kinetic and isothermal studies. Micro Nano Lett 14(8):823–827. https://doi.org/10.1049/mnl.2018.5775
112. Nemeş L, Bulgariu L (2016) Optimization of process parameters for heavy metals biosorption onto mustard waste biomass. Open Chem 14(1):175–187. https://doi.org/10.1515/chem-2016-0019
113. Nezami S, Ghaemi A, Yousefi T (2023) Application of titanium carbide/nitride (MXene)-based NPs in adsorption of radionuclides and heavy metal ions for wastewater remediation: a review. Case Stud Chem Environ Eng 7:1–22. https://doi.org/10.1016/j.cscee.2023.100326
114. Nur-E-Alam M, Abu Sayid Mia M, Ahmad F, Mafizur Rahman M (2018) Adsorption of chromium (Cr) from tannery wastewater using low-cost spent tea leaves adsorbent. Appl Water Sci 8(5):1–7. https://doi.org/10.1007/s13201-018-0774-y
115. Nyangi MJ (2021) Remediation of arsenic from water using iron and aluminum electrodes in electrocoagulation technology: adsorption isotherm and kinetic studies. Chem Afr 4(4):943–954. https://doi.org/10.1007/s42250-021-00268-2
116. Olu-Owolabi BI, Diagboya PN, Adebowale KO (2014) Evaluation of pyrene sorption-desorption on tropical soils. J Environ Manag 137:1–9. https://doi.org/10.1016/j.jenvman.2014.01.048
117. Onwordi CT, Uche CC, Ameh AE, Petrik LF (2019) Comparative study of the adsorption capacity of lead (II) ions onto bean husk and fish scale from aqueous solution. J Water Reuse Desalin 9(3):249–262. https://doi.org/10.2166/wrd.-2019.061
118. Pakzadeh B, Batista JR (2011) Chromium removal from ion-exchange waste brines with calcium polysulfide. Water Res 45(10):3055–3064. https://doi.org/10.1016/j.-watres.-2011.03.006
119. Pan J, Gao B, Wang S, Guo K, Xu X, Yue Q (2020) Waste-to-resources: Green preparation of magnetic biogas residues-based biochar for effective heavy metal removals. Sci Total Environ 737:1–10. https://doi.org/10.1016/j.scitotenv.2020.-140283
120. Park D, Yun Y, Park J (2014) The past, present, and future trends of biosorption. Biotechnol Bioprocess Eng 15:86–102

121. Parlayici Ş, Pehlivan E (2019) Comparative study of Cr(VI) removal by bio-waste adsorbents: equilibrium, kinetics, and thermodynamic. J Anal Sci Technol 10(1):1–8. https://doi.org/10.1186/s40543-019-0175-3
122. Pasricha S, Mathur V, Garg A, Lenka S, Verma K, Agarwal S (2021) Molecular mechanisms underlying heavy metal uptake, translocation and tolerance in hyperaccumulators-an analysis: heavy metal tolerance in hyperaccumulators. Environ Challenges 4:1–15. https://doi.org/10.1016/j.envc.2021.100197
123. Poonam, Bharti SK, Kumar N (2018) Kinetic study of lead (Pb^{2+}) removal from battery manufacturing wastewater using bagasse biochar as biosorbent. Appl Water Sci 8(4):1–13. https://doi.org/10.1007/s13201-018-0765-z
124. Pyrzynska K (2019) Removal of cadmium from wastewaters with low-cost adsorbents. J Environ Chem Eng 7(1):102795. https://doi.org/10.1016/j.-jece.2018.-11.040
125. Rafique M, Hajra S, Tahir MB, Gillani SSA, Irshad M (2022) A review on sources of heavy metals, their toxicity and removal technique using physico-chemical processes from wastewater. Environ Sci Pollut Res 29(11):16772–16781. https://doi.org/10.1007/s11356-022-18638-9
126. Rahdar S, Taghavi M, Khaksefidi R, Ahmadi S (2019) Adsorption of arsenic (V) from aqueous solution using modified saxaul ash: isotherm and thermodynamic study. Appl Water Sci 9(4):1–9. https://doi.org/10.1007/s13201-019-0974-0
127. Rahman, MM, Karmaker, SC, Pal, A, Eljamal, O, Saha, B. B (2021) Statistical techniques for the optimization of cesium removal from aqueous solutions onto iron-based nanoparticle-zeolite composites. Environ Sci Pollut Res 28(10):12918–12931. https://doi.org/10.1007/s11356-020-11258-1
128. Rahman DZ, Vijayaraghavan J, Thivya J (2023) A comprehensive review on zinc(II) sequestration from wastewater using various natural/modified low-cost agro-waste sorbents. Biomass Convers Biorefinery 13(7):5469–5499. https://doi.org/10.1007/s13399-021-01822-1
129. Rai R, Aryal RL, Paudyal H, Gautam SK, Ghimire KN, Pokhrel MR, Poudel BR (2023) Acid-treated pomegranate peel; an efficient biosorbent for the excision of hexavalent chromium from wastewater. Heliyon 9(5):1–17. https://doi.org/10.1016/j.-heliyon.2023.e15698
130. Raj D, Maiti SK (2019) Sources, toxicity, and remediation of mercury: an essence review. Environ Monit Assess 191(9):1–22. https://doi.org/10.1007/s10661-019-7743-2
131. Rani B, Trivedi A, Pathak RK, Maheshwari RK (2014) Fluoride ion: an inexplicable hydro toxicant. J Ind Pollut Control 30(1):23–32
132. Razzak A, Shafiquzzaman M, Haider H, Alresheedi M (2021) Arsenic removal by iron-oxidizing bacteria in a fixed-bed coconut husk column: experimental study and numerical modeling. Environ Pollut 272(xxxx):1–10. https://doi.org/10.1016/j.envpol.-2020.115977
133. Reza R, Singh G (2010) Heavy metal contamination and its indexing approach for river water. Int J Environ Sci Technol 7(4):785–792. https://doi.org/10.1007/BF03326187
134. Roy P, Dey U, Chattoraj S, Mukhopadhyay D, Mondal NK (2017) Modeling of the adsorptive removal of arsenic(III) using plant biomass: a bioremedial approach. Appl Water Sci 7(3):1307–1321. https://doi.org/10.1007/s13201-015-0339-2
135. Rwiza MJ, Oh SY, Kim KW, Kim SD (2018) Comparative sorption isotherms and removal studies for Pb(II) by physical and thermochemical modification of low-cost agro-wastes from Tanzania. Chemosphere 195(2):135–145. https://doi.org/10.1016/j.chemosphere.2017.12.043
136. Sabah H, Thouraya T, Melek H, Nadia M (2018) Application of response surface methodology for optimization of cadmium ion removal from an aqueous solution by eggshell powder. Chem Res Chin Univ 34(2):302–310. https://doi.org/10.1007/s40242-018-7163-9
137. Sahmoune MN, Louhab K, Boukhiar A (2011) Advanced biosorbents materials for removal of chromium from water and wastewaters. Environ Prog Sustain Energy 30(3):284–293. https://doi.org/10.1002/ep.10473
138. Salah AS (2015) Assessing the potential of laterite in adsorbing cadmium from mine leachate and surrogate cadmium solutions: a case study at Anglogold Ashanti Iduaperiem Gold Mine Ltd, Tarkwa. http://hdl.handle.net/123456789/6810

139. Samaniego JO, Tanchuling MAN (2019) Removal of heavy metals from an actual small scale gold mining wastewater by sorption onto cocopeat. ASEAN J Sci Technol Dev 36(1):1–7. https://doi.org/10.29037/ajstd.558
140. Samuel MS, Shang M, Klimchuk S, Niu J (2021) Novel regenerative hybrid composite adsorbent with improved removal capacity for lead ions in water. Ind Eng Chem Res 60(14):5124–5132. https://doi.org/10.1021/acs.iecr.-0c06277
141. Sarma GK, Sen Gupta S, Bhattacharyya KG (2019) Nanomaterials as versatile adsorbents for heavy metal ions in water: a review. Environ Sci Pollut Res 26(7):6245–6278. https://doi.org/10.1007/s11356-018-04093-y
142. Shahedi A, Darban AK, Taghipour F, Jamshidi-Zanjani A (2020) A review on industrial wastewater treatment via electrocoagulation processes. Curr Opin Electrochem 22:154–169. https://doi.org/10.1016/j.coelec.2020.05.009
143. Shahrin S, Lau WJ, Goh PS, Jaafar J, Ismail AF (2018) Adsorptive removal of As(V) ions from water using graphene Oxide-manganese ferrite and Titania nanotube-manganese ferrite hybrid nanomaterials. Chem Eng Technol 41(11):2250–2258. https://doi.org/10.1002/ceat.201800322
144. Sharma A, Anjana, Rana H, Goswami S (2022) A comprehensive review on the heavy metal removal for water remediation by the application of lignocellulosic biomass-derived nanocellulose. J Polym Environ 30(1):1–18. https://doi.org/10.1007/s10924-021-02185-4
145. Sharma A, Tomer A, Singh J, Chhikara BS (2019) Biosorption of metal toxicants and other water pollutants by Corn (Maize) plant: a comprehensive review. J Integr Sci Technol 7(2):19–29. http://pubs.iscience.in/journal/index.-php/jist/article/view/918
146. Siriwardena DP, James R, Dasu K, Thorn J, Iery RD, Pala F, Schumitz D, Eastwood S, Burkitt N (2021) Regeneration of per- and polyfluoroalkyl substance-laden granular activated carbon using a solvent based technology. J Environ Manage 289:112439. https://doi.org/10.1016/j.jenvman.2021.112439
147. Sivamani S, Prasad BSN, Nithya K, Sivarajasekar N, Hosseini-Bandegharaei A (2021) Back-propagation neural network: Box–Behnken design modelling for optimization of copper adsorption on orange zest biochar. Int J Environ Sci Technol 1–16. https://doi.org/10.1007/s13762-021-03411-1
148. Soliman NK, Moustafa AF (2020) Industrial solid waste for heavy metals adsorption features and challenges; a review. J Market Res 9(5):10235–10253. https://doi.org/10.1016/j.jmrt.2020.07.045
149. Sonone SS, Jadhav SV, Sankhla MS, Kumar R (2020) Water contamination by heavy metals and their toxic effect on aquaculture and human health through food chain. Lett Appl NanoBioSci 10(2):2148–2166. https://doi.org/10.33263/lianbs-102.-21482166
150. Steingräber LF, Ludolphy C, Metz J, Kierdorf H, Kierdorf U (2022) Uptake of lead and zinc from soil by blackberry plants (Rubus fruticosus L. agg.) and translocation from roots to leaves. Environ Adv 9:1–10. https://doi.org/10.1016/j.envadv.-2022.100313
151. Taha GM, Arifien AE, El-Nahas S (2011) Removal efficiency of potato peels as a new biosorbent material for uptake of Pb(II) Cd(II) and Zn(II) from their aqueous solutions. J Solid Waste Technol Manag 37(2):128–140. https://doi.org/10.5276/JSWTM.2011.128
152. Tang C, Shu Y, Zhang R, Li X, Song J, Li B, Zhang Y, Ou D (2017) Comparison of the removal and adsorption mechanisms of cadmium and lead from aqueous solution by activated carbons prepared from Typha angustifolia and Salix matsudana. RSC Adv 7(26):16092–16103. https://doi.org/10.1039/c6ra28035h
153. Tang J, Ptacek CJ, Blowes DW, Liu YY, Feng Y, Finfrock YZ, Liu P (2022) Mercury adsorption kinetics on sulfurized biochar and solid-phase digestion using aqua regia: a synchrotron-based study. Chem Eng J 428:1–9. https://doi.org/10.1016/j.cej.2021.131362
154. Tokay B, Akpınar I (2021) A comparative study of heavy metals removal using agricultural waste biosorbents. Bioresou Technol Rep 15:1–9. https://doi.org/10.1016/j.-biteb.2021.100719
155. Türkmen D, Özkaya Türkmen M, Akgönüllü S, Denizli A (2022) Development of ion imprinted based magnetic nanoparticles for selective removal of arsenic (III) and arsenic

(V) from wastewater. Sep Sci Technol (Philadelphia) 57(6):990–999. https://doi.org/10.1080/01496395.2021.1956972
156. Velusamy S, Roy A, Sundaram S, Kumar Mallick T (2021) A review on heavy metal ions and containing dyes removal through graphene oxide-based adsorption strategies for textile wastewater treatment. Chem Rec 21(7):1570–1610. https://doi.org/10.1002/tcr.202000153
157. Waghmare S (2015) Fluoride removal from water by various techniques—review. Int J Innovative Sci Eng Technol 2(9):560–571
158. Wahid SN, Maharaj R, Boodlal D, Smith JV (2022) The adsorption of phenol on granular activated carbon prepared from waste coconut shell in Trinidad. Environ Prog Sustain Energy 41(1):1–30. https://doi.org/10.1002/ep.13729
159. Wang B, Lan J, Bo C, Gong B, Ou J (2023) Adsorption of heavy metal onto biomass-derived activated carbon: review. RSC Adv 13(7):4275–4302. https://doi.org/10.1039/d2ra07911a
160. Wang LK, Hung YT, Lo HH, Yapijakis C (2005) Waste treatment in the process industries. In: Waste treatment in the process industries. Taylor and Francis Group, pp 1–638. https://doi.org/10.1201/9781420037159
161. Wang M, Bera G, Mitra K, Wade TL, Knap AH, Phillips TD (2021) Tight sorption of arsenic, cadmium, mercury, and lead by edible activated carbon and acid-processed montmorillonite clay. Environ Sci Pollut Res 28(6):6758–6770. https://doi.org/10.1007/s11356-020-10973-z
162. Wimalawansa SJ (2013) Purification of contaminated water with reverse osmosis: effective solution of providing clean water for human needs in developing countries. Int J Emerg Technol Adv Eng 3(12):15
163. Wu LK, Wu H, Zhang HB, Cao HZ, Hou GY, Tang YP, Zheng GQ (2018) Graphene oxide/$CuFe_2O_4$ foam as an efficient absorbent for arsenic removal from water. Chem Eng J 334:1808–1819. https://doi.org/10.1016/j.-cej.2017.11.096
164. Xia S, Song Z, Jeyakumar P, Bolan N, Wang H (2020) Characteristics and applications of biochar for remediating Cr(VI)-contaminated soils and wastewater. Environ Geochem Health 42(6):1543–1567. https://doi.org/10.1007/-s10653-019-00445-w
165. Xiao B, Jia J, Wang W, Zhang B, Ming H, Ma S, Kang Y, Zhao M (2023) A review on magnetic biochar for the removal of heavy metals from contaminated soils: preparation, application, and microbial response. J Hazard Mater Adv 10:1–13. https://doi.org/10.1016/j.hazadv.2023.100254
166. Xu Z, Dong J (2008) Synthesis, characterization, and application of magnetic nanocomposites for the removal of heavy metals from industrial effluents. Emerg Environ Technol 105–148. https://doi.org/10.1007/978-1-4020-8786-8_6
167. Yu XL, He Y (2018) Optimal ranges of variables for an effective adsorption of lead(II) by the agricultural waste pomelo (Citrus grandis) peels using Doehlert designs. Sci Rep 8(1):1–9. https://doi.org/10.1038/s41598-018-19227-y
168. Yusuff AS, Owolabi JO, Igbomezie CO (2021) Optimization of process parameters for adsorption of heavy metals from aqueous solutions by alumina-onion skin composite. Chem Eng Commun 208(1):14–28. https://doi.org/10.1080/0098-6445.-2019.1680371
169. Yusuff AS (2018) Optimization of adsorption of Cr(VI) from aqueous solution by Leucaena leucocephala seed shell activated carbon using design of experiment. Appl Water Sci 8(8):1–11. https://doi.org/10.1007/s13201-018-0850-3
170. Zhang C, Wang X, Jiang S, Zhou M, Li F, Bi X, Xie S, Liu J (2021) Heavy metal pollution caused by cyanide gold leaching: a case study of gold tailings in central China. Environ Sci Pollut Res 28(23):29231–29240. https://doi.org/10.1007/s11356-021-12728-w
171. Zhang H, Wu X, Li X (2012) Oxidation and coagulation removal of COD from landfill leachate by Fered-Fenton process. Chem Eng J 210:188–194. https://doi.org/10.1016/j.cej.2012.08.094
172. Zhang L, Tang S, He F, Liu Y, Mao W, Guan Y (2019) Highly efficient and selective capture of heavy metals by poly(acrylic acid) grafted chitosan and biochar composite for wastewater treatment. Chem Eng J 378:122215. https://doi.org/10.1016/j.cej.2019.122215
173. Zhang X, Li M, Yang H, Li X, Cui Z (2018) Physiological responses of Suaeda glauca and Arabidopsis thaliana in phytoremediation of heavy metals. J Environ Manage 223:132–139. https://doi.org/10.1016/j.jenvman.2018.-06.025

174. Zhang Y, Song X, Zhang P, Gao H, Ou C, Kong X (2020) Production of activated carbons from four wastes via one-step activation and their applications in Pb^{2+} adsorption: insight of ash content. Chemosphere 245:1–9. https://doi.org/10.1016/j.chemosphere.-2019.125587
175. Zhao J, Yu L, Ma H, Zhou F, Yang K, Wu G (2020) Corn stalk-based activated carbon synthesized by a novel activation method for high-performance adsorption of hexavalent chromium in aqueous solutions. J Colloid Interface Sci 578:650–659. https://doi.org/10.1016/j.jcis.2020.06.031

An Introduction to Heavy Metals and Oxidative Stress in Ageing and Cancer: A Role for Detoxification Through Diet

Chanda Siddoo-Atwal

Abstract Ever wonder why there is so much variation among adults of the same age, while there is much less variation among children of a similar age? The good news is that it's not just genetics—it's largely to do with lifestyle. Science has shown that your chronological age and your biological age are two separate things. So, what ages you? There are a number of factors that are involved in this mysterious process. Any kind of illness or disease ages you significantly because of the stress it puts on your body; naturally, *degenerative diseases* are the most destructive of all. *Obesity and a poor diet with high fat content* (which can contribute to heart disease and cause skin cell senescence); *dehydration* (which is associated with chronic disease, biological ageing, increased mortality); *alcohol consumption* (which accelerates biological ageing), *regular use of hard drugs* like LSD [lysergic acid diethylamide], cocaine & heroin (which cause a decline in the brain among other organ functions); *childbirth* (which is associated with shorter telomere length, a marker of cellular ageing); and, other internal or external stressors like *lack of physical exercise* (which increases your risk of chronic diseases) and *exposure to heavy metals* (which can cause oxidative stress and induce cell senescence), are all contributors to ageing. What can you do to slow down ageing? Some anti-ageing tips are eating a Mediterranean diet; including organic superfoods in your daily diet; taking Greek superherbs in teas; drinking matcha green tea; and taking resveratrol supplements derived from grapes. Superfoods are generally rich in amino acids, sugars, essential fatty acids, vitamins, minerals, and antioxidants. Superherbs are rich in phytochemicals with antioxidant properties. Antioxidants help the body to eliminate free radicals, which contribute to ageing by causing cellular damage and oxidative stress. They can also be involved in triggering cancer. Matcha green tea contains antioxidants (polyphenols) that reduce UV damage in your skin and resveratrol is an antioxidant that scavenges free radicals, which contribute to ageing in the skin. Both are known to have anti-cancer qualities, too, since the generation of free radicals can activate apoptotic pathways resulting in carcinogenesis.

C. Siddoo-Atwal (✉)
Moondust Cosmetics Ltd., West Vancouver, Canada
e-mail: moondustcosmetics@gmail.com

N. Kumar (ed.), *Heavy Metal Remediation*, Earth and Environmental Sciences Library,
https://doi.org/10.1007/978-3-031-53688-5_12

Keywords Ageing · Heavy metals · Oxidative stress · Cancer · Superfoods · Superherbs · Carcinogenesis · Apoptosis

1 Introduction

It has been stated in many different cultures from around the world that the body is your temple and you must take care of it in order to live an optimal life. Naturally, this entails feeling young enough to pursue an active lifestyle and enjoying good health while staying physically fit for as long as possible. It has been observed that chronological age is not an accurate indicator of the ageing process. Different individual rates of ageing lead to differences between chronological and biological age [1]. Biological age is an estimate of the functional status of an individual and how well they function in comparison with others of the same chronological age. This implies that it is something that can be measured or calculated for any age group. Various biomarkers or functional parameters exist to measure biological age such as obvious markers like grey hair, skin inelasticity, and dental health or biochemical markers like **telomere length**, **cross-linking of collagen**, and **glycosylated hemoglobin**. However, no universal consensus for a definitive test has been reached so far [2]. This is understandable since an indicator like grey hair may appear at any age, while hair may retain its natural colour as other degenerative bodily processes proceed. Therefore, even though, retaining one's natural hair colour is valued in society for cosmetic reasons, it cannot be truly considered an accurate predictor of age.

There are a number of factors that are involved in the mysterious process of ageing which scientists seem to be exploring more and more as people in affluent societies start to age and the proportion of the aged in relation to younger groups starts to increase. Ageing can be described as a progressive decline in tissue homeostasis with an increasing predilection for degenerative diseases and death. It is quite clear that any kind of illness or disease ages you significantly because of the stress it puts on your body and, naturally, ***degenerative diseases*** are the most destructive of all. ***Obesity and a poor diet with high fat content*** (which can contribute to heart disease and cause skin cell senescence); ***dehydration*** (which is associated with chronic disease, biological ageing, increased mortality); ***alcohol consumption*** (which accelerates biological ageing), ***regular use of hard drugs*** like LSD [lysergic acid diethylamide], cocaine and heroin (which cause a decline in brain activity among other organ functions); ***childbirth*** (which is associated with shorter telomere length, a marker of cellular ageing); and, other external or internal stressors like ***lack of physical exercise*** (which increases your risk of chronic diseases), are all contributors to ageing [3–11]. ***Heavy metal exposure*** can also cause cell senescence and premature ageing via oxidative stress, DNA damage, and mitochondrial dysfunction [12].

2 Chronological Versus Biological Age

As a general rule, people who function better than their contemporaries may be considered as "biologically younger", while people who do not function so well may be considered to be "biologically older". Several biological parameters that best represent the ageing process could be compiled into an index consisting of these biomarkers and is a priority in gerontology. Thus far, some biological markers that have been used as predictors of ageing and in statistical modelling studies include longevity, lack of physical decline, successful ageing, active life expectancy without disability, and health expectancy without certain key diseases like cardiovascular disease. At the same time, biochemical biomarkers that can be accurately measured may be the most scientific method available for determining biological age since other physical indicators may not be so consistent from person to person. In fact, a table of average values could be compiled for a panel of standard biochemical tests in each age range.

Interestingly, in the Framingham Offspring Study it was established that people who enjoy the best quality of life for the longest period of time and live the longest share certain inherited genetic traits like a **lack of cardiovascular disease**, which is the most prevalent cause of morbidity and mortality in older age [13]. Cardiovascular disease profiles tend to be better in centenarians, their siblings, and their offspring than in controls [14]. In another informative study with Ashkenazi Jews of exceptional longevity and their offspring, both **high density lipoprotein (HDL)** and **low density lipoprotein (LDL) particle sizes** were significantly higher than in controls. This lipid profile is associated with an absence of hypertension, metabolic syndrome, and cardiovascular disease [15]. Moreover, while psychosocial variables show the least association with age [16], measures of **immune function** show a strong association [17]. Anage-related decline was observed in human blood neutrophils and lymphocytes and mouse peritoneal macrophages and lymphocytes. The lowest values of these cell types were found in elderly humans and old mice. Other factors that have received less attention but also may affect ageing include **sex** of the subject since 2 X chromosomes seem to confer an advantage by increasing cell viability and proliferative capacity [18] and genetically inherent **ethnic** differences.

Other physiological markers of note include decreases in **muscle mass** and changes in skeletal markers and **bone formation** with age. In addition, there is a decline in the number of **brain cells** and, possibly, alterations in neuropychological **cognitive functions** such as memory impairment. Histological changes in the **lung parenchyma** and decreases in ventilatory muscle strength can also contribute to a lower vital capacity and respiratory function in elderly people.

Telomere length—Telomere length is recognized as one of the best biomarkers of ageing. Shortening of the telomere occurs during each consecutive cell division and is a well-established hallmark of cell senescence since this limits the number of times a cell can divide. It is also a common feature of age-related diseases [19, 20].

Collagen cross-linking—There is biochemical evidence from several species that there is a progressive age-related deterioration of the process that controls the collagen glycoxidation rate. Pentosidine is a marker of glycoxidative stress in skin collagen and the rate of increase correlated inversely with maximum lifespan. Interestingly, dietary restriction markedly inhibited glycoxidation rate in the rodent [21].

Glycosylated hemoglobin—The presence of free radicals and cross-linked proteins can cause loss of normal functioning of tissues as a result of the ageing process. In fact, it has been shown that there is a steady increase in glycated haemoglobin levels with age [22].

3 Oxidative Stress in Ageing and Cancer

There are several theories of ageing and many involve organelles called mitochondria which are often described as the "powerhouse" of the cell. They produce the basic energy unit of the body, adenosine triphosphate (ATP), via oxidative phosphorylation and the electron transport chain. **Reactive oxygen species (ROS)** are a natural by-product of ATP production. Free radicals and ROS, which are a subset of **free radicals** containing oxygen, contribute to **oxidative stress** in the cell. They can cause **mutations** and other **DNA damage** such as strand breaks by interacting with nucleic acids in mitochondria and the nucleus. ROS can also damage cellular structures by **oxidizing phospholipid membranes** thereby leading to premature cell death. Mitochondria are the main internal source of cellular oxidative stress in the body and this process is linked to ageing in many tissues including the skin [23]. Other external sources of oxidative stress include ultraviolet radiation (UVR) [24, 25], ionizing radiation (IR) [25, 26], electromagnetic radiation (EMR) [27], environmental pollutants such as heavy metals [28], smoke [29], and radioactive elements [30, 31]. A significant amount of oxidative stress is linked to cancer in all of these models.

Classically, experimental carcinogenesis is a complex, multi-stage process including initiation, promotion, and malignant progression in which the failure of DNA repair mechanisms and the subsequent clonal expansion of mutated cells play a pivotal role [32]. However, not all carcinogens are mutagens, which suggests there may be a different mechanism [33]. Therefore, more recently, it has become apparent that the pathogenesis of many cancers is closely connected with aberrantly regulated apoptotic cell death and the resulting deregulation of cell proliferation [34, 35]. In fact, scientific animal studies have shown that simply **increasing the basal frequency of apoptosis (cell death)** in murine skin cells can be linked to the development of squamous cell carcinomas in transgenic mice [36]. In the **Two Stage Model of Tumour Formation, Step I** exposure to a carcinogen, possibly facilitated by a genetic predisposition, results in an epigenetic or genetic event causing continuous apoptotic activation of cells in the target tissue. In **Step II**, when the carcinogen may or may not be present, resistance to apoptosis and continuous cell proliferation result due to another genetic or epigenetic event [28, 37].

UV radiation exerts some of its carcinogenic and cytotoxic effects via the formation of thymine dimers and pyrimidine photodimers in DNA, but UV exposure also results in the generation of reactive oxygen species (ROS) in the cell [24]. This is true of X-rays, as well, which cause the formation of ROS in addition to DNA double-strand damage [26]. The generation of reactive oxygen species in response to microwave radiation has been observed in human cells and polycyclic aromatic hydrocarbons, which are a prominent component of outdoor air pollution such as smoke, can cause a dramatic increase in cellular oxidative stress. Radioactive elements such as plutonium and uranium are also potential sources of ROS. Moreover, oxidative stress, which can signal cell death, has been reported in cell cultures in response to heavy metals like lead, mercury, and chromium [28]. Many genes and proteins that respond to conditions of oxidative stress within the cell, subsequently trigger apoptosis. Because mitochondria are important regulators of cellular redox status, the induction of **oxidative stress** exhibits its effects upon these organelles to trigger the **intrinsic apoptotic pathway** [35, 38] and, thus, to initiate **carcinogenesis** according to the two-step model described above under the right conditions [39–41].

4 Superfoods and Superherbs

Superfoods generally have a good selection of essential nutrients including **amino acids, sugars, essential fatty acids, vitamins, minerals, and antioxidants**. The reason for this is that, frequently, superfoods like beans, nuts, and seeds are "starter kits" for life on Earth and contain the building blocks for the citric acid cycle which manufactures the body's energy currency. They often have **energizing, anti-ageing**, and **anti-cancer qualities** as a result of being concentrated nutrition packages that also contain other important micronutrients such as resveratrol. As climate-change intensifies and agricultural crop yields decrease, the importance of superfoods that deliver so many nutrients to the human body at once can only increase.

Superherbs contain high concentrations of various **phytochemical antioxidants**. For example, certain herb extracts and other plant materials are rich in **phenolic compounds,** which exhibit considerable **free-radical scavenging antioxidant activity**. As such, they can reduce the oxidative degradation of lipids and chronic oxidative stress experienced by the human body, which is linked to ageing. In addition, they have the ability to act as **complexing agents for pro-oxidant metals** and reducing agents and **quenchers of singlet oxygen** formation. These plant-derived polyphenols can also display **antifungal** and **antimicrobial properties**. In Greece, which is known for cultivating specific superherbs for use in herbal "beauty" teas, local plants including sea buckthorn, hairy rockrose, and common hawthorn leaves were found to have the highest phenolic content and antioxidant capacity among various plant species tested [42, 43]. **Polyphenols** in green and black teas derived from the leaves of the *Camellia sinensis* plant have important anti-cancer activity as well [41].

5 Some Examples of Superfoods

Soya beans (dry) contain a host of essential and non-essential nutrients required by the human body. These are protein (36%); carbohydrates (35%); lipids (19%) including essential omega-3 and omega-6 fatty acids which are necessary for structural lipids in the brain and are required for hormone production; dietary fibre (17%); minerals (5%); and several other significant components including vitamins, antioxidants such as **isoflavones**, and phytochemicals like **phytoestrogens** [17, 44–46]. Soya bean consumption is associated with lowered risks for several cancers including breast, prostate, endometrial, lung, colon, liver, and bladder cancers. Studies have also reported a correlation with a lowered incidence of cardiovascular diseases and hypercholesterolemia among other metabolic effects [45]. Isoflavones and their related flavonoid compounds, particularly **genistein**, exert antiviral properties against a wide range of DNA and RNA viruses in vitro and in vivo. A variety of host cell signalling processes can also be affected by isoflavones including induction of gene transcription factors and secretion of cytokines. As a result, they have a unique effect on immune function and anti-inflammatory activity [47, 48]. **Miso, a fermented soy product,** deserves a special mention in relation to cancer prevention. There is a famous story from Nagasaki about a medical clinic 1.4 km from the hypocenter of the plutonium atomic bomb where the head physician (Dr. Akizuki) had all the patients drink miso soup daily prior to the event. Following the event, the health effects of the nuclear radiation in the vicinity were significant on the general populace [31]. Surprisingly, though, these particular patients from the clinic did not suffer from any acute radiation disease after the bombing. This has been ascribed to the radioprotective effects of bacterial fermentation products contained within the miso soup [49]. Laboratory experiments and chemical analysis suggested that **melanoidin** and **a melanoidin-like substance** extracted from miso were responsible for the protective properties observed against irradiation [50]. Melanoidins are anti-mutagenic compounds that are able to bind metal ions and have anti-tumorigenic activity [51]. Moreover, antioxidant activity has been attributed to various soluble melanoidins [52]. Important nutritional facts for some superfoods is tabulated in Table 1.

Peanuts contain lipids (49.2%) including omega-6 essential fatty acids, protein (25.8%), carbohydrate (16%), dietary fibre (8.5%), plus a number of micronutrients including a variety of vitamins, minerals, and important antioxidants such as **resveratrol** which has been associated with reducing risk of cancer, cardiovascular and Alzheimer's disease, and delaying ageing (US Department of Agriculture website; [53, 54]. Resveratrol is particularly promising in the prevention and treatment of skin cancer [41]. However, some people are dangerously allergic to peanuts or other nuts [55].

Table 1 Important nutritional facts for some superfoods

	Protein (%)	Carbohydrates (%)	Vitamins and minerals	Main antioxidants
Soya beans [fermented]	36	35	Trace amounts of B complex vitamins, vitamin E; Ca/K/Fe/Mg/Mn/Na/S/Cu/Zn/Se	Trace amounts of isoflavones (like genistein) and phytoestrogens **[melanoidins]**
Peanuts	25.8	16	Trace amounts of B complex vitamins, vitamin E; Ca/Mg/Fe/P/K/Na/Zn/Cu/Mn/Se	Trace amounts of resveratrol
Sunflower Seeds	20	20	Trace amounts of vitamins A, B complex, C, E; Ca/Mg/Fe/P/K/Na/Cu/Zn/Mg/Se	Trace amounts of alpha, beta, and gamma tocopherols
Honey	–	83	Trace amounts of vitamins B2, B3, B5, B6, B9, and C 0.17% Ca/Cu/Fe Mg, P	Trace amounts of flavonoids, phenolic acids, carotenoids
Bee propolis Bee pollen Royal jelly	< 40 22.7 18	– 30 12–15	Trace amounts of: vitamins/minerals vitamins/minerals vitamins/minerals	Trace phenolics, flavonoids, terpenes, polyphenols, etc.

Sunflower seeds contain lipids (51%) including omega-6 essential fatty acids (while being low in saturated fat), protein (20%), carbohydrate (20%), dietary fibre (8.6%), plus a wide range of essential vitamins and minerals including zinc and selenium [56]. A number of therapeutic benefits have been ascribed to sunflower seeds including the attenuation of some widespread chronic diseases like cardiovascular and inflammatory diseases [57]. Moreover, they are a good source of **selenium**, which has been identified as one of the most promising micronutrients for cancer prevention and therapy in clinical trials for lung, colorectal, and prostate cancers while cell culture studies have shown promising results in glioma cell lines [41]. **Zinc** may have a radioprotective effect against uranium toxicity which can be caused via the initiation of oxidative stress in animal tissues and its functional role in metallothioneins may also be relevant for heavy metal detoxification in humans [30].

Honey and honey bee products like bee pollen, propolis, and royal jelly can also be classified together as superfoods on the basis that they contain many essential nutrients required by the human body. Honey is mainly composed of sucrose [83%] (which is readily absorbed and utilized by the human body instead of amino acids for energy production in the citric acid cycle), but it also contains small quantities of vitamins, minerals, and antioxidants such as flavonoids [58]. In addition to other macronutrients and micronutrients commonly found in honey bee products, bee pollen, bee propolis, and royal jelly all contain some form of protein as well [59–63]. Two antioxidant compounds (**caffeic acid esters**) in Chinese propolis have been demonstrated to have anti-cancer properties [41] while bee pollen has been found to have both hepatoprotective and anti-cancer properties [61]. The unique protein and lipid components of royal jelly have been shown to promote healthy ageing and longevity [62].

6 Some Examples of Superherbs

Many plants are valued for their essential oil content which includes complex mixtures of volatile compounds. One of the major components of essential oils are terpenes, but minor components also include aldehydes, alcohols, and esters. Plants synthesize essential oils as protection against pests and microorganisms, to attract pollinating insects, and for signalling. Important antioxidants in some superherbs is tabulated in Table 2.

Basil was known as the herb of kings by the Greeks. It is also commonly used in Italian, Iranian, Indian, and Chinese cuisines and serves as an ingredient in traditional medicines. The predominant essential oils in basil are eugenol, chavicol, and terpenoids. Its main phenolic components are phenolic acids and flavonol-glycosides. The most important antioxidant compounds are **caffeic acid**, vanillic, rosmarinic,

Table 2 Important Antioxidants in Some Superherbs

	Main antioxidant components	General therapeutic effects
Basil	Caffeic, vanillic, rosmarinic, chlorogenic, and p-hydroxybenzoic acids; quercetin, rutin, apigenin	• Antibiotic, Antivira • Antifungal • Anti-inflammatory • Anti-diabetic • Anti-pyretic • Anti-cancer
Oregano	Terpenes—terpinene, carvacrol, thymol, cymene, etc.	• Antibiotic, Antiviral • Antifungal • Anti-inflammatory • Anti-diabetic • Anti-cancer
Fenugreek	Flavonoids Phenolics Alkaloids Saponins	• Antibiotic, Anti-viral • Anti-inflammatory • Anti-cancer • Antihyperlipidemic • Hypoglycemic
Garlic [fermented]	Phenolic compounds Organic sulphides Polysaccharides [melanoidins]	• Antibiotic • Anti-cancer • Anti-hypertensive • Cardiovascular disease prevention

chlorogenic, and p-hydroxybenzoic acids, quercetin, rutin, and apigenin. Basil has shown positive effects against some viral, fungal, and bacterial infections. It also displays anti-inflammatory, immunomodulatory, anti-diabetic, anti-pyretic and anti-cancer activity [64]. In particular, caffeic acid has been demonstrated to display some potent anti-cancer and neuroprotective effects [65].

Oregano is particularly known for containing **terpenes** which are important phytochemicals with a broad range of antioxidant effects. Essential oils of oregano have been demonstrated to display antimicrobial, antiviral, and antifungal properties. In addition to being potent antioxidants, new evidence suggests that these compounds are anti-inflammatory, anti-diabetic, and, possibly, anti-cancer agents with potential benefits for human health [66].

Fenugreek is known as Greek hay and has a long history for use as a culinary and medicinal herb in the ancient world. It has been used in Ayurvedic and Chinese traditional medicine from a digestive aid to a general health tonic [67]. Fenugreek is both a herb and seed and as such it is high in protein, fat, vitamins, and minerals [68]. It contains both flavonoids and phenolics like **gallic acid**. In addition, there is a fairly high quotient of alkaloids, saponins, and other antioxidants. Pharmacological properties include anti-carcinogenic, anti-viral, anti-microbial, and anti-inflammatory activity among many other medical uses [69]. It has possible hypoglycemic and antihyperlipidemic properties, as well [67].

Garlic is definitely one of the most impressive superherbs that has been used for medicinal purposes since ancient times. It was prized for its antibiotic properties during epidemics due to the **antibacterial agent allicin** that can be derived from crushed garlic. The other main bioactive compounds in garlic are **organic sulphides**, saponins, phenolic compounds, and polysaccharides. The therapeutic properties of garlic have been demonstrated against cancer, cardiac disease, high blood pressure, diabetes, bone and skin diseases, and other pathologies. Strong antioxidant activity has been ascribed to the **phenolic compounds** in garlic. Moreover, it contains a variety of vitamins and minerals including zinc and selenium [70]. Interestingly, like fermented soya beans or miso, fermented garlic also contains various **melanoidins** [52].

7 Some Examples of Supergreens

Another interesting subset of superfoods is comprised of algae, certain green vegetables, and seaweed, which may contain protein, vitamins, minerals, and/or very special antioxidants.

Spirulina is a blue-green algae that has a long history as a food supplement dating back to the Aztec civilization. It is particularly interesting because it has been used by NASA as a successful dietary supplement for astronauts on space missions. Spirulina has a surprisingly high protein content (up to 70%); contains polysaccharides such as rhamnose and glycogen; essential lipids like γ-linolenic acid (an essential omega-3 fatty acid); vitamins [especially pro-vitamin A {**β-carotenes**}, vitamin B-12, and vitamin E {**tocopherols**}] and minerals (especially iron). In addition, it is rich in antioxidants such as phenolic acids. Since spirulina lacks cellulose cell walls, it is easily digestible as well. It is **a high energy food** that increases healthy lacto-bacillus in the intestine and facilitates the production of vitamin B-6. It exhibits anti-inflammatory and anti-allergic effects and may also display anti-cancer and anti-viral activity [71].

Wheat grass is formed from sprouting wheat grains over a period of 6–10 days with its nutritional value reaching a peak on day 7 of germination. It contains an impressive array of nutrients including amino acids; vitamins (A, B complex, C, E, and K); minerals (sulfur, copper, calcium, iodine, phosphorus, magnesium, potassium, selenium, iron, zinc, boron and molybdenum); antioxidants including bioflavonoids such as **apigenin**, quercetin and lutein, **indole compounds**, and **chlorophyll**. Although scientific studies on wheat grass are limited, it has shown potential anti-inflammatory and anti-ageing qualities [72]. Apigenin is an antioxidant with anti-inflammatory activity and it may also enhance cognitive function [73]. Wheat grass juice may be an effective way to cleanse the body of environmental pollutants like toxic heavy metals due to its high levels of antioxidant enzymes such as superoxide dismutase, amino acids, and chlorophyll (which is similar in structure to hemoglobin) [72].

Broccoli is a cruciferous vegetable that belongs to the same family as cabbage and cauliflower. It is a good source of important nutrients including fiber; vitamins A, C, and K; minerals (calcium, potassium, and iron); and special antioxidants. These notable bioactive compounds include glucosinolates, sulforaphane, and indole-3-carbinol. In addition to antioxidant activity, they also have anti-cancer and anti-inflammatory properties. **Glucosinolates** act by the induction of antioxidant and detoxifying enzymes like glutathione-S-transferases and UDP- glucuronosyl transferase which help to bind carcinogens for expulsion, while flavonoids and other phenolic antioxidants act by direct free-radical scavenging [74, 75]. Antioxidant **indole compounds** protect both lipids and proteins from peroxidation [76]. **Sulforaphane** activates cellular antioxidant and anti-inflammatory responses and is implicated in increasing cellular lifespan and preventing neurodegeneration [77].

Chlorella is a green algae that is a type of freshwater seaweed. It gained prominence in the 1950s when its use as a food source was initiated by the Japanese amid a global food crisis. Chlorella contains a high quotient of protein (55–60%); carbohydrates including dietary fiber and **antioxidant polysaccharides** (17%); lipids (11%, including omega-6 essential fatty acids); vitamins (pro-vitamin A {α/β-carotenes}, B complex, C, D, E, and K); minerals (sodium, iron, calcium, potassium, magnesium, zinc, copper, phosphorus, manganese); and antioxidants including **antheraxanthin**, **zeaxanthin**, and **lutein**. Chlorella supplementation in humans and other mammals has been reported to have immunomodulatory, hepatoprotective, antidiabetic, antihypertensive, and antihyperlipidemic effects. It has also been proven to have a detoxifying effect on heavy metal compounds such as methylmercury [78, 79].

8 Conclusion

Ageing and cancer share certain commonalities—oxidative stress causes mutations and DNA damage in the cell as a result of normal daily mitochondrial function and environmental pollutants such as heavy metals including radioactive elements. Cellular oxidative stress in the body is linked to ageing in many tissues. It can also trigger apoptosis (cell death) in cells which is directly linked to carcinogenesis. Superfoods and superherbs often contain large concentrations of various antioxidants which scavenge free radicals and specific reactive oxygen species thereby slowing down ageing and preventing cancer. Greece is known for special herbal "beauty" teas which contain high levels of such phenolic antioxidant compounds. Similarly, India, China, and Japan have their own set of antioxidant superherbs which have been used to improve human health for centuries. Thus, depending on a person's diet and other health factors, which can enhance bodily functions, not everybody ages at the same rate. This difference is known as "chronological age versus biological age" and one's outcome can be improved through a healthy lifestyle and is not solely dependent upon inherited genetic characteristics. In addition, a healthy intake of natural anti-cancer compounds in one's diet can help in cancer prevention.

Acknowledgements I would like to thank Dr. Martin von Rosen (Co-Director of the Schlosspark Klinik in Germany) for introducing me to the importance of diet in detoxification during my first visit to the clinic. After attending a science conference in Bosnia & Herzogovina in 2017, I developed certain symptoms consistent with uranium/depleted uranium exposure and had to undergo a course of metal chelation therapy at the clinic accompanied by a detoxification diet rich in superfoods, superherbs, and supergreens.

References

1. Elliott ML, Caspi A, Houts RM et al (2021) Disparities in the pace of biological aging among midlife adults of the same chronological age have implications for future frailty risk and policy. Nat Aging 1:295 308. https://doi.org/10.1038/s43587 021 00044 4
2. Karasik D et al (2005) Disentangling the genetic determinants of human aging: biological age as an alternative to the use of survival measures. J Gerontol Series A 60(5):574–587. https://doi.org/10.1093/gerona/60.5.574
3. Bachi K, Sierra S, Volkow ND, Goldstein RZ, Alia-Klein N (2017) Is biological aging accelerated in drug addiction? Curr Opin Behav Sci 13:34-39. https://doi.org/10.1016/j.cobeha.2016.09.007. PMID: 27774503; PMCID: PMC5068223
4. Booth FW, Roberts CK, Laye MJ (2012) Lack of exercise is a major cause of chronic diseases. Compr Physiol 2(2):1143-211. https://doi.org/10.1002/cphy.c110025. PMID: 23798298; PMCID: PMC4241367
5. Dmitrieva NI, Gagarin A, Liu D, Wu CO, Boehm M (2023) Middle-age high normal serum sodium as a risk factor for accelerated biological aging, chronic diseases, and premature mortality. EBioMedicine. 87:104404. https://doi.org/10.1016/j.ebiom.2022.104404. Epub 2023 Jan 2. PMID: 36599719; PMCID: PMC9873684

6. Gu C, Coomans CP, Hu K, Scheer FAJL, Stanley HE, Meijer JH (2015) Lack of exercise leads to significant and reversible loss of scale invariance in both aged and young mice. Proc Natl Acad Sci 112(8):2320–2324. https://doi.org/10.1073/pnas.1424706112
7. Kennedy BK et al (2014) Geroscience: linking aging to chronic disease. Cell 159(4):709–713. https://doi.org/10.1016/j.cell.2014.10.039
8. Liang Y, Gu T, Peng S, Lin Y, Liu JB, Wang X, Huang X, Zhang X, Zhu J, Zhao L, Fan C, Wang G, Gu X, Lin JD (2022) p16^{INK4a} plays critical role in exacerbating inflammation in high fat diet induced skin. Oxidative Med Cell Longevity 3415528(20 pages). https://doi.org/10.1155/2022/3415528
9. Pollack AZ et al (2018) Parity associated with telomere length among US reproductive age women. Hum Reprod 33(4):736–744. https://doi.org/10.1093/humrep/dey024
10. Tam BT, Morais JA, Sylvia Santosa S (2020) Obesity and ageing: two sides of the same coin. Obesity Rev 21(4). https://doi.org/10.1111/obr.12991
11. Topiwala A, Taschler B, Ebmeier KP et al (2022) Alcohol consumption and telomere length: mendelian randomization clarifies alcohol's effects. Mol Psychiatry 27:4001–4008. https://doi.org/10.1038/s41380-022-01690-9
12. Cui H, Kong Y, Zhang H, 2012. Oxidative stress, mitochondrial dysfunction, and aging. J Signal Transduct. 2012, 646354. https://doi.org/10.1155/2012/646354. Epub 2011 Oct 2. PMID: 21977319; PMCID: PMC3184498
13. Brand FN, Kiely DK, Kannel WB, Myers RH (1992) Family patterns of coronary heart disease mortality: the Framingham longevity study. J Clin Epidemiol 4:169–174
14. Evert J, Lawler E, Bogan H, Perls T (2003) Morbidity profiles of centenarians: survivors, delayers, and escapers. J Gerontol A Biol Sci Med Sci 58:232–237
15. Barzilai N, Atzmon G, Schechter C et al (2003) Unique lipoprotein phenotype and genotype associated with exceptional longevity. JAMA 290:2030–2040
16. Anstey K, Lord S, Smith G (1996) Measuring human functional age: a review of empirical findings. Exp Aging Res 22:245–266
17. de Toda IM, Maté I, Vida C, Cruces J, De la Fuente M (2016) Immune function parameters as markers of biological age and predictors of longevity. Aging (Albany NY) 8(11):3110–3119. https://doi.org/10.18632/aging.101116. PMID: 27899767; PMCID: PMC5191888
18. Shephard R (2002) Constitution or environment? In: Shephard R (ed) Gender, physical activity, and aging. CRC Press, Boca Raton (FL), pp 151–174
19. Muraki K, Nyhan K, Han L, Murnane JP (2012) Mechanisms of telomere loss and their consequences for chromosome instability. Front Oncol 2:135. https://doi.org/10.3389/fonc.2012.00135. PMID: 23061048; PMCID: PMC3463808
20. Vaiserman A, Krasnienkov D (2021) Telomere length as a marker of biological age: state-of-the-art, open issues, and future perspectives. Front Genet 21(11):630186. https://doi.org/10.3389/fgene.2020.630186. PMID: 33552142; PMCID: PMC7859450
21. Sell DR, Lane MA, Johnson WA, Masoro EJ, Mock OB, Reiser KM, Fogarty JF, Cutler RG, Ingram DK, Roth GS, Monnier VM (1996) Longevity and the genetic determination of collagen glycoxidation kinetics in mammalian senescence. Proc Natl Acad Sci USA 93(1):485–490. https://doi.org/10.1073/pnas.93.1.485.PMID:8552666;PMCID:PMC40263
22. Suvarna HIS, Moodithaya S, Sharma R (2017) Metabolic and cardiovascular ageing indices in relation to glycated haemoglobin in healthy and diabetic subjects. Curr Aging Sci 10(3):201–210. https://doi.org/10.2174/1874609810666170216124039. PMID: 28215180
23. Birch-Machin MA, Bowman A (2016) Oxidative stress and ageing. Br J Dermatol 175(Suppl. 2):26–29. https://doi.org/10.1111/bjd.14906
24. Lu YP, Lou YR, Xie JG, Yen P, Huang MT, Conney AH (1997) Inhibitory effect of black tea on the growth of established skin tumours in mice: effects on tumour size, apoptosis, necrosis, and bromodeoxyuridine incorporation into DNA. Carcinogenesis 18:2163–2169
25. Siddoo-Atwal C (2009) AT, apoptosis, and cancer: a viewpoint. Indian J Ecol 36(2):103–110
26. Liao W-C, Haimovitz-Friedman A, Persaud RS, McLoughlin M, Ehleiter D, Zhang N, Gatei M, Lavin M, Kolesnick R, Fuks Z (1999) Ataxia telangiectasia-mutated gene product inhibits DNA damage-induced apoptosis via ceramide synthase. J Biol Chem 274(25):17908–17917

27. Siddoo-Atwal C (2018) Electromagnetic radiation from cellphone towers: a potential health hazard for birds, bees, and humans. In: Current understanding of apoptosis—programmed cell death. InTech. https://doi.org/10.5772/intechopen.76084
28. Siddoo-Atwal C (2017A) Heavy metal carcinogenesis: a possible mechanistic role for apoptosis. Vegetos 30(Special). https://doi.org/10.5958/2229-4473.2017.00046.5
29. Siddoo-Atwal C (2020A) A role for heavy metal toxicity and air pollution in respiratory tract cancers. In: Heavy metal toxicity in public health. IntechOpen
30. Siddoo-Atwal C (2019A) Chapter 17—biological effects of uranium and its decay products on soil microbes, plants, and humans. In: Varma A, Tripathi S, Prasad R (eds) Plant microbe interface. Springer. https://doi.org/10.1007/978-3-030-19831-2
31. Siddoo-Atwal C (2020B). Chapter 10—Sellafield, seascale, and scandinavia: a legacy of radioactive contamination with future implications for gene evolution in affected ecosystems. In: Chen Y-C, Chen S-J (eds) Gene expression and phenotypic traits, INTECHOPEN. https://doi.org/10.5772/intechopen.82955
32. El-Abaseri TB, Putta S, Hansen LA (2006) Ultraviolet irradiation induces keratinocyte proliferation and epidermal hyperplasia through the activation of the epidermal growth factor receptor. Carcinogenesis 27(2):225–231
33. Lijinsky W (1992) Chemistry and biology of N-nitroso compounds. Cambridge University Press
34. Denmeade SR, Isaacs JT (1996) Programmed cell death (apoptosis) and cancer chemotherapy. Cancer Control 3:303–309
35. Martin KR (2006) Targeting apoptosis with dietary bioactive agents. Exp Biol Med 231:117–129
36. Van Hogerlinden M, Rozell BL, Ahrlund-Richter L, ToftgArd R (1999) Squamous cell carcinomas and increased apoptosis in skin with inhibited rel/nuclear factor-κB signaling. Can Res 59:3299–3303
37. Siddoo-Atwal C (2017B) A new approach to cancer risk assessment: an overview. Lambert Academic Publishing
38. Sun S, Hail N, Lotan R (2004) Apoptosis as a novel target for cancer chemoprevention. J Natl Cancer Inst 96:662–678
39. Siddoo-Atwal C (2015) A case-study of abnormal growth formation and regression. Published online 22 Dec 2015. www.science20.com
40. Siddoo-Atwal C, 2019B. Chapter 7—an approach to cancer risk assessment and carcinogenic potential for three classes of agricultural pesticides. In: Peshin R, Dhawan AK (eds) Natural resource management: ecological perspectives. Springer
41. Siddoo-Atwal C (2023) Chapter 59—a potential role for resveratrol and other phytochemicals in cancer chemoprevention and therapy; in "comprehensive clinical oncology: current practices", 3rd edn. Libromed Panama (edited by Restrepo CJ)
42. Parsaeimehr A, Sargsyan E, Javidnia KA (2010) Comparative study of the antibacterial, antifungal and antioxidant activity and total content of phenolic compounds of cell cultures and wild plants of three endemic species of Ephedra. Molecules 15:1668–1678. https://doi.org/10.3390/molecules15031668
43. Roidaki A, Kollia E, Panagopoulou E, Chiou A, Varzakas T, Markaki P, Proestos C (2016) Super foods and Super herbs: antioxidant and antifungal activity. Curr Res Nutr Food Sci. 4(Special Issue Conference October 2016). https://doi.org/10.12944/CRNFSJ.4.Special-Issue-October.19
44. Prabakaran M, Lee KJ, An Y, Kwon C, Kim S, Yang Y, Ahmad A, Kim SH, Chung IM (2018) Changes in Soybean (*Glycine max* L.) flour fatty-acid content based on storage temperature and duration. Molecules. 23(10):2713. https://doi.org/10.3390/molecules23102713. PMID: 30347888; PMCID: PMC6222662
45. Sherif M (2013) Soybean, nutrition and health. Soybean—Bio-Active Compounds. InTech. https://doi.org/10.5772/54545
46. Simopoulos AP (2016) An increase in the Omega-6/Omega-3 fatty acid ratio increases the risk for obesity. Nutrients 8(3):128. https://doi.org/10.3390/nu8030128. PMID: 26950145; PMCID: PMC4808858

47. Andres A, Donovan SM, Kuhlenschmidt MS (2009) Soy isoflavones and virus infections. J Nutr Biochem 20(8):563–569. https://doi.org/10.1016/jnutbio.2009.04.004
48. Sakai T, Kogiso M (2008) Soy Isoflavones and Immunity. J Med Invest 55(3–4):167–173. https://doi.org/10.2152/jmi.55.167
49. Watanabe H (2013) Beneficial biological effects of miso with reference to radiation injury, cancer and hypertension. J Toxicol Pathol 26(2):91–103. https://doi.org/10.1293/tox.26.91. Epub 2013 Jul 10. PMID: 23914051; PMCID: PMC3695331
50. Watanabe H, Shiono T, Kawamura D, Kawano K, Kajimura J, Kashimoto N, Kamiya K (2006) Radioprotective effects of melanoidine against irradiation in B6C3F1 mice. J Nagasaki Med Ass 81:296–298 (in Japanese)
51. Langner E, Rzeski W (2014) Biological properties of melanoidins: a review. Int J Food Prop 17(2):344–353. https://doi.org/10.1080/10942912.2011.631253
52. Kim JS (2020) Antioxidant activity of various soluble melanoidins isolated from black garlic after different thermal processing steps. Prev Nutr Food Sci 25(3):301–309. https://doi.org/10.3746/pnf.2020.25.3.301. PMID: 33083380; PMCID: PMC7541931
53. Brenna JT, Akomo P, Bahwere P, Berkley JA, Calder PC, Jones KD, Liu L, Manary M, Trehan I, Briend A (2015) Balancing omega-6 and omega-3 fatty acids in ready-to-use therapeutic foods (RUTF). BMC Med 13:17. https://doi.org/10.1186/s12916-015-0352-1. PMID: 25980919; PMCID: PMC4433071
54. Sales JM, Resurreccion AV (2014) Resveratrol in peanuts. Crit Rev Food Sci Nutr 54(6):734-770. https://doi.org/10.1080/10408398.2011.606928. PMID: 24345046
55. Al-Muhsen S, Clarke AE, Kagan RS (2003) Peanut allergy: an overview. CMAJ 168(10):1279–1285. Erratum in: CMAJ. 2003 Jun 10; 168(12), 1529. PMID: 12743075; PMCID: PMC154188
56. Djuricic I, Calder PC (2021) Beneficial outcomes of omega-6 and omega-3 polyunsaturated fatty acids on human health: an update for 2021. Nutrients 13(7):2421. https://doi.org/10.3390/nu13072421. PMID: 34371930; PMCID: PMC8308533
57. Khurana S, Singh R (2021) Sunflower (*Helianthus annuus*) seed. In: Tanwar, B, Goyal A (eds) Oilseeds: health attributes and food applications. Springer, Singapore. https://doi.org/10.1007/978-981-15-4194-0_5
58. Siddoo-Atwal C, Atwal AS (2012) A possible role for honey bee products in the detoxification of mycotoxins. In: ISHS Acta Horticultura 963: I international symposium on mycotoxins in nuts and dried fruits
59. Gabrys J, Konecki J, Krol W, Scheller S, Shani J (1986) Free amino acids in bee hive product (propolis) as identified and quantified by gas-liquid chromatography. Pharmacol Res Commun 18(6):513–518. https://doi.org/10.1016/0031-6989(86)90146-3. PMID: 3749241
60. Khalifa SAM, Elashal MH, Yosri N, Du M, Musharraf SG, Nahar L, Sarker SD, Guo Z, Cao W, Zou X, Abd El-Wahed AA, Xiao J, Omar HA, Hegazy MF, El-Seedi HR (2012) Bee pollen: current status and therapeutic potential. Nutrients. 13(6):1876. https://doi.org/10.3390/nu1306 1876. PMID: 34072636; PMCID: PMC8230257
61. Komosinska-Vassev K, Olczyk P, Kaźmierczak J, Mencner L, Olczyk K (2015) Bee pollen: chemical composition and therapeutic application. Evid Based Complement Alternat Med 297425. https://doi.org/10.1155/2015/297425. Epub 2015 Mar 11. PMID: 25861358; PMCID: PMC4377380
62. Kunugi H, Ali AM (2019) Royal jelly and its components promote healthy aging and longevity: from animal models to humans. Int J Mol Sci. 20(19):4662. https://doi.org/10.3390/ijms20 194662. PMID: 31547049; PMCID: PMC6802361
63. Pasupuleti VR, Sammugam L, Ramesh N, Gan SH (2017) Honey, propolis, and royal jelly: a comprehensive review of their biological actions and health benefits. Oxid Med Cell Longev 1259510. https://doi.org/10.1155/2017/1259510. Epub 2017 Jul 26. PMID: 28814983; PMCID: PMC5549483
64. Shahrajabian MH, Sun W, Cheng Q (2020) Chemical components and pharmacological benefits of Basil (Ocimum basilicum): a review. Int J Food Prop 23(1):1961–1970. https://doi.org/10.1080/10942912.2020.1828456

65. Alam M, Ahmed S, Elasbali AM, Adnan M, Alam S, Hassan MI, Pasupuleti VR (2022) Therapeutic implications of caffeic acid in cancer and neurological diseases. Front Oncol. 12:860508. https://doi.org/10.3389/fonc.2022.860508. PMID: 35359383; PMCID: PMC8960963
66. Leyva-López N, Gutiérrez-Grijalva EP, Vazquez-Olivo G, Heredia JB (2017) Essential oils of oregano: biological activity beyond their antimicrobial properties. Molecules 22(6):989. https://doi.org/10.3390/molecules22060989.PMID:28613267;PMCID:PMC6152729
67. Basch E, Ulbricht C, Kuo G, Szapary P, Smith M (2003) Therapeutic applications of fenugreek. Altern Med Rev 8(1):20–27
68. Altuntaş E, Özgöz E, Taşer OF (2005) Some physical properties of fenugreek (Trigonella foenum-graceum L.) seeds. J Food Eng 71(1):37–43. ISSN 0260-8774. https://doi.org/10.1016/j.jfoodeng.2004.10.015
69. Ahmad A, Alghamdi SS, Mahmood K, Afzal M (2016) Fenugreek a multipurpose crop: potentialities and improvements. Saudi J Biol Sci. 23(2):300–310. https://doi.org/10.1016/j.sjbs.2015.09.015. Epub 2015 Sep 14. PMID: 27307778; PMCID: PMC4894452
70. Ansary J, Forbes-Hernández TY, Gil E, Cianciosi D, Zhang J, Elexpuru-Zabaleta M, Simal-Gandara J, Giampieri F, Battino M (2020) Potential health benefit of garlic based on human intervention studies: A brief overview. Antioxidants (Basel) 9(7):619. https://doi.org/10.3390/antiox9070619. PMID: 32679751; PMCID: PMC7402177
71. Karkos PD, Leong SC, Karkos CD, Sivaji N, Assimakopoulos DA (2011) Spirulina in clinical practice: evidence-based human applications. Evid Based Complement Alternat Med 2011:531053. https://doi.org/10.1093/ecam/nen058. Epub 2010 Oct 19. PMID: 18955364; PMCID: PMC3136577
72. Hattarki S, Bogar C (2017) Triticum aestivum (wheat grass); a power house plant—a review. Dent J Adv Stud 05(01):025–029. https://doi.org/10.1055/s-0038-1672077
73. Salehi B, Venditti A, Sharifi-Rad M, Kręgiel D, Sharifi-Rad J, Durazzo A, Lucarini M, Santini A, Souto EB, Novellino E, Antolak H, Azzini E, Setzer WN, Martins N (2019) The therapeutic potential of apigenin. Int J Mol Sci 20(6):1305. https://doi.org/10.3390/ijms20061305. PMID: 30875872; PMCID: PMC6472148
74. Taymour S, Alreshidi KSM, Elhassan Taha MM, Mohan S (2023) Broccoli: a multi-faceted vegetable for health: an in-depth review of its nutritional attributes, antimicrobial abilities, and anti-inflammatory properties. Antibiotics (Basel) 12(7):1157. https://doi.org/10.3390/antibiotics12071157. PMID: 37508253; PMCID: PMC10376324
75. Williamson G, Faulkner K, Plumb GW (1998) Glucosinolates and phenolics as antioxidants from plant foods. Eur J Cancer Prev 7(1):17–21. PMID: 9511848
76. Süzen S (2007) Antioxidant activities of synthetic indole derivatives and possible activity mechanisms. In: Khan MTH (eds) Bioactive Heterocycles V. topics in heterocyclic chemistry, vol 11. Springer, Berlin, Heidelberg. https://doi.org/10.1007/7081_2007_074
77. Santín-Márquez R, Alarcón-Aguilar A, López-Diazguerrero NE, Chondrogianni N, Königsberg M (2019) Sulforaphane—role in aging and neurodegeneration. Geroscience. 41(5):655–670. https://doi.org/10.1007/s11357-019-00061-7. Epub 2019 Apr 2. PMID: 30941620; PMCID: PMC6885086
78. Bito T, Okumura E, Fujishima M, Watanabe F (2020) Potential of *Chlorella* as a dietary supplement to promote human health. Nutrients. 12(9):2524. https://doi.org/10.3390/nu12092524. PMID: 32825362; PMCID: PMC7551956
79. Peñalver R, Lorenzo JM, Ros G, Amarowicz R, Pateiro M, Nieto G (2020) Seaweeds as a functional ingredient for a healthy diet. Marine Drugs 18(6):301. https://doi.org/10.3390/md18060301. PMID: 32517092; PMCID: PMC7345263

Propolis as a Bioindicator of Contamination with Toxic Metals

Ana Paola Balderrama-Carmona, Víctor Ramos-García, Liliana Ruiz-López, and Ximena Felipe-Ortega-Fonseca

Abstract Bioindicators are living organisms used to predict the level of environmental pollution in a certain area. These organisms and their products are sensitive to surrounding changes in the concentrations of toxic components that, whether due to geographical or anthropogenic origin, can increase their presence. High agricultural, mining, and industrial activity can be sources of chemical contaminants, among these, toxic metals are the most common. Contamination with heavy metals is a public health concern because they can trigger different pathologies associated with their exposure. This is why constant monitoring is important to maintain nce doses below the limits recommended by international regulations. In recent years, bee products have been suggested as bioindicators of contamination of toxic compounds, among them, the most susceptible is propolis. This research aims to analyze the main studies that propose propolis as a bioindicator of contamination with toxic metals.

Keywords Industrial areas · Cadmium · Lead · Principal component analysis

1 Introduction

Propolis is produced by bees, in all their species, to cover the gaps in the hives, and protect them from the weather and invading organisms, which is why it is also called "bee glue" [1–3]. The bee leaves the hive and, from exudates collected from the surrounding plants [4, 5], produces propolis through chewing, where salivary enzymes modify the composition of the gum and produce a resinous substance in

A. P. Balderrama-Carmona (✉) · L. Ruiz-López · X. Felipe-Ortega-Fonseca
Department of Chemical Biological and Agricultural Sciences, University of Sonora, Navojoa, Sonora, Mexico
e-mail: paola.balderrama@unison.mx

V. Ramos-García
Department of Physics, Mathematics and Engineering, Universidad de Sonora, Navojoa, Sonora, Mexico

N. Kumar (ed.), *Heavy Metal Remediation*, Earth and Environmental Sciences Library,
https://doi.org/10.1007/978-3-031-53688-5_13

its highest percentage, which also contains wax, essential oils, pollen and bioactive compounds [6] which have given it antioxidant [7, 8], antibacterial [9], antifungal [10], antiviral [11], and antiproliferative activity [12], this is the reason for its utilization since ancient times due to its therapeutic properties [13].

The bioactive substances of propolis, among which flavonoids and terpenes are distinguished [14–16], depending on their quantity and quality on the flora that was used to collect the exudates, lead to saying that the components of propolis have an exclusive environmental origin of the geographical area where they are produced and confers the capacity as a biomarker of environmental quality [15, 17, 18]. Bees are insects that are very sensitive to climate changes [19, 20], additionally to having constant exposure to pollutants, when they go out to collect nectar, they are in contact with chemical residues from plants, air, water, and soil, furthermore, they can travel an average distance of 5.5 km from the hive [21], and taking into account that each hive can house 10,000–80,000 bees [19, 20], the sweep of the area is exhaustive, which means that the bees when exposed to toxic elements, take them, to the hive and the products it produces such as honey, wax, royal jelly, and propolis [22], this is why its use as biomarkers or bioindicators is suggested because these products serve to predict the damage caused by some type of contaminant in the environment [23].

Environmental contaminants can be chemical or biological and of natural or anthropogenic origin [24]. Among the chemical pollutants of anthropogenic origin are the emerging ones represented mainly by pharmaceuticals, pesticides, organic pollutants, industrial substances, resistant microorganisms [25], and microplastics, on the other hand, there are heavy metals that can be of natural or anthropogenic origin; natural contamination with minerals is called geogenic [24]. Heavy metals are chemical elements, and although they are essential for life in small quantities, they can be highly toxic when they accumulate in excess in the environment and living organisms, which is why they are classified as persistent and bioaccumulative pollutants in the environment [22]. Because it not only includes metals but also metalloids, the term heavy metal is out of use and is currently defined as a toxic metal or potentially toxic element [26], these metals include lead, cadmium, mercury, and arsenic, are known for their harmful effects on human health and the ecosystem in general. Heavy metal pollution is a global problem that poses significant challenges to public health and environmental conservation; this is because they produce harmful effects on the organisms that consume them, with those with greater sensitivity and exposure being the most affected.

The evaluation of environmental pollution is necessary to monitor and detect the presence of toxic metals in nature, so the search for effective methods to achieve this has become essential [27]. Bioindicators can accumulate and reflect the concentration of these metals, among these bioindicators, propolis has gained recognition in recent years due to its ability to accumulate heavy metals [28, 29]. In this review, we will explore in detail the ability of propolis as a bioindicator of toxic metals, analyzing its chemical structure, its accumulation properties, and the methods used for its analysis. Furthermore, we will examine the benefits and limitations of its application in environmental monitoring studies and its relevance in the protection of human health and the conservation of biodiversity. The objective of this research is to analyze the potential of propolis as a tool for the detection and evaluation of contamination with toxic metals, contributing to the understanding and mitigation of one of the most urgent environmental challenges of our time.

2 Bioindicators of Toxic Metals

A bioindicator is a living organism integrated within a broader ecological system or community, capable of providing pertinent data concerning the quantitative dimensions of the ecological integrity within its habitat [30]. Bioindicators are creatures that endure prolonged exposure to a variety of toxic substances, which may include but are not limited to heavy metals, metalloids, nitrogen, phosphorus, pesticides, radionuclides, emerging pollutants, and more [31]. This exposure to contaminants may be due to pollution related to industrial activity, mining, agriculture, or even natural disasters. Examples of these bioindicators encompass a wide array of living organisms, including animals, plants, microbial systems, fungi, and humans, and can be used as bioindicators of environmental pollution by toxic metals due to their capacity to absorb these substances from air, water, sediment, ground and the trophic chain [30].

According to various authors [32–34] bioindicating organisms must exhibit several key characteristics. These include having a prolonged life expectancy that accommodates various maturity stages, the ability to accumulate toxic substances or pollutants in significant concentrations without fatally harming the host, occupying the uppermost levels of the trophic chain, inhabiting distinct and well-defined geographical regions to enable straightforward localization and convenient sampling, and existing in abundant quantities or at least in sufficient numbers that facilitate multiple samplings for research purposes. Another desirable characteristic is the size of the organism, which should be substantial enough to permit dissection for cellular-level analysis. Additionally, the ability to cultivate these organisms for research purposes is another sought-after feature.

3 Contamination with Metals and Metalloids

3.1 Geographic Origin

Propolis is an antiseptic that prevents the growth of microorganisms in the hive [35, 36]. Bees are found in dry tropical areas, humid forests, and some subtropical regions around the world [37], such as in Brazil [38], Bulgaria [39], India [40], Indonesia [41], Malaysia [42], Mexico [43], Thailand [44], Colombia [45]. Turkey [46], Croatia [47], China [48], Brunei [49], Argentina [50] and Vietnam [51].

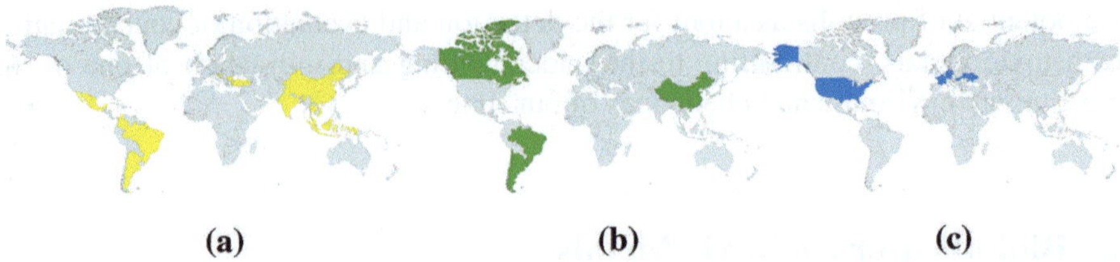

Fig. 1 Globally: **a** regions with the highest bee population in tropical areas, **b** major global propolis producers, and **c** primary consumers of propolis

Presently, Brazil stands as one of the primary producers of propolis [52]. Nevertheless, other notable countries engaged in the production of this substance encompass China, Argentina, Chile, Cuba, Uruguay, and Canada. Conversely, in terms of consumption, France, Germany, Ukraine, Denmark, Hungary, and the United States emerged as the most significant consumers of propolis, as illustrated in Fig. 1.

Propolis characteristics may vary, depending on the bee species, chemical composition, temperature and humidity of each region, and the state of maturity of the vegetation [53, 54], propolis can be classified into five different types [53]: Type I propolis (resinous exudate), it appears in warm climates (west of Asia, Europe and North America), type II also called green propolis comes predominantly from Brazil. Type III is found in Venezuela and Cuba, where the principal source of resins comes from cusia flowers. Type IV is in the Pacific specifically in places like Taiwan and Oceania. Lastly, Type V is in the Baltic peninsula like in Turkey, Greece and, Crete [52].

Conversely, Mexico holds one of the foremost positions on the global stage in honey production [55]. However, it lags in the development of propolis, as the chemical composition and biological activity of this substance in the country are limited and, at times, non-existent for its production [52].

3.2 Anthropogenic Activities that Cause Metal Contamination in Propolis

Each organic entity within a biological system, such as honey-producing bees, helps indicate the health of their environment. Therefore, these species are relevant bioindicators that allow predicting the degree of pollution in each region [56]. Therefore, bees have become a suitable marker for application as a bioindicator due to their vast patrol area and their advantageous body characteristics by which the metal content of their habitat can be identified. These bees involuntarily accumulate contaminants (such as heavy metals) that are present in flowers, along with pollen, nectar, water, and propolis in search of food [57].

In multiple regions, soil and water reservoirs are contaminated with heavy metals, so the need to search for food is significantly influenced by bioaccumulation [58].

These toxic metals are contaminants that can affect not only human health but also reduce the diversity of plant and animal species [59, 60]. The quantification of the concentration of toxic metals in the air, soil, and water is important to understand the biogeochemical processes and monitor the health of the ecosystem, due to different anthropogenic and geogenic factors, which help increase the metal content in the air, soil, the water, plants, and other elements worldwide [57, 61–63]. Therefore, metal accumulation in bee hives is a relevant opportunity to monitor the environmental conditions of a certain region [19, 20].

The concentration of heavy metals can be identified in different environmental behaviors, such as volcanic [64], the agricultural sector [65], urban areas [66], natural reserves [67], Islands [68] and mountains [69]. Furthermore, in the mining industry [70] and industries in the manufacturing sector [25], such as manufacturing, combustion, and pesticide processes, they release toxic byproducts into the environment that affect ecosystems. and humans [29], and are the main sources responsible for generating considerable concentrations of heavy metals that are discharged into bodies of water, which is why they are biomagnified [71, 72], in turn, these metals can sediment and be stored in riverbeds, they can also infiltrate into groundwater [60, 73]. It has been reported in the scientific literature that, worldwide, sites located near mines contain metals at levels that exceed the criteria for drinking and surface water [60, 73–75]. High concentrations of metals in the bodies of living beings affect biochemical and physiological processes, producing various pathologies as chronic exposure to these types of heavy metals can cause skin lesions, neurological disorders, diabetes, cardiovascular problems, and cancer [76].

In modern agriculture, it has been considered one of the largest sources of heavy metals, due to the use of fertilizers in the soil, especially phosphorus and nitrogen fertilizers, pesticides, compost derived from waste, and sludge from sewage treatment plants, water, and manure [77, 78]. Therefore, these metals can migrate upward by absorption in the vegetation (Barisic et al. 2002). Therefore, the persistence of contaminants in the soil is much longer than in other compartments of the biosphere, which is permanent [79].

In pollution in the beekeeping sector, bees can capture pollutants diffused in the air, deposited on the surface of plants, or absorbed by them from the soil, or they can assimilate pollutants located in water [80]. In the case of bee honey, honeybee honey is generated from foraging nectar in flowers and is processed by said insects, consequently, it contains the lowest concentrations of contaminants [81, 82]. Pollen, on the other hand, has proven to be a good bioindicator of contamination by heavy metals and pesticides in vegetation [83]. Contamination in the food chain is the most important exposure route for people [84], so the elimination of heavy metals in food and water is one of the main challenges for food safety [73].

3.3 *Toxicity of Metals and Metalloids*

Metals are components of minerals found in soil; they are significant due to their biological activity in the development of plants and living beings. Some heavy metals are essential such as zinc, selenium, copper, and manganese, and have important biological functions in living beings. However, they can be toxic when present at prominent levels or when homeostasis is affected [85]. The toxic metals or heavy metals as they are commonly called include Pb, Cd, Hg, As, Ni, and Cr. Once the metals are emitted, they are easily absorbed through biological membranes, due to their high affinity. Chemical to proteins, which facilitates their bioaccumulation and makes their degradation and elimination difficult, increasing the risk of toxicity in humans [75].

Metals are found at low levels in water and food, these being the main contact with metals; However, the main source of toxicity is through work activities such as agriculture, industry, and mining. In some jobs there are specific exposure risks to certain elements such as lead in the paint, battery, and electrical cables industry; exposure to mercury in electrolysis operations and dental mechanics due to exposure to beryllium, mercury, and nickel [86].

In biological organisms, the toxicity of heavy metals occurs through enzymatic inactivation due to the formation of bonds between the metal and sulfhydryl groups (-SH), as well as other functional groups of proteins and enzymes, blocking biological activities that cannot be reversed. Other routes of heavy metal toxicity are the displacement reactions of different metal ions or the modification of the active structure of biological molecules [87].

The WHO considers toxic metals a threat to health due to the harmful and pathological effects they cause, such as genetic changes that determine congenital malformations, and fertility problems, among others; that occur due to their high toxicity in the body, causing the induction of oxidative stress, inflammation, apoptosis and pathological alterations in signaling pathways [88].

3.3.1 Mechanisms of Lead Toxicity

Lead (Pb) is one of the toxic metals with the greatest environmental pollution, due to its non-biodegradable character that makes it persistent in the environment [75]. The main routes of lead absorption are the gastrointestinal, respiratory system, and to a lesser extent through the skin. Some studies indicate that prolonged exposure to small doses of lead greater than 5 μg/dL is associated with subclinical effects such as hypertension, damage to kidney function, cognitive dysfunction, and reproductive disorders [88]. Likewise, these doses have toxic effects on the liver, nervous system, and hematopoietic system in addition to being carcinogenic.

The harmful effects of Pb occur after the induction of oxidative stress an increase in reactive oxygen species (ROS), and a decrease in antioxidant levels. It has also been observed that Pb affects biological activities regulated by calcium and the generation of genotoxic compounds; and alters cellular processes such as ion transport, protein folding, and maturation. Likewise, it causes irregularities that affect the nervous system and cell differentiation and maturation [85].

The Codex Alimentarius reports a maximum limit for lead and cadmium of 0.1 mg/kg and established a provisional tolerable monthly intake of 25 μg/kg of body weight since lead is associated with a decrease of at least three points in the intelligence quotient in children and an increase in blood pressure in adults [89].

3.3.2 Mechanisms of Mercury Toxicity

Among toxic metals, mercury (Hg), like lead, is an environmental contaminant generated from natural and anthropogenic sources. In the atmosphere, it is found in elemental form (Hg0) and accumulates through terrestrial and aquatic food webs as an organic form (methylmercury, the most toxic form). Suspended Hg particles are transported by air in the atmosphere and are considered an indicator of urban and industrial pollution [86].

The most important pathways of mercury toxicity are Hg-induced ROS generation and the onset of oxidative stress. Most of it is absorbed in the gastrointestinal tract and some in the brain and blood. The main organ of inorganic Hg accumulation is the kidneys (Famurewa, 2022), as well as the brain, but it can affect other organs such as nerves, and muscles. It alters membrane potential by disrupting intracellular calcium homeostasis. Mercury binding to selenohydryl and sulfhydryl groups when reacting with methylmercury also damages protein structure and alters cellular function. Methylmercury is a neurotoxic compound that is responsible for microtubule destruction, mitochondrial damage, lipid peroxidation, and accumulation of neurotoxic molecules such as serotonin, aspartate, and glutamate [90].

The initial manifestations of poisoning are lung irritation, eye irritation, skin rashes, vomiting, and diarrhea. In the long term, it can cause chronic kidney disease, alterations in behavior, mood, and sleep; and reproductive problems (abortions, infertility, and stillbirth). Dysfunction of glands such as the thyroid and adrenal glands can also be observed; anemia, heart, and blood vessel diseases (heart attacks, high cholesterol levels). Respiratory difficulty, tooth decay, pneumonia, kidney failure, and death [91].

Hg (organic) compounds generate a host of conditions in important biological systems such as the central nervous system (CNS), immune, cardiovascular, and reproductive systems, affecting kidney function, structural bone deformity, and developmental processes. He has been found to interfere with the functioning of enzymes, altering protein synthesis and disrupting energy production [75].

3.3.3 Mechanisms of Cadmium Toxicity

Cadmium (Cd) is a byproduct of zinc and copper processing; as well as the production of nickel and cadmium batteries and, the manufacturing of cement and phosphate fertilizers; Cadmium reacts forming oxides, chlorides, and sulfides, thus being one of the main soil contaminants [91].

In humans and other mammals, the absorption of cadmium is equal to that of essential metals such as iron and zinc; caused by calcium and iron deficiencies in the diet or low protein diets. The entry of Cd into cells is different in all systems and can be mediated by passive or active transport, or by calcium channels. One of the mechanisms is due to reactive oxygen species (ROS) that act as second messengers and alter different cell signaling pathways and anti-apoptotic responses. This is of medical importance related to several types of cancer, which is why Cd is classified as category I carcinogenic [92].

Cd-induced oxidative stress is an important mechanism causing kidney and liver diseases. It has also been discovered that Cd causes damage to mitochondria since they are responsible for the formation of ROS causing an alteration in the redox balance. Epidemiological studies have reported that exposure to toxic metals, such as Cd, can promote the development of musculoskeletal diseases, such as osteoporosis, rheumatoid arthritis (RA), and osteoarthritis (OA). Cadmium can increase bone resorption, affect osteoclast activity and the absorption of calcium (Ca), and impair kidney function, which favors the development of osteoporosis. In the case of RA, Cd interferes with the activity of antioxidant proteins, such as superoxide dismutase (SOD) and catalase (CAT). It also promotes an inflammatory state, inducing the citrullination process, which affects the proteins of the immune response [93].

The different conditions of the organisms occur in various organs and tissues such as the kidney (producing tubular renal dysfunction, proteinuria, and chronic renal failure), and heart (producing aortic and coronary atherosclerosis, increase in cholesterol and fatty acids); Likewise, the bones, testicles, placenta, central and peripheral nervous system are affected. The lung is a very susceptible organ to exposure to cadmium; subacute chronic inhalation can produce bronchitis with progressive alveolar damage, secondary fibrosis, and emphysema [94].

3.3.4 Mechanisms of Arsenic Toxicity

Arsenic (As) is a natural element with oxidation $-3, 0, +3$, and $+5$, it is distributed in the Earth's crust, and is released into the environment; air, water, and land; as well as anthropogenic contamination. Most arsenical compounds in organisms and foods are in the pentavalent oxidation state; in food [95]. The highest concentrations of this element are found in fish and seafood, followed by meats and grains. In fruits, vegetables, and dairy products they are present in lower concentrations. On the other hand, we can see that rice (contains monomethyl arsenic acid ($MMA^{+5)}$, dimethylarsinic acid ($DMA^{+5)}$, arsenic tri-glutathione (ATG), monomethyl arsenic acid di-glutathione (MADG) and dimethyl also—glutathione (DMAG) and seaweed (rich in arsenosugars) have the highest arsenic concentrations in the world [95].

The toxicity of As depends on its form and oxidation state, in addition to other factors, such as dose, duration, and frequency of exposure. In the human body, a series of successive reductions and oxidative reactions, as well as methylation reactions transform inorganic arsenic to produce different metabolites that depending on the valence will be the reactivity, half-life, and distribution in the body. Therefore, each species in which As is present has different harmful effects, such as stomach irritation, skin damage, decrease in red blood cells, in addition, to being a carcinogenic agent. It is related to lung cancer, in this sense there is the possibility of synergy with smoking [91].

The toxicity mechanisms that have been most reported at the subcellular and cellular level are the generation of reactive oxygen species (ROS) and nitrogen species (RNS) related to the differentiation and proliferation of keratinocytes, endothelial dysfunction, resistance to insulin, with the induction of lipid peroxidation in the liver, renal tubular necrosis and with changes in the expression of the estrogen receptor. As can cause alterations in molecular and cellular processes, which are reflected in different pathologies. It has interference with multiple enzyme systems. Binding to sulfhydryl groups (proteins, glutathione, cysteine) affects enzymes of cellular respiration, gluconeogenesis, glucose uptake, and glutathione (As $^{+3)}$ metabolism [96].

4 Propolis as a Bioindicator of Toxic Metals

Propolis is resinous and sticky, its chemical composition varies depending on its geographical location, the season, and the bee species that produce it, and it has a complex mixture of compounds [97, 98]. The main characteristics that make it effective as a bioindicator of toxic metals are its organic components and their chemical functionality, adsorption, and bioaccumulation.

4.1 Organic Components of Propolis

Propolis contains a wide variety of organic compounds, such as flavonoids, terpenoids, and phenols [14–16, 99]. These organic compounds can form complexes with toxic metals [100, 101], which facilitates the accumulation of these elements in propolis. Flavonoids (quercetin, kaempferol, luteolin naringenin, rutin, galantine, apigenin, pinobanksin 5-methyl ether) and other phenols (p- coumaric coffee and trans ferulic acid, catechin) [102] They contain functional groups such as hydroxyls (–OH) and carbonyls (C = O) that can act as ligands (Fig. 2). These groups have lone pairs of electrons that can interact with metal ions [103].

4.1.1 Chemical Structure

Functional groups can interact and bind to toxic metals through chelation processes. This chelation ability allows propolis to trap and retain toxic metals in its matrix

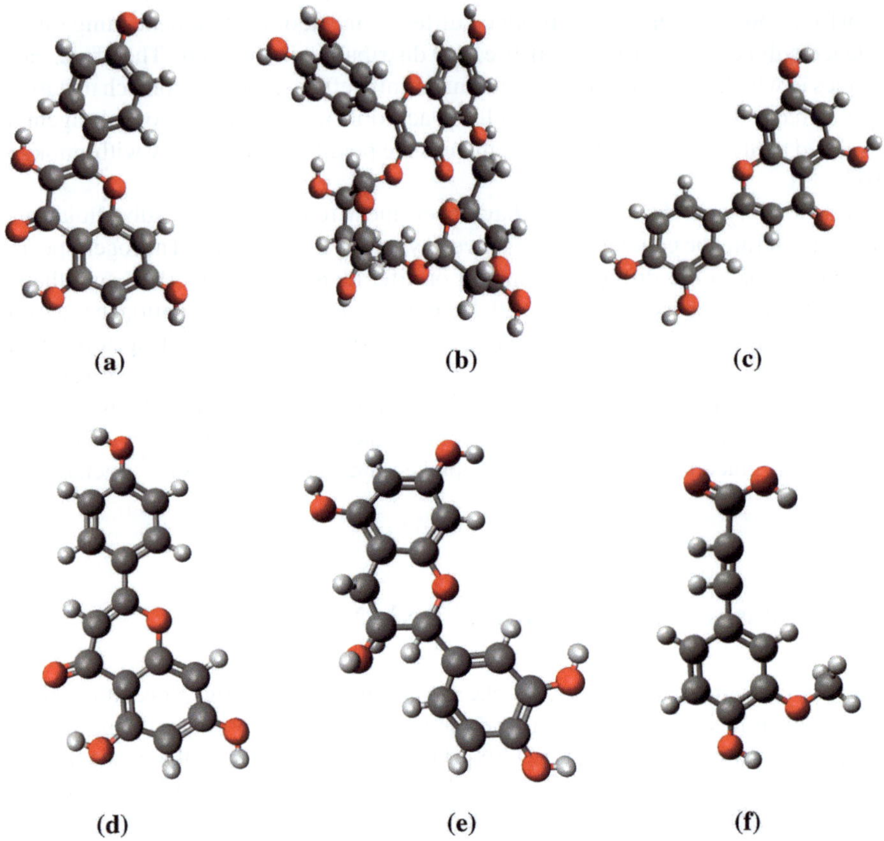

Fig. 2 Main chemical components of propolis: **a** Kaempferol, **b** Rutin, **c** Luteolin, **d** Apigenin, **e** Cathechin, **f** Ferulic acid

[104]. When toxic metal ions, such as Pb^{2+} [105, 106], Cd^{2+} [107], and Hg^{2+} [108], are present in the environment, the ligands in propolis can interact with these ions. Ligands donate electron pairs to metal ions, creating coordinated chemical bonds between the metal and the ligand [103]. This results in the formation of a table metal–ligand complex (Fig. 3). The complexes formed are more stable than free metal ions in the environment [101]. This means that the metal is "trapped" within the complex, making it difficult to release and transport in the body or environment [109]. The formation of these metal–ligand complexes in propolis is what allows it to function as a bioindicator of toxic metals. The number of complexes formed is related to the concentration of toxic metals in the environment, making it easier to detect and measure heavy metal contamination.

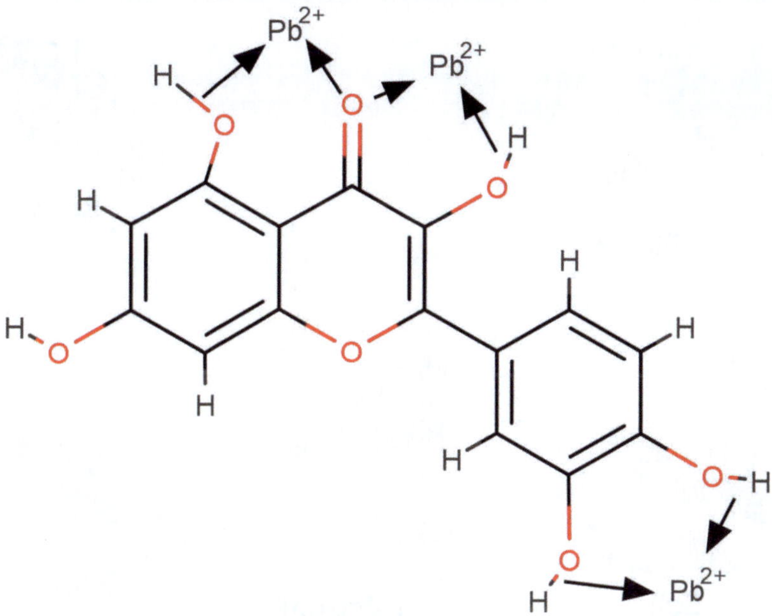

Fig. 3 Anionic lead coordination in the possible chelating sites of quercetin

4.2 *Adsorption and Absorption of Metals in Propolis*

Propolis has a porous structure that allows it to retain particles suspended in air or water [110]. This is known as physical adsorption, and it is a process in which particles adhere to the surface of propolis due to physical forces, such as Van der Waals forces or electrostatic forces [111]. In addition to adsorption on the surface, some metal ions can be absorbed by propolis and diffuse into its interior. This means that propolis can effectively accumulate toxic metals from its environment.

4.3 *Bioaccumulation of Metals in Propolis*

Bees collect plant resins from various sources, such as trees and plants, to produce propolis [112]. These resins contain lesser amounts of toxic metals that may have been absorbed from the soil or surrounding air, subsequently, the bees process these collected plant resins, mixing them with salivary enzymes and other components to create propolis [113]. During this process, some of the toxic metals present in the resins may be released and dispersed in the hive. They can transfer the toxic metals accumulated in the propolis to the hive and to hive products such as honey and wax [114]. Bees use propolis to seal and protect the hive; as propolis meets the environment, it may encounter more toxic metals present in the air, water, or

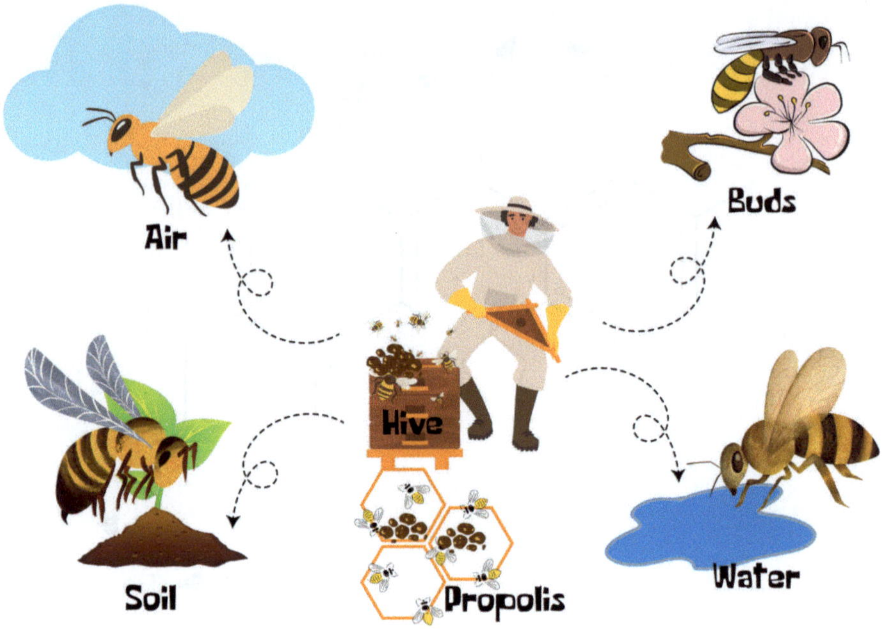

Fig. 4 Exposure of bees to the different environmental matrices polluted potentially with toxic metal

surrounding surfaces [29]. As bees continue to use propolis in the hive, the toxic metals accumulated in propolis can be transferred to other products of the hive, such as honey or wax (Fig. 4. This means that hive products can also function as indicators of the presence of toxic metals in the environment [18, 115]. Propolis can selectively accumulate some toxic metals, such as Pb, Cd, or Hg, through the chelation process, trapping them in its chemical structure [100, 103, 116]. This makes propolis an indirect bioindicator of the presence of toxic metals in each area.

4.4 Limitations of Propolis as a Bioindicator of Heavy Metals

The chemical composition of propolis can vary significantly depending on geographical location, season, source of resins, and other factors, in addition, different bee colonies can produce propolis with different chemical compositions, and during the collection of propolis by bees, there is the possibility of contamination with heavy metals from the environment occurs, which can lead to biased results and makes it difficult to standardize and compare data between studies, this can make the results inconsistent and make a comparison between studies difficult [2, 98, 102, 112, 117].

Improper storage and processing of propolis can alter its chemical composition and affect the accuracy of the results [117]. It is important to follow strict protocols to minimize contamination and deterioration of propolis.

The presence of heavy metals in propolis does not necessarily indicate the concentration of these metals in the surrounding environment. There may be other factors that affect the accumulation of metals in propolis, such as the bees' diet or interaction with other chemicals in the environment [28]. Additionally, propolis can accumulate a variety of chemical compounds, not just heavy metals. Therefore, it is important to perform additional chemical analyses to identify and quantify the specific metals present. Unlike other bioindicators, such as lichens or mosses, which have been widely used in heavy metal studies and have well-established sampling and analysis methods [118–120], the use of propolis as a bioindicator is not as standardized.

5 Assessment of Health Risks from Exposure to Metals in Propolis

Assessing health risks from exposure to metals in propolis involves identifying, evaluating, and managing potential adverse effects that may arise when people consume beehive products, such as honey or propolis, that contain high concentrations of toxic metals [22, 121, 122]. Figure 5 shows the steps to evaluate health risks from exposure to metals.

5.1 *Exposure Identification*

The first step in a risk assessment is to determine the source of exposure, in this case, identifying which hive products, such as honey or propolis, may contain toxic metals due to accumulation in the hive [22]. The study of [28], analyzes the use of bees to analyze toxic metals, however, the use of pollen, wax, and propolis is an alternative so as not to have to kill the bee to carry out the analysis. In this step it is also

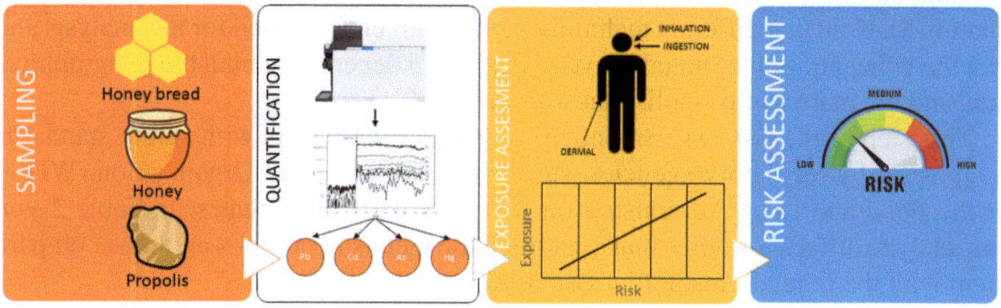

Fig. 5 Risk assessment of toxic metal in bees' products

necessary to characterize who is at risk of exposure, including vulnerable population groups such as children, pregnant women, or people with chronic diseases [27, 123] reported that people with chronic kidney diseases have higher values of metals in urine than healthy patients. Although the mechanisms are unknown Specific factors, it is important in identifying the exposure to include the factors of age and gender since they influence the distribution and consumption of metals. Other factors such as exercise, socioeconomic level, smoking, and genetic hypersusceptibility can also be considered [124].

5.2 Determination of Metal Concentration

The second step is measuring the concentration of toxic metals in propolis samples. This is typically done using chemical analysis techniques. This involves taking an appropriate amount and homogenizing it to ensure that the sample is representative of the overall composition of propolis. To extract the metals from the propolis matrix, it is necessary to perform acid digestion. The sample is heated in the presence of these acids to dissolve the components and release the metals and after digestion, the solution is diluted with deionized water or appropriate solvents to obtain a concentration suitable for analysis. Before performing the analysis, the analysis instrument must be calibrated with standard solutions of the metals of interest. This allows instrument readings to be converted into accurate concentrations. It is important to carry out quality checks throughout the analysis process to ensure that the results are accurate and dependable.

Concentration can be determined using various analysis techniques, depending on the precision required and available resources, some examples are atomic absorption spectroscopy (AAS, GFAAS, FAAS), used for the quantitative determination of individual metals and allow high sensitivity and precision; Inductively coupled plasma spectroscopy (ICP-OES or ICP-MS), these techniques are very sensitive and can detect multiple metals simultaneously; X-ray fluorescence allows up to 72 elements to be analyzed without requiring prior digestion; spectrophotometry, it is possible to use specific spectrophotometry methods to determine the concentration of certain metals and electrochemistry, also for specific metals, an example of this method is voltammetry [28]. In Table 1 you can consult some studies in which metals have been quantified. Note that different techniques and treatments are used for the analysis and that the concentration of metals is very varied and depends on both the geographical area and the matrix of propolis that was analyzed.

This step of the risk assessment also compares the results with safety standards established by regulatory agencies, such as the World Health Organization (WHO) or the Food and Drug Administration (FDA) in the United States. In Table 2 you can analyze different studies that quantified the metals in propolis and then compare them with the regulations of your country.

Table 1 Quantification of Toxic metals in different propolis matrices

Equipment	Mineralization treatment	Average or range of metals elements and concentration (mg/kg)		Propolis matrix	Location	References
AAS	Microwave system	Pb (1.06–4.32) Cd (0.62–6.59)	Cr (1.83–7.03)	Raw propolis	Italy	[125]
AAS	Microwave system	Ace (0.039–3.02) CD (0)	Mn (0.840–8.232) Pb (0.314–64.02)	Raw propolis	Croatia	[47]
		Ace (0) CD (0)	Mn (0.031–0.806) Pb (0–0.16)	Propolis tinctures		
FAAS	Wet digestion	CD (1.48–10.77) Pb (1.22–3.21) Faith (0.021–0.84)	Nor (1.9–5.07) Mg (0.94–4.01) Zn (4.2–44.3)	Raw propolis	Poland	[81]
ICP-OES	Microwave system	Ca (1773–6683) K (735–4790) Mg (301–1405) Faith (312–1270) Al (308–582) P (171–611)	Min 6.42–27.1 Cu 2.08–4.72 Nor (0.64–3.65) Pb (< 0.070–3.75) Cd (0.038–0.11) Ace (< 0.075–0.13)	Raw propolis	Spain	[126]
FAAS/GFAAS	Wet ashing	Al (608) ca (1660) K (7590) Mg (1270) Mn (80)	Na (580) Zn (20) Cd (0.13) Cr (5.53) Pb (9.85)	Raw propolis	Brazil	[127]
FAAS/GFAAS	Wet digestion	Pb (0.04585–0.961.04) Ace (0.00899–0.578.46)	Cd (1695–19,388) Hg (0)	Raw propolis	Turkey	[110]

(continued)

Table 1 (continued)

Equipment	Mineralization treatment	Average or range of metals elements and concentration (mg/kg)		Propolis matrix	Location	References
ICP-OES	Microwave system	Al (43.0–833.9) Ca (219.1–5173.0) Faith (46.1–1538.0) K (267.0–4428.3) Mg (63.5–1398.0) P (116.0–729.0)	Cr (0.8–48.9) Cu (Nd–33.4) Ni (Nd–29.9) Pb (Nd–73.9) Zn (5.5–460.7)	Raw propolis	Chile and Spain	[48]
AAS	Wet digestion	Mg (1545.5–1021) Faith (413.0 -343.8) Na (44.8–396.0)	Ca (3985.9–2066.7) Cu (2.24–11.5) Zn (87.43–43.3)	Raw propolis	Brazil	[128]
AAS	Wet digestion	Pb (0.62- 6.88) Ni (0.1–42.5) Hg (0.1–1.5)	Cr (0.1–23.85) Cd (0.25–1.22) Sn (1.8–63.3)	Raw propolis	Brazil	[129]
		Pb (0.1–0.83) Nor (0) Hg (0.3–0.6)	Cr (0.16) Cd (0.12–0.2) Sn (0.1–7.98)	Ethanolic extracts		
AAS	Wet ashing	Cu (0.025) Pb (0.040) Cd (0.21)	Cr (0.024) Zn (0.025)	Raw propolis	Macedonia	[130]
ICP-OES	Wet ashing	Ca (1773–6683) K (735–4790) Mg (301–1364) P (171–611) Na (93–225) Faith (116–284)	Zn (0.004–3.23) Cr (0.8) Min 6.51 Pb (2.0–9.7) Cd (0.067 -0.31) Ace (0) Hg (0)	Raw propolis	Croatia	[78]

(continued)

Table 1 (continued)

Equipment	Mineralization treatment	Average or range of metals elements and concentration (mg/kg)		Propolis matrix	Location	References
ICP-OES	Microwave system	Ba (54.83) Cd (0.07) Cr (1.99) Cu (3.69) Mn (9.61) Na (73.54) Nor (17.87) Pb (1.53)	Zn (21.84) At (259.84) Ca (1505.21) Faith (2.42) K (876.56) Mg (378.88)	Raw propolis	Turkey	[131]
ICP-MS	Microwave system	Cr (0.10–17.7) Ni (0.01–7.01) Cu (0.01–6.44)	Zn (0.01–6.44) Pb (0.03–7.21)	Propolis in commercial products	Spain, Portugal, and Chile	[132]
ICP-OES	microwave system	At (48.59) Ca (550.05) Cd (0.664) Co (0.774) Cr (2.08) Cu (24.25) Faith (69.64) K (728.50)	Mg (378.23) Min (106.03) Na (273.26) Nor (2.20) Pb (1.00) Zn (2.97) Ace (4.45)	Raw propolis	Brunei	[133]
FAAS/GFAAS	microwave system	Ace (< 0.048–8.47) Pb (< 0.006–0.72) Se (< 0.041–0.54)	Cu (1110–1150) K (230–4980)	Raw propolis	Brazil	[134]

(continued)

Table 1 (continued)

Equipment	Mineralization treatment	Average or range of metals elements and concentration (mg/kg)		Propolis matrix	Location	References
ICP-MS	microwave system	Ag (0.01) Al (106.5) Ace (0.07) Ba (2.4) Cd (0.04) Co (0.12) Cr (0.48) Cu (1.65)	Mn (7.2) Mo (0.07) Nor (0.48) Pb (0.66) Sb (0.04) It (0.05) V (0.25) Zn (13.67)	Raw propolis	Poland	[135]
ICP-OES	microwave system	Cu (2.76) Pb (1.20) Zn (36.3) Faith (390) Ace (0.68) Al (249)	Nor (5.61) Co (0.15) Cr (1.60) Mn (10.6) V (2.00) Sn (7.47)	Raw propolis	Turkey	[122]
AAS	Wet digestion	Cd (0.016 ± 0.004) Cu (0.402 ± 0.059) Zn (0.947 ± 0.090) Cr (1.868 ± 0.065)	Pb (0.160 ± 0.044) Ni (0.876 ± 0.082) Mn (1.026 ± 0.086)	Raw propolis	Romania	[37]

AAS Atomic absorption spectroscopy, *FAAS* Flame Atomic Absorption Spectrometry, *GFAAS* Graphite Furnace Atomic Absorption Spectrometry, *ICP-OES* Inductively coupled plasma-optical emission spectrometry, *ICP-MS* Inductively coupled plasma mass spectrometry

Table 2 Comparison between metal concentration detected in propolis and allowable concentration indicated in the regulations

Propolis type	Toxic metal/ Concentration in ppm	Regulation/allowable concentration	Observation	References
Raw propolis	Pb 2.94 ± 0.27 Cd 54.4 ± 22.2	Regulation of the Minister of Health from Polish	• Do not exceed the allowable concentration	[81]
Raw propolis	Pb < 0.070 -3.75	Japan Propolis Conference 20 mg Pb/kg	• Lower than the upper permissible limit	[126]
Raw propolis	Pb 9.85 CD 0.13	• FAO-WHO/ PTWI 0.25 mg Pb/kg and 0.07 mg Cd/kg • ANVISA Pb 0.3 ppm, Cd 0.1 ppm	• Lower than tolerable levels • Greater than Brazilian honey regulation	[127]
Raw propolis	Pb 2.0–9.7	Official Gazette of the Republic of Serbia 3 ppm	• Greater than official regulation	[69]
Products with propolis	Pb 0.03–7.21	*Codex Alimentarius* / MrL 0.1 ppm of Pb	• 42% of products are greater than *Codex*	[132]
Raw propolis	As < 0.048–8.47 Pb < 0.006–0.72	FAO-WHO/ PTWI 0.25 mg Pb/kg and 0.15 mg Ask	• Lower than tolerable levels	[134]
Raw propolis	Cr 0.48 Pb 0.66	Polish regulation (PN-P-77627:1997)	• Levels of Cr and Pb were above the regulation	[135]
Raw propolis	Cd 0.08 Pb 0.651	Commission Regulation (EC) no. 1881/2006 *Codex Alimentarius*	• Exceed allowable concentration	[121]

MrL Maximum residue Level, *Rf* Daily reference dose, PTWI Provisional Tolerable Weekly Intake, *ANVISA* Brazilian Health Regulatory Agency

5.3 Exposure Assessment

The third step in calculating the estimated daily intake is to determine how much propolis you typically consume per day estimate the number of toxic metals ingested and identify the routes of exposure, oral, dermal, or inhalation in the study [136], a consumption of 250 mg of propolis is proposed, twice a day for three months as a supplement for breast cancer; In another study reported by Salehi-Sahlabadi et al. [137], reviewed several controlled trials in which a treatment of 906 mg of propolis per day for 10 days was given to diabetic people. In addition to oral exposure, there is also dermal contact with propolis, especially by beekeepers who collect it and by pharmacists who process it into tinctures for consumption as herbal medicine.

5.4 Risk Assessment

It consists of establishing a tolerable exposure limit by comparing it with the tolerable exposure limits established by regulatory agencies to determine if a health risk exists and reviewing the scientific literature to identify possible adverse health effects associated with the consumption of propolis for this purpose. It is considered that the concentration of toxic carcinogenic metals is measured [138]. In this step, it is necessary to determine whether exposure to toxic metals in propolis represents a significant risk to the health of the population and recognize the limitations in the available data and any uncertainties in the assessment. The Delphi classification of risks is categorized into seven levels (Orosun et al. 2020), with these values we can continue with the necessary actions to resolve the present risk.

5.5 Risk Management

Risk management consists of implementing measures to reduce exposure, such as establishing maximum allowable limits for metals in beehive products or providing guidance on safe consumption and communicating risks, informing the population about the risks, and recommending mitigation measures. [139], propose the genetic engineering of microorganisms as a novel tool for the removal of toxic metals in environmental matrices. After management, the concentration of toxic metals in hive products and the exposure of the population must continue to be periodically evaluated to ensure that it remain within safe levels.

6 Principal Component Analysis to Correlate Levels of Toxic Metal Contamination

Multivariate statistical tools are the tools currently most used to define the behavior of the different variables to be studied concerning a certain problem. One of these methods that supports the researcher in knowing these variables in more detail is Principal Component Analysis (PCA). This type of statistical analysis is widely used in environmental research [140]. Its purpose is to reduce a series of observable variables within one or more dimensions to have those items that can explain as much as possible and that have a relationship with said dimension.

Furthermore, this is a projection technique that allows easy visualization of all the information contained in a data set and, therefore, helps to find differences between one sample and another [50]. Therefore, this manages to eliminate a great variability of those variables that do not contribute to the strengthening of said dimensions.

[50], used PCA to observe similarities between 96 samples of raw propolis obtained from hives in the central region of Argentina. They were reducing a lower dimensional space, keeping most of the information from the original data set in the observable variables.

Principal Component Analysis (PCA) is widely used in environmental research [140]. Its purpose is to measure the original variables [141] and reduce a set of observable variables into one or more dimensions by eliminating items that have no relationship. At the same time, it reduces the variability in the study of variables that do not contribute to the enhancement of these constructs, aiming to provide a more precise explanation of these dimensions for predictive purposes. The PCA procedure involves transforming the data into variables that account for a substantial portion of the fluctuations in the original variables and their interrelationships [142].

Furthermore, all multivariate tools rely on Principal Component Analysis (PCA), such as Multiple Factor Analysis, Correspondence Analysis, Canonical Analysis, STATIS [143], and Structural Equation Modeling, among others. Additionally, it is possible to achieve synthetic representations of a set of quantitative data by describing similarities (distances) between the objects of study, the correlations between variables, and dimension reduction, typically in the first factorial plane [143, 144]. One of the advantages of this technique is that it can be applied to a wide variety of questions, having immense importance in the development of diagnostics or prediction.

References

1. Alvarenga L, Cardozo LFMF, Borges NA, Chermut TR, Ribeiro M, Leite M (2021) To be or not to be? The bee extract propolis is a bioactive compound in the burden of lifestyle diseases. Nutrition. https://doi.org/10.1016/j.nut.2020.111094
2. Hodel KVS, MacHado BAS, Santos NR, Costa RG, Menezes-Filho JA, Umsza-Guez MA (2020) Metal content of nutritional and toxic value in different types of Brazilian propolis. Sci World J. https://doi.org/10.1155/2020/4395496
3. Lima B, Tapia A, Luna L, Fabani MP, Schmeda-Hlrschmann G, Podio NS et al (2009) Main flavonoids, DPPH activity, and metal content allow the determination of the geographical origin of propolis from the Province of San Juan (Argentina). J Agric Food Chem 57(7). https://doi.org/10.1021/jf803866t
4. Bhargava P, Mahanta D, Kaul A, Ishida Y, Terao K, Wadhwa R, Kaul SC (2021) Experimental evidence for therapeutic potentials of propolis. Nutrients 13(8). https://doi.org/10.3390/nu13082528
5. Sforcin JM, Bankova V (2011) Propolis: Is there a potential for the development of new drugs? J Ethnopharmacol 133:253–260
6. Salatino A, Salatino MLF (2021) Scientific note: often quoted, but not factual data about propolis composition. Apidologie 52(2). https://doi.org/10.1007/s13592-020-00821-x

7. Freitas AS, Cunha A, Oliveira R, Almeida-Aguiar C (2022) Propolis antibacterial and antioxidant synergisms with gentamicin and honey. J Appl Microbiol 132(4). https://doi.org/10.1111/jam.15440
8. Matin G, Kargar N, Buyukisik HB (2016) Biomonitoring of cadmium, lead, arsenic, and mercury in industrial districts of Izmir, Turkey by using honeybees, propolis, and pine tree leaves. Ecol Eng 90. https://doi.org/10.1016/j.ecoleng.2016.01.035
9. Almuhayawi MS (2020) Propolis is a novel antibacterial agent. Saudi J Biol Sci. https://doi.org/10.1016/j.sjbs.2020.09.016
10. Parmar TK, Rawtani D, Agrawal YK (2016) Bioindicators: the natural indicator of environmental pollution. Front Life Sci 9:110–118. https://doi.org/10.1080/21553769.2016.1162753
11. Rodríguez-Eugenio N, McLaughlin M, Pennock D (2019) La contaminación del suelo: una realidad oculta. Organizacion de las Naciones Unidas para la alimentacion y la agricultura FAO
12. Alday E, Valencia D, Garibay-Escobar A, Domínguez-Esquivel Z, Piccinelli AL, Rastrelli L (2019) Plant origin authentication of Sonoran Desert propolis: an antiproliferative propolis from a semi-arid region. Sci Nat 106(5–6). https://doi.org/10.1007/s00114-019-1620-2
13. Lambert O, Piroux M, Puyo S et al (2012) Bees, honey and pollen as sentinels for lead environmental contamination. Environ Pollut 170:254–259. https://doi.org/10.1016/j.envpol.2012.07.012
14. Abdelrazeg S, Hussin H, Salih M, Shaharuddin B (2020) Propolis composition and applications in medicine and health. Int Med J 25(3)
15. Kasote DM, Pawar MV, Gundu SS, Bhatia R, Nandre VS, Jagtap SD, Mahajan SG, Kulkarni MV (2019) Chemical profiling, antioxidant, and antimicrobial activities of Indian stingless bees propolis samples. J Apic Res 58:617–625
16. Liu S, Tian S, Li K, Wang L, Liang T (2018) Heavy metal bioaccessibility and health risks in the contaminated soil of an abandoned, small-scale lead and zinc mine. Environ Sci Pollut Res 25:15044–15056. https://doi.org/10.1007/s11356-018-1660-8
17. Borg D, Attard, E. 2020. Honeybees and their products as bioindicators for heavy metal pollution in Malta. Acta Brasiliensis, 4(1). https://doi.org/10.22571/2526-4338282
18. Girotti S, Ghini S, Ferri E, Bolelli L, Colombo R, Serra G et al (2020[a]) Bioindicators and biomonitoring: honeybees and hive products as pollution impact assessment tools for the Mediterranean area. Euro-Mediterr J Environ Integr 5. https://doi.org/10.1007/s41207-020-00204-9
19. Cunningham MM, Tran L, McKee CG, Ortega Polo R, Newman T, Lansing L et al (2022) Honeybees as biomonitors of environmental contaminants, pathogens, and climate change. Ecol Ind. https://doi.org/10.1016/j.ecolind.2021.108457
20. Cunningham MM, Tran L, McKee CG et al (2022) Honeybees as biomonitors of environmental contaminants, pathogens, and climate change. Ecol Ind 134:108457. https://doi.org/10.1016/j.ecolind.2021.108457
21. Beekman M, Ratnieks FLW (2000) Long-range foraging by the honeybee, Apis mellifera L. Funct Ecol 14(4). https://doi.org/10.1046/j.1365-2435.2000.00443.x
22. Vera Chang JF, Morejon Lucio RN, Salgado Tello IP, Flores Mancheno CI, Morejón Centeno MR (2022) Use of enzymes and leavening agents as a strategy to reduce the presence of cadmium in the fermentation process of Theobroma cacao L. Almonds. J Pharm Negative Results (13):604–614. https://doi.org/10.47750/pnr.2022.13.03.089
23. Asif N, Malik MF, Chaudhry MF (2018) A review of environmental pollution bioindicators. Pollution 4(1)
24. Babuji P, Thirumalaisamy S, Duraisamy K, Periyasamy G (2023) Human health risks due to exposure to water pollution: a review. Water (Switzerland). Multidisciplinary Digital Publishing Institute (MDPI). https://doi.org/10.3390/w15142532
25. Moroń D, Grześ IM, Skórka P, Szentgyorgyi H, Laskowski R, Potts SG, Woyciechowsk M (2012) Abundance and diversity of wild bees along gradients of heavy metal pollution. J Appl Ecol 49:118–125. https://doi.org/10.1111/j.1365-2664.2011.02079.x

26. Predes L, Hidalgo V, Vargas T, Molinet A (2003) Diagnósticos estructurales en los sistemas de producción de ganadería doble propósito en el municipio Alberto Arvelo Torrealba del estado Barinas Venezuela. Zootecnia Tropical 21(1):87–108
27. Herrera LMG, Calvo ILM, Peña RLM (2019) El propóleo y su potencial económico como producto de la industria apícola. Centro de Investigación Científica de Yucatán, A.C. 11:190–194. ISSN: 2395-8790
28. Farias RA, Nunes CN, Quináia SP (2023) Bees reflect better on their ecosystem health than their products. Environ Sci Pollut Res. https://doi.org/10.1007/s11356-023-28141-4
29. Van Der Perck M (2014) Soil and water contamination, 2nd edn. CRC Press
30. Szczepaniak K, Biziuk M (2003) Aspects of the biomonitoring studies using mosses and lichens as indicators of metal pollution. Environ Res. https://doi.org/10.1016/S0013-9351(03)00141-5
31. Rodríguez-Arce E, Saldías M (2021) Antioxidant properties of flavonoid metal complexes and their potential inclusion in the development of novel strategies for the treatment against neurodegenerative diseases. Biomed Pharmacother. https://doi.org/10.1016/j.biopha.2021.112236
32. Dmowski K (1999) Birds as bioindicators of heavy metal pollution: review and examples concerning European species. Acta Ornithologica
33. Tiller KG (1989) Heavy metals in soils and their environmental significance. Adv Soil Sci 9:113–142
34. Zingue S, Maxeiner S, Rutz J, Ndinteh DT, Chun FKH, Fohouo FNT et al (2020) Ethanol-extracted Cameroonian propolis: antiproliferative effects and potential mechanism of action in prostate cancer. Andrologia 52(9). https://doi.org/10.1111/and.13698
35. Ibibia E, Bankole TA, Ogunjimi DV, Mmuo AI, Amuda MO (2023) GC-MS Profile, synthesis and spectral characterization of flavonoid-metal (cadmium, cobalt, copper, nickel) complexes of jatropha curious leaves. Int J Chem Res. https://doi.org/10.22159/ijcr.2023v7i2.217
36. Stankovic S, Stankovic AR (2013) Bioindicators of toxic metals. . https://doi.org/10.1007/978-94-007-6836-9_5
37. Mititelu M, Udeanu DI, Nedelescu M, Neacsu SM, Nicoara AC, Oprea E, Ghica M (2022) Quality control of different types of honey and propolis collected from Romanian accredited beekeepers and consumer's risk assessment. Crystals 12(1). https://doi.org/10.3390/cryst12010087
38. Sforcin JM, Fernandes A, Lopes CAM, Bankova V, Funari SRC (2000) Seasonal effect on Brazilian propolis antibacterial activity. J Ethnopharmacol 73:243–249. https://doi.org/10.1016/s0378-8741(00)00320-2
39. Bankova V, Popova M, Trusheva B (2014) Propolis volatile compounds: chemical diversity and biological activity: a review. Chem Cent J 8:1–8
40. Kasprzak MM, Erxleben A, Ochocki J (2015) Properties and applications of flavonoid metal complexes. RSC Adv. https://doi.org/10.1039/c5ra05069c
41. Tutun H, Aluç Y, Kahraman HA, Sevin S, Yipel M, Ekici H (2022) The content and health risk assessment of selected elements in bee pollen and propolis from Turkey. J Food Comp Anal 105. https://doi.org/10.1016/j.jfca.2021.104234
42. INEGI (2007) La apicultura en la Penín-sula de Yucatán. Censo agropecuario 2007, VIII Censo Agrícola, Ganadero y Forestal. Aguascalientes, Ags
43. Hassaan MA, El Nemr A, Madkour FF (2016) Environment assessment of heavy metal pollution and human health risk. Am J Water Sci Eng 2(3). https://doi.org/10.11648/j.ajwse.20160203.11
44. Santos LM, Fonseca MS, Sokolonski AR, Deegan KR, Araújo RPC, Umsza-Guez MA et al (2020) Propolis: types, composition, biological activities, and veterinary product patent prospecting. J Sci Food Agric.https://doi.org/10.1002/jsfa.10024
45. Virolia BJD, Gil GJH, Durango RDL, García CM (2012) Physicochemical characterization of propolis from the region of Bajo Cauca Antioqueño (Antioquia, Colombia). Biotecnología en el Sector Agropecuario y Agroindustrial. 10(1):77–86

46. Kumpiene J, Lagerkvist A, Maurice C (2008) Stabilization of As, Cr, Cu, Pb and Zn in soil using amend-ments—a review. Waste Manage 28(1):215–225
47. Cvek J, Medic-Saric M, Vitali D, Vedrina DI, Smit Z, Tomic S (2008) The content of essential and toxic elements in Croatian propolis simples and their tinctures. J Apic Res Bee World 47(1):35–45
48. González-Martín MI, Escuredo O, Revilla I, Vivar-Quintana AM, Carmen Coello M, Riocerezo CP, Moncada GW (2015) Determination of the mineral composition and toxic element contents of propolis by near-infrared spectroscopy. Sensors (Switzerland), 15(11):27854–27868. https://doi.org/10.3390/s151127854
49. Olivares B (2014) Aplicación del Análisis de Componentes Principales (ACP) en el diagnóstico socioambiental. Caso: sector Campo Alegre, municipio Simón Rodríguez de Anzoátegui Multiciencias, vol. 14, núm. 4, octubre-diciembre, 2014, pp. 364–374 Universidad del Zulia Punto Fijo, Ven
50. Cantarelli MA, Caiña JM, Pettenati EM, Marchevsky EJ (2010) Trace mineral content of Argentinean raw propolis by neutron activation análisis (NAA): assessment of geographical provenance by chemometrics. LWT Food Sci Technol 44:256e260
51. Georgieva K, Popova M, Dimitrova L, Trusheva B, Thanh LN, Phuong DTL, Lien NTP, Najdenski H, Bankova V (2019) PhytoChem. Analysis of Vietnamese propolis produced by the stingless bee lisotrigona cacciae. PLoS One 14:e0216074
52. Herrera-López MG, Richomme P, Peña-Rodríguez LM, Calvo-Irabien LM (2023) Bee species, botanical sources and the chemical composition of propolis from yucatan, Mexico. J Chem Ecol. https://doi.org/10.1007/s10886-023-01429-y
53. Salehi-Sahlabadi A, Chhabra M, Rahmani J, Momeni A, Karam G, Nattagh-Eshtivani E et al (2020) The effect of propolis on anthropometric indices and lipid profile: a systematic review and meta-analysis of randomized controlled trials. J Diabetes Metabolic Disord. https://doi.org/10.1007/s40200-020-00604-2
54. Serra Bonvehí J, Orantes Bermejo FJ (2013) Element content of propolis collected from different areas of South Spain. Environ Monit Assess 185(7):6035–6047. https://doi.org/10.1007/s10661-012-3004-3
55. Jaishankar M, Tseten T, Anbalagan N, Mathew BB, Beeregowda KN (2014) Toxicity mechanism and health effects of some heavy metals. Interdiscip Toxicol Epub 7(2):60–72. PMID: 26109881; PMCID: PMC4427717. https://doi.org/10.2478/intox-2014-0009
56. Perugini M, Manera M, Grotta L et al (2011) Heavy metal (Hg, Cr, Cd, and Pb) contamination in urban areas and wildlife reserves: honeybees as bioindicators. Biol Trace Elem Res 140:170–176. https://doi.org/10.1007/s12011-010-8688-z
57. Almeida FR, Nogueira NC, Pércio QS (2023) Bees reflect better on their ecosystem health than their products. Environ Sci Pol Res 30:79617–79626. https://doi.org/10.1007/s11356-023-28141-4
58. Butt A, Qurat-ul-Ain RK et al (2018) Bioaccumulation of cadmium, lead, and zinc in agriculture-based insect food chains. Environ Monit Assess 190:1–12. https://doi.org/10.1007/s10661-018-7051-2
59. Burden CM, Morgan MO, Hladun KR et al (2019) Acute subletal exposure to toxic heavy metals alters honeybee (Apis mellifera) feeding behavior. Sci Rep 9:1–10. https://doi.org/10.1038/s41598-019-40396-x
60. Yushin N, Chaligava O, Zinicovscaia I, Grozdov D, Vergel K (2020) Mosses as bioindicators of heavy metal air pollution in the lockdown period adopted to cope with the covid-19 pandemic. Atmosphere 11(11). https://doi.org/10.3390/atmos11111194
61. Diels L, Van Der Lelie N, Bastiaens L (2002) New developments in treatment of heavy metal contami-nated soils. Rev Environ Sci Bio/Technol 1:75–82
62. Kuropatnicki AK, Szliszka E, Krol W (2013) Historical aspects of propolis research in modern times. Evid-Based Complement Altern Med. https://doi.org/10.1155/2013/964149
63. WHO (World Health Organization) (2013) Guidelines for drinking-water quality, 4th edn.
64. Bastías JM, Jambon P, Munoz O et al (2013) Honey as a bioindicator of arsenic contamination due to volcanic and mining activities in Chile. Chil J Agric Res 73:147–153. https://doi.org/10.4067/S0718-58392-01300-02000-10

65. Ćirić J, Spirić D, Baltić T, Branković Lazić I, Trbović D, Parunović N, Petronijević R, Đorđević V (2021) Honey bees and their products as indicators of environmental element deposition. Biol Trace Elem Res 199:2312–2319. https://doi.org/10.1007/s12011-020-02321-6
66. Sousa J, Furtado N, Jorge R, Soares A, Bastos J (2007) Perfis fisicoquímico y cromatográfico de amostras de própolis produzidas nas microrregiões de Franca (SP) e Passos (MG) Brasil. Brazilian J Pharmacognosy 17(1):85–93
67. Perugini M, Manera M, Grotta L, Abete MC, Tarasco R, Amorena M (2021) Heavy metal (Hg, Cr, Cd, and Pb) contamination in urban areas and wildlife reserves: honeybees as bioindicators. Biol Trace Elem Res 140:170–176. https://doi.org/10.1007/s12011-010-8688-z
68. Lebart LMA, Piron M (2006) Statistique exploratoire multidimensionnelle. Dunod, Paris
69. Tosic S, Stojanovic G, Mitic S, Pavlovic A, Alagic S (2017) Mineral composition of selected serbian propolis samples. J Apic Sci 61(1). https://doi.org/10.1515/JAS-2017-0001
70. Ayuso EÁ, Abad-Valle P (2017) Trace element levels in an área impacted by old mining operations and their relationship with beehive products. Sci Total Environ. 599–600:671–678. https://doi.org/10.1016/j.scitotenv.2017.05.030
71. Martínez Flores K, Souza Arroyo V, Ortiz B, Gómez Quiroz LE, Gutiérrez Ruiz MC, (2013) Cadmio: efectos sobre la salud. Respuesta celular y molecular. Acta Toxicol Argent 21(1):33–49
72. Xia M, Chen Z, Li Y, Li C, Ahmad NM, Cheema WA, Zhu S (2019) Removal of Hg(ii) in aqueous solutions through physical and chemical adsorption principles. RSC Adv 9(36). https://doi.org/10.1039/c9ra01924c
73. Wongsasuluk P, Chotpantarat S, Siriwong W, Robson M (2013) Heavy metal contamination and human health risk assessment in drinking water from shallow groundwater wells in an agricultural area in Ubon Ratchathani province Thailand. Environ Geochem Health 36:169–182. https://doi.org/10.1007/s10653-013-9537-8
74. Calla LH, Cabrera CC (2010) Calidad del agua en la cuenca del río Rímac, sector de San Mateo, afectado por las actividades mineras. Revista del Instituto de Investigaciones de la facultad de Geología, Minas, Metalurgia y ciencias Geográficas 13(25):87–94
75. Modoi OC, Roba C, Török Z, Ozunu A (2014) Environmental risks due to heavy metal pollution of water resulted from mining wastes in NW Romania. Env Eng Manage J 13(9):2325–2336
76. Tellez M, Merchant M (2015) Biomonitoring heavy metal pollution using an aquatic apex predator, the American Alligator, and its Parasites. PLoS One 10(11). https://doi.org/10.1371/J.pone.0142522
77. Adriano DC (2001) Trace elements in terrestrial environments: biogeochemistry, bioavailability and risks of metals, 2nd edn. Springer, New York, pp 867. https://doi.org/10.1007/978-0-387-21510-5
78. Tosic S, Stojanovic G, Mitic S et al (2017) Mineral composition of selected Serbian propolis samples. J Apic Sci 61:5–15. https://doi.org/10.1515/JAS-2017-0001
79. Beltrán J, Aguilar JM (2022) Análisis de la contaminación por metales pesados en la ciudad de Cuenca mediante el uso de bioindicadores. Avances en Ciencias e Ingeniería. 14(2). https://doi.org/10.18272/aci.v14i2.2600
80. Pourret O, Bollinger JC, Hursthouse A (2021) Heavy metal: a misused term? Acta Geochimica 40(3). https://doi.org/10.1007/s11631-021-00468-0
81. Formicki G, Greń A, Stawarz R, Zyśk B Gał A (2013) Metal content in honey, propolis, wax, and bee pollen and implications for metal pollution monitoring. Polish J Environ Stud 22(1)
82. Dżugan M, Tomczyk M, Sowa P, Grabek-Lejko D (2018) Antioxidant activity as biomarker of honey variety. Molecules 23(8):2069. https://doi.org/10.3390/molecules23082069
83. Cabrera de Oliveira R, do Nascimento Queiroz SC, Fernandes Pinto da Luz C, Silveira Porto R, Rath S (2016) Bee pollen as a bioindicator of environmental pesticide contamination. Chemosphere 163:525–534. https://doi.org/10.1016/j.chemosphere.2016.08.022
84. Reyes-Hinojosa D, Lozada-Pérez CA, Zamudio-Cuevas Y, López-Reyes A, Martínez-Nava G, Fernández-Torres J, Olivos-Meza A, Landa-Solis C, Gutiérrez-Ruiz MC, Rojas del Castillo E, Martínez-Flores K (2019) Toxicity of cadmium in musculoskeletal diseases. Environ Toxicol Pharmacol. Epub (72):103219. PMID: 31494513. https://doi.org/10.1016/j.etap.2019.103219

85. Famurewa AC, Renu K, Elad MA, Chakraborty R, Myakala H, El-Sherbiny M, Elsherbini DMA, Vellingiri B, Madhyastha H, Wanjari UR, Mukherjee AG, Gopalakrishnan AV (2022) Hesperidina y hesperetina contra la toxicidad de metales pesados: información sobre el mecanismo molecular de mitigación. Biomed Pharmacother 149(2022):112914. https://doi.org/10.1016/j.biopha.2022.112914
86. Astolfi ML, Conti ME, Ristorini M, Frezzini MA, Papi M, Massimi L, Canepari S (2021) An analytical method for the biomonitoring of mercury in bees and beehive products by cold vapor atomic fluorescence spectrometry. Molecules 26:4878. https://doi.org/10.3390/molecules26164878
87. Beltrán-Pineda ME, Gómez-Rodríguez AM (2016) Biorremediación de metales pesados cadmio (Cd), cromo (Cr) y mercurio (Hg), mecanismos bioquímicos e ingeniería genética: una revisión. Revista Facultad de Ciencias Básicas 12(2):172–197. https://doi.org/10.18359/rfcb.2027
88. Porrini C, Ghini S, Girotti S, Sabatini A, Gattavecchia E, Celli G (2002) Use of honeybees as bioindicators of environmental pollution in Italy. In: Devillers J, Pham-Delègue M-H (eds) Honeybees: estimating the environmental impact of chemicals. CRC Press, pp 188–224. https://doi.org/10.1201/9780203218655
89. Morin-Crini N, Lichtfouse E, Liu G, Balaram V, Ribeiro ARL, Lu Z et al (2022) Worldwide cases of water pollution by emerging contaminants: a review. Environ Chem Lett. https://doi.org/10.1007/s10311-022-01447-4
90. Kasote D, Bankova V, Viljoen AM (2022) Propolis: chemical diversity and challenges in quality control. Phytochemistry Rev. https://doi.org/10.1007/s11101-022-09816-1
91. Covarrubias SA, Peña Cabriales JJ (2017) Contaminación ambiental por metales pesados en México: problemática y estrategias de fitorremediación. Revista Internacional de Contaminación Ambiental (33):7–21. https://doi.org/10.20937/RICA.2017.33.esp01.01
92. Masek A, Chrzescijanska E, Latos M, Kosmalska A (2019) Electrochemical and spectrophotometric characterization of the propolis antioxidants properties. Int J Electro Chem Sci 14(2). https://doi.org/10.20964/2019.02.66
93. Ripari N, Sartori AA, Honorio MDS, Conte FL, Tasca KI, Santiago KB, Sforcin JM (2021) Propolis antiviral and immunomodulatory activity: a review and perspectives for COVID-19 treatment. J Pharm Pharmacol. https://doi.org/10.1093/JPP/RGAA067
94. Villanueva C, Muhammad I, Harvey C, Esquivel H (2003) Tipologías de fincas con ganadería bovina y cobertura arbórea en pasturas en el trópico seco de Costa Rica. Agroforestería en las Américas 10(39):9–16
95. Michener CD (2013) The Meliponini. In: Vit P, Pedro SRM, Roubik D (eds) Pothoney: a legacy of stingless bees. Springer, New York, pp 3–17
96. Orosun MM, Adewuyi AD, Salawu NB, Isinkaye MO, Orosun OR, Oniku AS (2020) Monte Carlo approach to risks assessment of heavy metals at automobile spare part and recycling market in Ilorin, Nigeria. Sci Rep 10(1). https://doi.org/10.1038/s41598-020-79141-0
97. Keskin N, Hazir S, Baser KHC, Kürkçüoglu M (2001) Antibacterial activity and chemical composition of Turkish propolis. Z. Naturforsch 56c:1112–1115
98. Saravanan A, Kumar PS, Ramesh B, Srinivasan S (2022) Removal of toxic heavy metals using genetically engineered microbes: molecular tools, risk assessment and management strategies. Chemosphere 298. https://doi.org/10.1016/j.chemosphere.2022.134341
99. do Nascimento TG, dos Santos Arruda RE, da Cruz Almeida ET, dos Santos Oliveira JM, Basílio-Júnior ID, Celerino de Moraes Porto IC et al (2019) Comprehensive multivariate correlations between climatic effect, metabolite profile, antioxidant capacity, and antibacterial activity of Brazilian red propolis metabolites during the seasonal study. Sci Rep 9(1). https://doi.org/10.1038/s41598-019-54591-3
100. de Souza RFV, de Giovani WF (2004) Antioxidant properties of complexes of flavonoids with metal ions. Redox Rep 9(2). https://doi.org/10.1179/135100004225003897
101. Şahinler N, Gül A, Şahinler S (2017) Heavy metals and trace elements content of different propolis types. Eur Int J Sci Technol 6(3)

102. Smith KE, Weis D (2020) Evaluating spatiotemporal resolution of trace element concentrations and Pb isotopic compositions of honeybees and hive products as biomonitors for urban metal distribution. GeoHealth 4:e2020GH000264. https://doi.org/10.1029/2020GH000264
103. Kasprzak MM, Erxleben A, Ochocki J (2015b) ChemInform abstract: properties and applications of flavonoid metal complexes. ChemInform 46(29). https://doi.org/10.1002/chin.201529267
104. Kekecoglu M, Keskin M, Birici C, Birici E, Kolaili S (2021) Effects of honeybee race and season on propolis composition. Tarım Bilimleri Dergisi. https://doi.org/10.15832/ankutbd.619996
105. Adhikari A, Darbar S, Chatterjee T, Das M, Polley N, Bhattacharyya M (2018) Spectroscopic studies on dual role of natural flavonoids in detoxification of lead poisoning: bench-to-bedside preclinical trial. ACS Omega 3(11). https://doi.org/10.1021/acsomega.8b02046
106. Wang YM, Chen TC, Yeh KJ, Shue MF (2001) Stabilization of an elevated heavy metal contaminated site. J Hazard Mater. 88(1):63–74
107. Ibrahim N, Zakaria AJ, Ismail Z, Mohd KS (2016) Antibacterial and phenolic content of propolis produced by two Malaysian stingless bees, Heterotrigona itama and Geniotrigona thoracica. Int J Pharmacogn Phytochem Res 8:156–161
108. Alasady SA, Muhamad YH, Ahmed RS (2020) Theoretical and thermodynamics studies of complexes formation between natural flavonoids and Hg (II) Ion. Syst Rev Pharm 11(12)
109. Silva-Beltrán NP, Umsza-Guez MA, Rodrigues DMR, Gálvez-Ruiz JC, de Paula Castro TL, Balderrama-Carmona AP (2021) Comparison of the biological potential and chemical composition of Brazilian and Mexican propolis. Appl Sci (Switzerland) 11(23). https://doi.org/10.3390/app112311417
110. Matuszewska E, Klupczynska A, Maciołek K, Kokot ZJ, Matysiak J (2021) Multielemental analysis of bee pollen, propolis, and royal jelly collected in west-central Poland. Molecules 26(9). https://doi.org/10.3390/molecules26092415
111. Yengle RC (2012) Application of principal component analysis as a technique to obtain synthetic indices of environmental quality. Scientia 4(2):145–153
112. Dezmirean DS, Paşca C, Moise AR, Bobiş O (2021) Plant sources are responsible for the chemical composition and main bioactive properties of poplar-type propolis. Plants. https://doi.org/10.3390/plants10010022
113. Salatino A, Fernandes-Silva CC, Righi AA, Salatino ML (2011) Propolis research and the chemistry of plant products. Nat Prod Reports 28(5):925–936
114. Navarro CJM, Casa CGM, González RE (2010) Principal component and regression analysis for categorical data. Application to arterial hypertension. Rev Mat Teoría Appl 17(2):199–230. ISSN: 1409-2433
115. Conti ME, Astolfi ML, Finoia MG, Massimi L, Canepari S (2022) Biomonitoring of element contamination in bees and beehive products in the Rome province (Italy). Environ Sci Pollut Res 29(24). https://doi.org/10.1007/s11356-021-18072-3
116. da Cruz FB, de Martins DHN, Juliana Ferreira F, De Oliveira Magalhães P, Silveira DY, Fonseca Bazzo M (2022) Antioxidant activity of apis mellifera bee propolis: a review. J Nat Prod Discov 1(1)
117. Ibarra EJI, Carmona GCA, Escalante VF, Avila RF (2020) Efecto del propóleo y aceite de orégano sobre parámetros productivos, leucocitos, metabolitos y estabilidad oxidativa de la pechuga de pollo. Rev Mex Cienc Pecu 11(1):153–166. https://doi.org/10.22319/rmcp.v11i1.4882
118. Ramos CG, Gómez Á, De Ascencao A (2004) Caracteres morfológicos determinantes en dos poblaciones de cacao criollo del occidente de Venezuela. Agronomía Trop 54(1):45–62
119. Tchounwou PB, Yedjou CG, Patlolla AK, Sutton DJ (2012) Heavy metal toxicity and the environment. In: Luch A (eds) Molecular, Clinical and Env. Toxicology. Experientia Supplementum, vol 101, pp 133–164. https://doi.org/10.1007/978-3-7643-8340-4_6
120. Zamorano A, Demey J, Gonzalez V (1995) Grupos homogéneos de crecimiento y manipulación in vitro de seis cultivares comerciales de caña de azúcar en Venezuela. Agron Trop Ven 45(1):51–72

121. Mititelu M, Udeanu D, Docea A, Tsatsakis A, Calina D, Arsene AL, Nedelescu M, Neacsu S, Velescu BS, Ghica M (2023) New method for risk assessment in environmental health: the paradigm of heavy metals in honey. Env Res 236(1):115194, ISSN 0013-9351. https://doi.org/10.1016/j.envres.2022.115194
122. Vakhonina EA, Lapynina EP, Lizunova AS (2021) Study of toxic elements in propolis. IOP Conf Ser Earth Environ Sci. https://doi.org/10.1088/1755-1315/845/1/012122
123. Danziger J, Dodge LE, Hu H, Mukamal KJ (2022) Susceptibility to environmental heavy metal toxicity among Americans with Kidney Disease. Kidney 360:3(7). https://doi.org/10.34067/KID.0006782021
124. Gong S, Luo L, Gong W, Gao Y, Xie M (2012) Multivariate analyses of element concentrations revealed the groupings of propolis from different regions in China. Food Chem 134:583
125. Conti ME, Botrè F (2001) Honeybees and their products as potential bioindicators of heavy metals contamination. Environ Monit Assess 69(3). https://doi.org/10.1023/A:1010719107006
126. Sforcin JM (2016) Biological properties and therapeutic applications of propolis. Phytotherapy Res. https://doi.org/10.1002/ptr.5605
127. Finger D, Filho IK, Torres YR, Quináia SP (2014) Propolis is an indicator of environmental contamination by metals. Bull Environ Contam Toxicol 92(3). https://doi.org/10.1007/s00128-014-1199-4
128. Souza EAR, Veiga ZN, Orsi RO (2016) Effects of seasonal variations and collection methods on the mineral composition of propolis from Apis mellifera Linnaeus Beehives. Braz J Biol 76(2):396-401. https://doi.org/10.1590/1519-6984.16714
129. Papp Z, Bouchelaghem S, Szekeres A, Meszéna R, Gyöngyi Z, Papp G (2021) The scent of antifungal propolis. Sensors 21(7). https://doi.org/10.3390/s21072334
130. Bogdanova Popov B, Karapetkovska Hristova V, Presilski S, Ali Shariati M, Najman S (2017) Assessment of heavy metals in propolis and soil from the Pelagonia region, Republic of Macedonia. Maced J Chem Chem Eng 36(1). https://doi.org/10.20450/mjcce.2017.1004
131. Salatino A (2022) Perspectives for uses of propolis in therapy against infectious diseases. Molecules. https://doi.org/10.3390/molecules27144594
132. Guzmán-Gutiérrez SL, Nieto-Camacho A, Castillo-Arellano JI, Huerta-Salazar E, Hernández-Pasteur G, Silva-Miranda M, Argüello-Nájera O, Sepúlveda- Robles O, Espitia CI, Reyes-Chilpa R (2018) Mexican propolis: a source of antioxidants and anti-inflammatory compounds, and isolation of a novel chalcone and e-caprolactone derivative. Molecules 23:334
133. Abdullah NA, Zullkiflee N, Zahirah Zaini SN, Taha H, Hashim F, Usman A (2020) Phytochemicals, mineral contents, antioxidants, and antimicrobial activities of propolis produced by Brunei stingless bees *Geniotrigona thoracica, Heterotrigona itama*, and *Tetrigona binghami*. Saudi J Biol Sci 27(11):2902–2911. https://doi.org/10.1016/j.sjbs.2020.09.014
134. Hossain R, Quispe C, Khan RA, Saikat ASM, Ray P, Ongalbek D et al (2022) Propolis: an update on its chemistry and pharmacological applications. Chin Med (United Kingdom). https://doi.org/10.1186/s13020-022-00651-2
135. Medina-Pizzali M, Robles P, Mendoza M, Torres C (2018) Ingesta de arsénico: el impacto en la alimentación y la salud humana. [Arsenic Intake: Impact in Human Nutrition and Health]. Revista Perú Medica Exp Salud Publica 35(1):93–102. PMID: 29924286 Spanish. https://doi.org/10.17843/rpmesp.2018.351.3604
136. Davoodi SH, Yousefinejad V, Ghaderi B, Akbari ME, Darvishi S, Mehrabi Y, Darvishi N (2022) Oral propolis, nutritional status and quality of life with chemotherapy for breast cancer: a randomized, double-blind clinical trial. Nutr Cancer 74(6). https://doi.org/10.1080/01635581.2021.1988118
137. Sanpa S, Popova M, Bankova V, Tunkasiri T, Eitssayeam S, Chantawannakul P (2015) Antibacterial compounds from propolis of Tetragonula laeviceps and Tetrigona melanoleuca (Hymenoptera: Apidae) from Thailand. PLoS One 10:e0126886
138. Di Vaio P, Magli E, Caliendo G, Corvino A, Fiorino F, Frecentese F et al (2018) Heavy metals size distribution in PM10 and environmental-sanitary risk analysis in Acerra (Italy). Atmosphere 9(2). https://doi.org/10.3390/atmos9020058

139. Sawaya ACHF, Cunha IBS, Mar-cucci MC, de Oliveira RF, Eberlin MN (2006) Brazilian propolis of Tetragonisca angustula and Apis mellifera. Apidologie 37(3):398–407
140. Polo Bravo C, Quispe LS (2019) Metales pesados: fuentes y su toxicidad sobre la salud humana. Ciencias. Revista de la facultad de ciencias. Universidad Nacional Jorge Basadre Grohmann 2(1):20–36. https://doi.org/10.33326/27066320.2018.1.842
141. Nurul AA, Nadzirah Z, Siti NZ, Hussein T, Fatimah H, Anwar U (2020) Phytochemicals, mineral contents, antioxidants, and antimicrobial activities of propolis produced by Brunei stingless bees Geniotrigona thoracica, Heterotrigona itama, and Tetrigona binghami. Saudi J Biol Sci 27:2902–2911
142. Abdi HW (2010) Principal component analysis. WIREs Comp Stats 2(4):433–459
143. León GH, Llinás S, Tilano J (2008) Análisis multivariado aplicando componentes principales al caso de los desplazados. Ing Desarro 23:119–142
144. Aluja T (1999) Aprender de los Datos: El Análisis de Componentes Principales. EUB, Barcelona

Ecotechnology Approach for Remediation of Heavy Metals

Shaik Riyazuddin, Daka Gopamma, Kolli Suresh Kumar, Nyayapathi Priyanka Priyadarshini, and Namuduri Srinivas

Abstract Ecotechnologies are crucial in reducing water, soil, and air pollution. With increasing concern over environmental degradation and its detrimental effects on human health, adopting sustainable practices that mitigate pollution and promote a healthier ecosystem has become imperative. Ecotechnologies offer innovative solutions that address pollution issues and contribute to our planet's sustainability. These technologies encompass a range of approaches, such as filtration and treatment with natural coagulants that remove contaminants and provide clean water for reuse, enhancing economic efficiency. Additionally, soil remediation techniques can help restore polluted land by eliminating toxins and promoting healthy vegetation growth. Remediating soil and water can be done using physical, chemical, and biological technologies. Implementing these ecotechnologies can pave the way for a greener future and ensure the environment's and humanity's long-term well-being. Ecotechnologies such as filtration with nano-ceramic oxide utilization of natural coagulants can help reduce GHGS (Green House Gases) to synergetic climate change. Renewable options such as phyto and bioremediation techniques can also aid in better resource management. Furthermore, adopting ecotechnologies can lead to a more sustainable and resilient society, as people sustainably utilize the resources and risks related to the environment. Overall, ecotechnologies are crucial for achieving sustainability and prosperity.

Keywords Ecotechnologies · Sustainability · Environmental remediation · Heavy metals · Waste management

S. Riyazuddin · D. Gopamma · K. Suresh Kumar · N. Priyanka Priyadarshini · N. Srinivas (✉)
Department of Environmental Science, GITAM School of Science, GITAM (Deemed to be University), Visakhapatnam, Andhra Pradesh, India
e-mail: snamudur@gitam.edu

N. Kumar (ed.), *Heavy Metal Remediation*, Earth and Environmental Sciences Library,
https://doi.org/10.1007/978-3-031-53688-5_14

1 Introduction

Ecotechnology was first introduced in the late twentieth century when environmental issues were beginning to garner international attention. Governments and organizations started looking for alternate solutions as the detrimental effects of industrialization and unsustainable practices became evident. To overcome these obstacles, the idea of ecotechnology—which combines ecological concepts with technical advancements—appears. This strategy sought to create cutting-edge techniques and tools to reduce adverse environmental effects while advancing sustainable development [1]. "Ecotechnology" refers to various methods, including waste management plans, renewable energy systems, and green building architecture. Ecotechnology presents a viable route toward a more ecologically sensitive and sustainable future by combining ecological knowledge with technical know-how [2].

Ecotechnology is defined as "the use of technological means for ecosystem management based on a deep understanding of principles on which natural ecological systems are built and on the transfer of such principles into ecosystem management in a way to minimize the costs of the measures and their harm to the global environment" [1]. By incorporating ecotechnology into ecosystem management practices, we can strive toward a more sustainable and harmonious coexistence with the natural world. This approach emphasizes the importance of understanding and respecting the intricate web of relationships within ecosystems while seeking innovative solutions that minimize negative environmental impacts. In essence, ecotechnology aims to harness the power of technology to work in tandem with nature, rather than against it, to achieve long-term ecological balance and resilience [3]. Since the early 1970s, a combination of environmental practices and technological intervention has been referred to as "ecotechnology." Although it is frequently used, there doesn't appear to be much agreement regarding its meaning [1].

Renewable energy systems are one area in which ecotechnology has achieved notable advancements. Developing solar panels, wind turbines, and hydropower generators has made using clean, renewable energy sources feasible [4]. These innovations offer a dependable and sustainable power supply in addition to lowering greenhouse gas emissions. Furthermore, ecotechnology has transformed methods for managing trash. Recycling, composting, and anaerobic digestion are more ecologically friendly alternatives to conventional waste disposal techniques like landfilling and incineration [5]. These methods assist in resource conservation and pollution reduction, as well as lowering the quantity of garbage in landfills. Recycling, for instance, lessens the need to mine raw materials and create new goods, conserving energy and reducing carbon emissions [6].

However, composting enables organic waste to break down and enrich the soil organically, encouraging the growth of healthier plants and lowering the need for synthetic fertilizers [7]. Finally, anaerobic digestion may convert organic waste into biogas, a renewable energy source. Using these ecotechnologies, we may considerably mitigate our ecological footprint and progress toward a more sustainable future [8].

2 Definition of Ecotechnologies

All technologies less harmful to the environment than standard technologies for the same purposes are called ecotechnologies. Their line of work is making products and services that are intended to quantify, cease, restrict, or correct environmental violations (air, water, and ground) and issues related to waste, noise, and ecosystems. As a result, they have an interest in sustainable development [9].

Ecotechnology is the use of technological instruments for ecosystem management based on a deep understanding of the principles that underpin the construction of natural ecological systems as well as the transference of these tenets into ecosystem management in a manner that cuts down on the costs of the actions and their adverse effects on the environment at large [10]. These concepts of ecosystems serve as the foundation for ecotechnology. The application of these ecotechnological guidelines for ecosystem management is provided because they are rather broad. Specific applications of the general regulations are provided, especially in relation to regional and watershed management. The following list includes the eco-techniques used in lake/reservoir and watershed management. It is demonstrated that in order for ecotechnological approaches to be successful, they must be applied in specific situations using statistical and mathematical models. These models are designed to choose a set of choices that, when combined, produce the desired ecosystem condition at the lowest possible cost and with the least amount of secondary environmental damage [11].

3 Importance of Ecotechnologies

Figuring out ways to improve the economy, technological innovation, and human well-being while remaining mindful of the environment and available resources is a global sustainable development goal in the modern world. The conversion of ecological consumption into economic gains, technological advancement, and human well-being depends on research on eco-efficiency, ecotechnology innovation, and eco-well-being performance. Between 1960 and 2018, the world economy grew from 1.37 to 85.93 trillion dollars [12]. However, this quick economic expansion has also brought about resource overuse and environmental issues [13]. Researchers' report shows a significant increase in the total primary energy supply over the past thirty years, from 8.8 to 13.9 billion tons [14]. Nitrous oxide emissions increased 1.5 times between 1970 and 2012, forest area decreased by 1.3 million km^2, global greenhouse gas emissions increased nearly two times.

GDP, generated by human economic activities, is largely fueled by resources, labor, and capital, but these actions also contribute to environmental pollution in air, water, and solid waste [15]. Examining the mutually beneficial equilibrium between environmental preservation and global economic growth is crucial and significant when it comes to sustainable development. Since eco-efficiency (EE) directly correlates to the effects on the economy, resources, and environment, it has been suggested by Schaltegger and Sturm [16] as a valuable metric for measuring sustainable development. Furthermore, eco-efficiency offers a way to approach sustainability because it measures the ability to generate more desired results, like GDP, with less natural resource consumption and ecological destruction [17].

4 Application in Ecotechnologies

Ensuring that labor, capital, and resource inputs are effectively translated into technological innovation and varying degrees of human well-being across nations is imperative in the interim. The goals of eco-well-being performance (EWP) and ecotechnology innovation (ETI) are to increase technological output and enhance human well-being while reducing resource consumption and adverse effects on the environment [18, 19]. Sustainable development requires continuous improvements in eco-efficiency, technology innovation, and well-being performance. Examining EE, ETI, and EWP (3E) in various nations can provide a fresh perspective on research for global sustainable development [12].

Ecotechnologies have been developed in many nations to address the deterioration and unsustainability of ecosystems. We should assess the technology to determine whether it is appropriate for our circumstances before learning about ecotechnology from other areas. In ecologically degraded areas, there is a strong desire to learn advanced ecotechnology in order to restore damaged ecosystems [20]. Ecotechnologies and their combinations can impact ecological governance projects, so it's crucial to assess and optimize them before implementing [21].

5 Merits and demerits of Ecotechnologies

Ecotechnologies for heavy metal remediation offer sustainable solutions for environmental restoration. Based on existing research studies, here is an overview of their merits and demerits of Ecotechnologies (Table 1).

Research in this field continues to evolve, focusing on optimizing ecotechnologies, enhancing their efficiency, and addressing their limitations. Integrating these approaches with traditional methods might offer a balanced and practical approach in remediation of heavy metals [36].

Table 1 Merits and demerits of ecotechnologies of early years researchers studies

S. No.	Merits	Demerits
1	**Environmental Friendliness**: Ecotechnologies often utilize natural processes, plants, or microorganisms, reducing the use of harsh chemicals and minimizing ecological disturbances [22, 23]	**Time Intensive Processes**: Ecotechnologies might take longer to show significant results than conventional methods due to the reliance on natural processes, potentially delaying remediation efforts [24]
2	**Cost-effectiveness**: Compared to traditional remediation methods, ecotechnologies can be more cost-effective due to lower operational costs and the utilization of natural resources [25]	**Variable Efficiency**: The efficiency of ecotechnologies can vary depending on factors like environmental conditions, contaminant concentration, and the chosen remediation strategy [26]
3	**Versatility**: These technologies can be adapted to various environments and contaminants, offering flexibility in addressing different heavy metal pollutants [27]	**Limited Scale of Application**: Some ecotechnologies might not be suitable for large-scale remediation due to constraints related to space, time, or resource availability [28]
4	**Long-Term Efficacy**: Some ecotechnologies promote sustainable remediation by improving soil or water quality over time, leading to long-lasting effects compared to conventional methods [29]	**Regulatory Challenges**: Regulatory frameworks may not always be well-equipped to accommodate or validate these newer, eco-friendly approaches, posing challenges in their widespread adoption [30]
5	**Reduced Secondary Pollution**: Ecotechnologies often result in minimal secondary pollution compared to conventional methods, reducing the risk of creating additional environmental issues during the remediation process [31]	**Geographical Limitations**: Certain ecotechnologies might be more effective or applicable in specific geographical regions or climates, limiting their universal applicability [2]
6	**Biodiversity Enhancement**: Certain ecotechnologies, such as phytoremediation (using plants), can promote biodiversity by providing habitats for various organisms, contributing to ecological restoration [32]	**Resource Intensiveness**: Implementing certain ecotechnologies might require significant resources such as specific plant species, microbial cultures, or specialized equipment, posing challenges in their widespread adoption [22, 23]
7	**Community Acceptance**: These methods often garner public acceptance due to their environmentally friendly nature, fostering community involvement and support [33]	**Knowledge Gaps**: There might be gaps in understanding the long-term effects or potential unintended consequences of implementing ecotechnologies, requiring further research [34]
8	**Synergistic Effects**: When combined with other remediation approaches or technologies, some ecotechnologies showcase synergistic effects that enhance overall efficiency [35]	

6 Types of Ecotechnologies (Physical, Chemical, Biological)

A variety of techniques such as physical chemical and biological approaches have been employed for removal of heavy metals from water and soil media (Fig 1). Research studies on various eco-technological approaches for removal of heavy metals from water and soils are given in Table 2.

6.1 Physical Remediation Techniques

Over the years, several innovative approaches have been developed, amongst which Filtration and Ion exchange are predominant. In addition to these physical techniques of remediation Electrokinetic remediation, thermal desorption, sorting, vapor stripping, soil washing, verification, encapsulation, and permeable barrier are some other methods opted for the removal of contaminants specifically heavy metals.

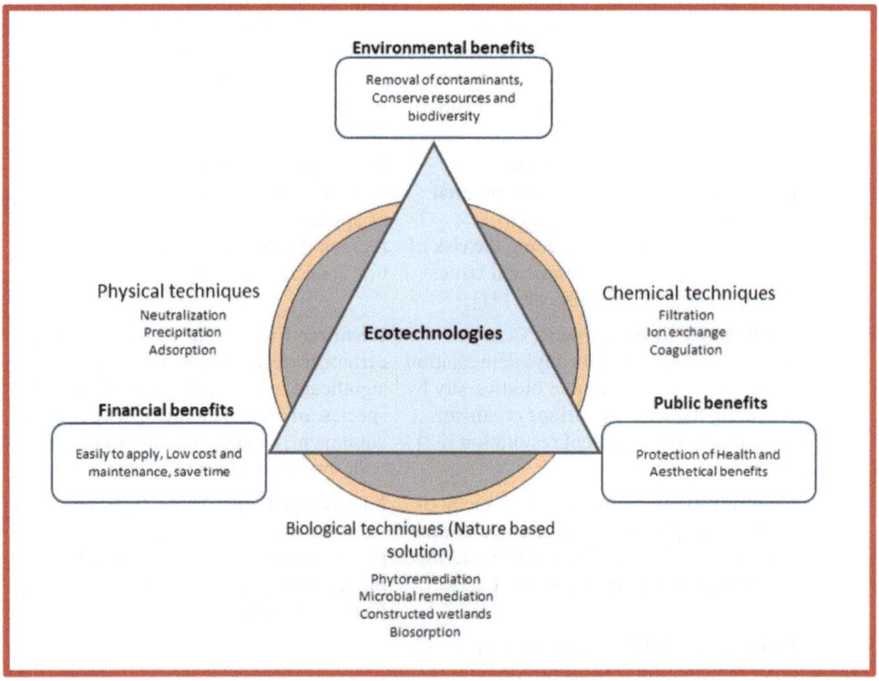

Fig. 1 Types of eco technologies and benefits of techniques to environmental, financial, and public

Table 2 Research studies on various eco-technological approaches to heavy metal removal from water and soils

S. No.	Segment	Article type	Origin	Method	Key highlights/results	Reference
1	Electro remediation coupled with phytoremediation techniques	Review	India	Electric charge alters waste characteristics, supporting plant physiology and microflora in the rhizosphere, thereby aiding in metal removal through phytoremediation	Phytoremediation, in combination with the electrokinetic method, enhances the availability of metals in soil solution, which enhances the remediation potential of plants through rhizosphere	[72]
2	Metals adsorption	Review paper	Sub-Saharan Africa	Sugarcane bagasse exhibits high heavy metals adsorption capacity. Kinetic and isotherm studies reveal cooperative adsorption as the primary mechanism	Sugarcane bagasse exhibits exceptional selectivity for HMs adsorption removals, indicating potential for future use in developing countries	[73]
3	Phytoremediation	Review	India	Phytoremediation is a non-invasive, cost-effective, and aesthetically pleasing technology. It utilizes metal-binding proteins from plants like metallothioneins, phytochelatins, and metal-activated enzymes	Model crops have provided insights into the biosynthesis, function, expression, and regulation of phytochelatins and metallothioneins, highlighting the importance of plants' tolerance to heavy metal stress	[71]
4	Heavy metal adsorption by cryogels	Review	Turkey	This manuscript reviews research on cryogels, a unique bio-adsorbent used for HM removal from wastewater treatment—the advantages of cryogels, including low cost, eco-friendliness, and high removal capacity	The research presented the structure, synthesis, and composition of cryogels, which are synthesized by bulk polymerization under semifrozen conditions. They have a high removal capacity of over 90% rate	[74]

(continued)

Table 2 (continued)

S. No.	Segment	Article type	Origin	Method	Key highlights/results	Reference
5	Recovering heavy metals	Book chapter	Canada	This paper reviewed advanced technologies for recovering metal pollutants from wastewater, focusing on their remediation mechanisms within the circular economy management framework	This paper reviewed research on remediating heavy metals (HMs) contaminated wastewater, focusing on recovering heavy metals as value-added products for recycling and reuse	[75]
6	Polymer-based adsorbents	Book chapter	Nepal and the Republic of Korea	Highlighted technological advancements in removing and recovering heavy metals from industrial wastewater and evaluated their efficiency	Adjusting pH, temperature, and pressure in membrane technology parameters can lower costs and enhance efficiency, while polymer-based adsorbents can effectively remove heavy metal ions from wastewater	[76]
7	Phytoremediation	Review	Algeria	Soil toxic metal contamination poses a significant environmental risk and costly chemical methods like heat treatment	Phytoremediation is a promising strategy that uses hyper-accumulator plant species to remove, degrade, or detoxify toxic metals from the environment	[77]
8	Constructed wetland	Research article	Bangladesh	Applied two-hybrid subsurface flow constructed vertical flow (VF) and horizontal flow (HF) wetland systems for removal of selected metals—zinc (Zn), chromium (Cr), nickel (Ni), and lead (Pb)—were removed from landfill leachate	The two-hybrid wetlands' overall percentages removal ranges from Zn (20–97%), Cr (95–99%), Ni (55–73%), and Pb (69–83%)	[78]

(continued)

Table 2 (continued)

S. No.	Segment	Article type	Origin	Method	Key highlights/results	Reference
9	Electrochemical	Research article	China	The study presents a simple synthesis of N, S-co-doped carbon materials (NS-C) from PVC plastic wastes as CDI electrode materials to remove heavy metal ions efficiently	NS-C-fabricated CDI electrodes demonstrated high removal efficiency, capacity, and regeneration capability for the adsorption of low-concentration heavy metal ions (Fe, Co, Ni, Cu, Pb, Cd, and PVC plastic wastes)	[79]
10	Biosensor microorganisms	Review	Ethiopia	Using biosensor microorganisms for hazardous metal balance is eco-friendly and cost-effective, offering various metal sequestration mechanisms and microbial tools for removal and recovery from solutions	Microorganisms like bacteria, fungi, and algae effectively remediate heavy metal-contaminated environments using metal-resistant strains in single, consortium, and immobilized forms, with immobilized forms potentially having more chemisorption sites	[80]
11	Integrated processes for heavy metal removal	Review	India, China, UK	Integrated processes effectively remove heavy metals from various environmental matrices, overcoming challenges of large-scale implementation	Chemical-biological integrated treatment offers an economical and eco-friendly alternative for treating wastewater containing heavy metals, with significant results reported by researchers worldwide	[36]
12	Vetiver grass to stabilize soil	Research article	USA	The study evaluated the potential of vetiver grass to stabilize soil contaminated with copper and its impact on its lignocellulosic composition and bioethanol production	The study shows that Cu-rich soil improves the lignocellulosic composition of vetiver, leading to increased hemicellulose and cellulose content, higher ethanol yield, and volumetric productivity	[81]

Filtration

It is a physical separation process separation fluids and solids from mixture of filter medium are in practice, among clay and ceramic filtration are effective in removal of heavy metals with low cost. According to studies of Robbins et al. [37], Yang et al. [38], ceramic water filters are developed using clay and agricultural waste. The removal of heavy metals may be accomplished with the use of clay filters alone [39].

Heavy metals in contaminated water may not be completely removed by conventional ceramic water filters, according to research. The potential of this method for removing arsenate was investigated in a study that used rice husks and ceramic water filters coated with nano-cerium oxide. The findings demonstrated that coating ceramic filters with nano-cerium oxide enhanced the As (V) ion removal from 64.04 to 84.58% in contrast to the 61.93% reduction of an uncoated filter. According to the study's findings, the addition of nano-cerium oxide to ceramic water filters increased their ability to remove heavy metals, suggesting that these elements working together could have a positive effect on heavy metal removal [38, 40].

Ion Exchange

Using a resin bed as a semipermeable membrane in ion exchange technology is a promising method for removing heavy metal ions from water bodies. The exchange of charged ions embedded in the resin bed can remove heavy metal ions via sulfonic acid and quaternary amine groups [41, 42]. Electro-deionization regenerates ion exchange layer, suitable for membrane use. Methods for heavy metal removal include biological oxidation, ion exchange, chemical precipitation, and membrane separation [43]. Ion exchange materials, on the other hand, can quickly reach adsorption saturation, necessitating frequent replacement or metal ion desorption cycles [44]. Biological ion exchange and nanofiltration are promising methods for wastewater treatment, generating biofilms without chemical regeneration and removing heavy metal ions with high energy efficiency and low pollution emissions [45, 46].

6.2 Chemical Remediation Techniques

Neutralization, Coagulation, Precipitation and Adsorption are methods used for extracting heavy metals at low cost in eco-friendly manner [47, 48].

Neutralization

Soil neutralization can be achieved through chemical methods like lime, hydrated lime, soda-ash, or caustic soda, which can accompany physical processes like coagulation, flocculation, and sedimentation. Alkali compounds like lime, limestone, chemical compounds, and industrial waste can also be used for AAT neutralization, which is easy and simple and requires a short time to remediate [49, 50].

Compost is a crucial carbon source for BPS growth because it can accumulate heavy metals and contain dissolved organic compounds. Organic husks, like coconut, rice, and hay, have a lower rate of dissolved organic carbon, resulting in higher removal efficiency. Organic carbon improves the reduction process, removes Fe^3+, and increases pH levels. Biological processing of acid pollutants is a promising alternative due to its low cost and environmental friendliness. This bioremediation method uses naturally living microbes in soil and water without harming humans or other living things. Passive processing alternatives, such as in situ or around polluted locations, are generally conducted in situ. The BPS consortium effectively optimizes the bioremediation process in acid-polluted environments [51, 52].

Coagulation

Coagulation is a physical–chemical process that removes metal ions from water and wastewater. It involves three steps:

- The vigorous stirring of a coagulation agent, which destabilizes the substance.
- The gentle stirring of a flocculant that forms large, easily separable flocculants.
- The separation of sludge and clarified effluents through settling, flotation, or filtration

However, this process cannot achieve 100% heavy metal removal efficiency, so it must be combined with other treatment processes. The process is essential for removing heavy metals from water and wastewater [53].

Coagulation can be classified into chemical and natural coagulation. Chemical coagulants, like Fe_2So_4, Fe_2So_3, and Al_2So_3, have high operational and disposal costs. Natural coagulants, like Moringa oleifera seed, Cicer arietinum seed, Nirmali seeds, cactus, tannin, and eggshells, are renewable, non-toxic, biodegradable, cost-effective, and efficient in removing turbidity from wastewater. However, chemical coagulants have limitations, such as large sludge quantities and disposal costs [53–55].

Precipitation

Precipitation is a process where contaminants are separated from a solution as sediment, which can be filtered, centrifuged, or separated from the liquid part. This occurs due to a complex formed between the precipitation agent and heavy metal ions, reducing their bioavailability [56].

Chemical precipitants like hydrogen and sulfide are widely used in the industry to remove pollutants from wastewater, soften water, remove heavy metals, and remove oil and grease from emulsified solutions. They are simple and inexpensive to operate but can also cause odor and toxic H_2S gas formation, harming the environment.

However, their versatility and cost-effectiveness make them a valuable resource for wastewater treatment [56, 57].

Adsorption

Adsorption is a surface phenomenon where an adsorbate solution is adsorbed on an adsorbent's surface. It can be physisorption, where the adsorbate binds to the adsorbent through van der Waals forces, and chemisorption, which occurs due to chemical reactions. Physiosorption is reversible, weak, and endothermic, while chemisorption is irreversible, selective, and exothermic [58, 59]. Adsorbents are classified as natural or synthetic, with natural sources including clays, minerals, charcoal, ores, and zeolites, and synthetic sources from industrial, agricultural, and waste sludge [58].

Activated carbons are widely used adsorbents, but their high cost and low regeneration limit their use. Low-cost adsorbents, such as carbon nanotubes, fullerenes, and nanocomposites, are needed to improve wastewater treatment efficiency. These nonconventional, readily available materials, mainly agricultural and industrial waste, are crucial for faster and more effective wastewater treatment [58]. Agricultural wastes such as rice shells, wheat shells, eggshells, coconut husks, palm fruit, bagasse, groundnut shells, fruit peels, and biochar have functional groups that increase their adsorption capacity. They can be utilized immediately by cleaning, pulverizing, and straining these waste materials to the proper size for adsorption testing. They can also be converted into chars and then further activated to increase the number of adsorption sites [60].

Heavy metals from wastewater can be effectively removed using biochar, a charred solid made by pyrolysis. It works well as an adsorbent because of its low ash content, diverse functional groups, high surface area, and mesoporous structure. Due to their high adsorption capacity, industrial wastes like fly ash, red mud, and slag can be used to solve waste treatment issues. Adsorbents made from industrial waste have been extensively used to remove Zn^{2+} from wastewater, indicating their potential for wastewater remediation. In waste treatment, these waste products have proven to be effective [58, 61, 62].

6.3 *Biological Remediation Techniques*

Bioremediation involves the use of organisms like algae, fungi, bacteria, protozoa, worms, insects, and enzymes to remove contaminants from wastewater. This process involves biosorption, biotransformation, activated sludge, anaerobic digestion, and biofiltration [47, 48, 63]. Biosorption occurs when contaminants are affixed to the cell surface of active, inactive, dead, or microbial biomass cells. The activated sludge process has a 95–98% Chemical Oxygen Demand (COD) removal efficiency. Extracellular polymeric substances extracted from activated sludge are crucial for this procedure [64]. Phytoremediation is another promising strategy for heavy metal removal from polluted areas, using hyper-accumulator plant species that are highly tolerant to HMs in the environment or soil. Green plants are used to remove, decompose, or detoxify hazardous metals from contaminated soils and water bodies [65].

Bioremediation

Microorganisms significantly influence metal transformations in biogeochemical cycles, with interactions influencing beneficial or harmful effects. Environmental factors like pH, moisture, temperature, and ions also influence microbial colonization and biofilm formation [66].

Bioremediation is a method that converts harmful contaminants like heavy metals into less harmful substances, eliminating toxic elements from contaminated environments. It degrades organic substances and mineralizes them into carbon dioxide, water, and nitrogen gas using dead or alive biomass. Bioremediation techniques involve bioaccumulation, biotransformation, and bioleaching, making them eco-friendly and cost-effective for removing heavy and toxic metals, such as Thiobacillus ferroxidases, from soil and water bodies [67].

Biosorbent Materials

Choosing a cost-effective, economical, and selective biosorbent is essential for treating wastewater. The pretreatment and experimental requirements of the bioagent determine the biosorbent's efficiency. Agro-waste is one of the many biological agents that have been employed in remediation [68]. Microorganisms can be very effective biosorbents for treating wastewater, including bacteria, fungus, yeast, and algae. The removal of heavy metals is facilitated by functional groups such as amide, amine, carbonyl, and carboxyl, which allow these organisms to absorb them at deficient concentrations. Nonetheless, the biosorption process is influenced by steric and conformational factors as well as the quantity and accessibility of reactive sites. Fungi, like Pleurotus sp. and Klebsiella oxytoca, display metal-binding capacity and are effective in lead-contaminated soil [69].

Algal Bioremediation (Phytoremediation)

Marine ecosystems are home to various species of algae, which are autotrophic organisms with low nutritional requirements and large biomass. Brown algae, among the three algal groups, have better biosorption capacity for metal ion biosorption. The type and structure of algal biomass and the charge and chemical constitution of heavy metal ions influence metal ion biosorption. Different algae, in live or dead forms, have been used for in-situ remediation. Algal proteins contain potential metal sites, which operate through complex formation methods during heavy metal remediation [69].

Phytoremediation

The phytoremediation of heavy metal-contaminated soil involves various mechanisms, including phytoextraction, phytostabilization, phytovolatilization, and rhizofiltration. Plants contain metal-binding proteins (MBPs) such as metallothioneins, phytochelatins, channel proteins, and metal-activated and metal-enzyme complexes. The significant presence of these chelating molecule in plant cells is believed to be responsible for hyperaccumulating heavy metals in various parts of the plant and induce greater tolerance or resistance [70]. The selection criteria for plants used in phytoremediation include being highly metal-tolerant, having a short life cycle, broad distribution, large biomass, and a translocation factor greater than one. The economic burden of soil remediation by physical or chemical methods are high whereas phytoremediation of soil in one acre costs approximately 60,000–1,000,000 US dollars only [71].

7 Conclusions

Ecotechnologies are necessary to address environmental challenges to minimizing our impact on the planet while maximizing resource efficiency. The integration of ecotechnology with mechanization has the potential to revolutionize industries and enhance productivity while reducing environmental harm. More sustainable and resilient future can be created by incorporating sustainable practices and technologies into various sectors, such as agriculture, energy, and waste management. Heavy metal contamination is one such issue that impacts all the systems of the environment. Numerous chemical, biological, and physical technologies have been developed due to the threat that these heavy metals impose. In order to have an efficient outcome in the long run, it is always suggested to adopt environmentally viable and economically sustainable approaches. One such effective option is adopting Ecotechnologies. By implementing these ecotechnologies, we can create a more sustainable and resilient agricultural system that meets the demands of a growing population while safeguarding the health of both humans and the planet.

Acknowledgements The support and resources provided by the Department of Environmental Science at GITAM School of Science, GITAM (Deemed to be University), Visakhapatnam, have been instrumental in enabling the authors to conduct their research and contribute to advancing this field.

References

1. Schwarz A (2022) Ecotechnology. In: Oxford research encyclopedia of environmental science
2. Haddaway NR, McConville J, Piniewski M (2018) How is the term 'ecotechnology' used in the research literature? a systematic review with thematic synthesis. Ecohydrol Hydrobiol 18(3):247–261. https://doi.org/10.1016/j.ecohyd.2018.06.008
3. Kataki S, Chatterjee S, Vairale MG, Dwivedi SK, Gupta DK (2021) Constructed wetland, an ecotechnology for wastewater treatment: a review on types of wastewater treated and components of the technology (macrophyte, biolfilm and substrate). J Environ Manage 283:111986
4. Paraschiv LS, Paraschiv S (2023) Contribution of renewable energy (hydro, wind, solar and biomass) to decarbonization and transformation of the electricity generation sector for sustainable development. Energy Rep 9:535–544
5. Abdel-Shafy HI, Mansour MS (2018) Solid waste issue: sources, composition, disposal, recycling, and valorization. Egypt J Pet 27(4):1275–1290
6. Amaral-Zettler LA, Zettler ER, Mincer TJ (2020) Ecology of the plastisphere. Nat Rev Microbiol 18(3):139–151
7. Sayara T, Basheer-Salimia R, Hawamde F, Sánchez A (2020) Recycling of organic wastes through composting: process performance and compost application in agriculture. Agronomy 10(11):1838
8. Roopnarain A, Rama H, Ndaba B, Bello-Akinosho M, Bamuza-Pemu E, Adeleke R (2021) Unravelling the anaerobic digestion 'black box': biotechnological approaches for process optimization. Renew Sustain Energy Rev 152:111717
9. Umme A, Ritambhara B, Shiv KS, Anagbogu FC (2023) Biofuel production via phyto and microbial power systems: adaptation of ecotechnology as a step toward sustainable energy. IA in OF for SD 161182
10. Gann GD, McDonald T, Walder B, Aronson J, Nelson CR, Jonson J, Hallett JG, Eisenberg C, Guariguata MR, Liu J, Hua F (2019) International principles and standards for the practice of ecological restoration. Restor Ecol 27(S1):S1–S46
11. Straškraba M (1993) Ecotechnology as a new means for environmental management. Ecol Eng 2(4):311–331. https://doi.org/10.1016/0925-8574(93)90001-V
12. Zhang Y, Mao Y, Jiao L, Shuai C, Zhang H (2021) Eco-efficiency, ecotechnology innovation and eco-well-being performance to improve global sustainable development. Environ Impact Assess Rev 89:106580. https://doi.org/10.1016/j.eiar.2021.106580
13. Huang Z, Fan H, Shen L, Du X (2021) Policy instruments for addressing construction equipment emission—a research review from a global perspective. Environ Impact Assess Rev 86:106486. https://doi.org/10.1016/j.eiar.2020.106486
14. IEA (2017). Data and statistics of international energy agency. Retrieved from https://www.iea.org/dataanstatistics?country=WORLD&fuel=Energy%20supply&indicator=Total%20primary%20energy%20supply%20(TPES)%20by%20source
15. Caiado RGG, de Freitas Dias R, Mattos LV, Quelhas OLG, Leal Filho W (2017) Towards sustainable development through the perspective of eco-efficiency-a systematic literature review. J Clean Prod 165:890–904. https://doi.org/10.1016/j.jclepro.2017.07.166
16. Schaltegger S, Sturm A (1990) Öologische rationalität (German/in English: environmental rationality). Die Unternehmung 4(4):117–131
17. Yang L, Zhang X (2018) Assessing regional eco-efficiency from the perspective of resource, environmental and economic performance in China: a bootstrapping approach in global data envelopment analysis. J Clean Prod 173:100–111. https://doi.org/10.1016/j.jclepro.2016.07.166
18. Mavi RK, Saen RF, Goh M (2019) Joint analysis of eco-efficiency and eco-innovation with common weights in two-stage network DEA: a big data approach. Technol Forecast Soc Chang 144:553–562. https://doi.org/10.1016/j.techfore.2018.01.035

19. Zhang S, Zhu D, Shi Q, Cheng M (2018) Which countries are more ecologically efficient in improving human well-being? an application of the index of ecological well-being performance. Resour Conserv Recycl 129:112–119. https://doi.org/10.1016/j.resconrec.2017.10.015
20. Lin Z, Huimin Y, Yunfeng H, Zhichao X, Yu X, Gaodi X, Jianxia M, Jijun W (2017) Overview of ecological restoration technologies and evaluation systems. J Resour Ecol 8(4):315–324. https://doi.org/10.5814/j.issn.1674-764x.2017.04.002
21. Hu X, Luo H, Guo M, Wang J (2022) Ecological technology evaluation model and its application based on logistic regression. Ecol Ind 136:108641. https://doi.org/10.1016/j.ecolind.2022.108641
22. Kumar M, Seth A, Singh AK, Rajput MS, Sikandar M (2021) Remediation strategies for heavy metals contaminated ecosystem: a review. Environ Sustain Indicat 12:100155. https://doi.org/10.1016/j.indic.2021.100155
23. Kumar S, Sindhu SS, Kumar R (2022) Biofertilizers: an ecofriendly technology for nutrient recycling and environmental sustainability. Curr Res Microb Sci 3:100094. https://doi.org/10.1016/j.crmicr.2021.100094
24. Jaiswal KK, Chowdhury CR, Yadav D, Verma R, Dutta S, Jaiswal KS, Karuppasamy KSK (2022) Renewable and sustainable clean energy development and impact on social, economic, and environmental health. Energy Nexus 7:100118. https://doi.org/10.1016/j.nexus.2022.100118
25. Rajasulochana P, Preethy V (2016) Comparison on efficiency of various techniques in the treatment of waste and sewage water–a comprehensive review. Resour Eff Technol 2(4):175–184. https://doi.org/10.1016/j.reffit.2016.09.004
26. Bala S, Garg D, Thirumalesh BV, Sharma M, Sridhar K, Inbaraj BS, Tripathi M (2022) Recent strategies for Bioremediation of emerging pollutants: a review for a green and sustainable environment. Toxics 10(8):484. https://doi.org/10.3390/toxics10080484
27. Ayangbenro AS, Babalola OO (2017) A new strategy for heavy metal polluted environments: a review of microbial biosorbents. Int J Environ Res Public Health 14(1):94. https://doi.org/10.3390/ijerph14010094
28. Lv H, Yang Z, Pan H, Wu R (2022) Electromagnetic absorption materials: current progress and new frontiers. Prog Mater Sci 127:100946
29. Elbasiouny H, Darwesh M, Elbeltagy H, Abo-Alhamd FG, Amer AA, Elsegaiy MA, Khattab IA, Elsharawy EA, Ebehiry F, El-Ramady H, Brevik EC (2021) Ecofriendly remediation technologies for wastewater contaminated with heavy metals with special focus on using water hyacinth and black tea wastes: a review. Environ Monit Assess 193(7):449. https://doi.org/10.1007/s10661-021-09236-2
30. Inês C, Guilherme PL, Esther MG, Swantje G, Stephen H, Lars H (2020) Regulatory challenges and opportunities for collective renewable energy prosumers in the EU. Energy Policy 138:111212
31. Anjum M, Miandad R, Waqas M, Gehany F, Barakat MA (2019) Remediation of wastewater using various nano-materials. Arab J Chem 12(8):4897–4919. https://doi.org/10.1016/j.arabjc.2016.10.004
32. Knapp S, Schmauck S, Zehnsdorf A (2019) Biodiversity impact of green roofs and constructed wetlands as progressive ecotechnologies in urban areas. Sustainability 11(20):5846. https://doi.org/10.3390/su11205846
33. Anderson CC, Renaud FG, Hanscomb S, Gonzalez-Ollauri A (2022) Green, hybrid, or grey disaster risk reduction measures: what shapes public preferences for nature-based solutions? J Environ Manage 310:114727. https://doi.org/10.1016/j.jenvman.2022.114727
34. Amon DJ, Gollner S, Morato T, Smith CR, Chen C, Christiansen S, Currie B, Drazen JC, Fukushima T, Gianni M, Gjerde KM (2022) Assessment of scientific gaps related to the effective environmental management of deep-seabed mining. Mar Policy 138:105006
35. Mi N, Peng Y, Qin Z, Fan B, Chen Q, Gai J (2022) Synergistic effects of microorganisms and passivation materials on the growth and Cd uptake of coriander (Coriandrum sativum L.) in Cd-contaminated soils. Rhizosphere 24:100604. https://doi.org/10.1016/j.rhisph.2022.100604

36. Selvi A, Rajasekar A, Theerthagiri J, Ananthaselvam A, Sathishkumar K, Madhavan J, Rahman PK (2019) Integrated remediation processes toward heavy metal removal/recovery from various environments-a review. Front Environ Sci 7:66. https://doi.org/10.3389/fenvs.2019.00066
37. Robbins EC, Guo J, Adams CD (2014) Removal of As (III) and As (V) in surface modified ceramic filters. J Water Sanit Hygiene Develop 4(2):214–222. https://doi.org/10.2166/washdev.2014.114
38. Yang X, Huang G, An C, Chen X, Shen J, Yin J, Song P, Xu Z, Li Y (2021) Removal of arsenic from water through ceramic filter modified by nano-CeO2: a cost-effective approach for remote areas. Sci Total Environ 750:141510. https://doi.org/10.1016/j.scitotenv.2020.141510
39. Farrow C, McBean E, Huang G, Yang A, Wu Y, Liu Z, Dai Z, Fu HY, Cawte T, Li YP (2018) Ceramic water filters: a point-of-use water treatment technology to remove bacteria from drinking water in Longhai City, Fujian Province. J Environ Inf 32(2):63–68
40. Pérez-Vidal A, Silva-Leal JA, Diaz-Gómez J, Meneses-Torres CJ, Arias-Vallejo JE, Giraldo-Tenorio GA, Bacca-Jordán MC (2021) Performance evaluation of ceramic pot filters combined with adsorption processes for the removal of heavy metals and phenolic compounds. J Water Health 19(5):750–761. https://doi.org/10.2166/wh.2021.052
41. Boussouga YA, Mohankumar MB, Gopalakrishnan A, Welle A, Schäfer AI (2021) Removal of arsenic (III) via nanofiltration: contribution of organic matter interactions. Water Res 201:117315. https://doi.org/10.1016/j.watres.2021.117315
42. Fu ZJ, Jiang SK, Chao XY, Zhang CX, Shi Q, Wang ZY, Liu ML, Sun SP (2022) Removing miscellaneous heavy metals by all-in-one ion exchange-nanofiltration membrane. Water Res 222:118888. https://doi.org/10.1016/j.watres.2022.118888
43. Peydayesh M, Mezzenga R (2021) Protein nanofibrils for next generation sustainable water purification. Nat Commun 12(1):3248
44. Edgar M, Boyer TH (2021) Removal of natural organic matter by ion exchange: comparing regenerated and non-regenerated columns. Water Res 189:116661. https://doi.org/10.1016/j.watres.2020.116661
45. Peng P, Barnes M, Wang C, Wang W, Li S, Swanson HL, Dardick W, Tao S (2018) A meta-analysis on the relation between reading and working memory. Psychol Bull 144(1):48
46. Su S, Du L, Jiang S (2021) Learning from the past: development of safe and effective COVID-19 vaccines. Nat Rev Microbiol 19(3):211–219
47. Kulkarni SJ, Dhokpande SR, Kaware JP (2013) Studies on flyash as an adsorbent for removal of various pollutants from wastewater. Int J Eng Res Technol (IJERT) 2(5):1190–1195
48. Kulkarni SJ, Dhokpande SR, Kaware JP (2013) Studies on flyash as an adsorbent for removal of various pollutants from wastewater. Int J Eng Res Technol (IJERT) 2(5):1190–1195
49. Acharya BS, Kharel G (2020) Acid mine drainage from coal mining in the United States–an overview. J Hydrol 588:125061. https://doi.org/10.1016/j.jhydrol.2020.125061
50. Sánchez-Andrea I, Sanz JL, Bijmans MF, Stams AJ (2014) Sulfate reduction at low pH to remediate acid mine drainage. J Hazard Mater 269:98–109. https://doi.org/10.1016/j.jhazmat.2013.12.032
51. Imaniar KR, Tazkiaturrizki T, Rinanti A (2021) Fipronil removal at various temperature and pollutant concentration by using Pseudomonas aeruginosa and Brevibacterium sp. in liquid media. In: IOP conference series: materials science and engineering, vol 1098, no 5, IOP Publishing, pp 052071. https://doi.org/10.1088/1757-899X/1098/5/052071
52. Sihotang MSM, Rinanti A, Fachrul MF (2021) Heavy metal removal and acid mine drainage neutralization with bioremediation approach. In: IOP conference series: earth and environmental science, vol 894, no 1, IOP Publishing, pp 012041. https://doi.org/10.1088/1755-1315/894/1/012041
53. Lupa L, Cocheci L (2023). Heavy metals removal from water and wastewater. https://doi.org/10.5772/intechopen.110228
54. Al Kindi GY, Gomaa GF, Abd Ulkareem FA (2020) Combined adsorbent is the best (chemical and natural) coagulation process for removing some heavy metals from wastewater. Int J Environ Sci Technol 17(7):3431–3448. https://doi.org/10.1007/s13762-020-02702-3

55. Alazaiza MY, Albahnasawi A, Ali GA, Bashir MJ, Nassani DE, Al Maskari T, Amr SSA, Abujazar MSS (2022) Application of natural coagulants for pharmaceutical removal from water and wastewater: a review. Water 14(2):140. https://doi.org/10.3390/w14020140
56. Fu F, Wang Q (2011) Removal of heavy metal ions from wastewaters: a review. J Environ Manage 92(3):407–418. https://doi.org/10.1016/j.jenvman.2010.11.011
57. Vishali S, Kavitha E (2021) Application of membrane-based hybrid process on paint industry wastewater treatment. In: Membrane-based hybrid processes for wastewater treatment, Elsevier, pp 97–117
58. Hussain A, Madan S, Madan R (2021) Removal of heavy metals from wastewater by adsorption. Heavy Metals—Their Environmental Impacts and Mitigation
59. Singh N, Gupta SK (2016) Adsorption of heavy metals: a review. Int J Innov Res Sci Eng Technol 5(2):2267–2281
60. Bhatnagar A, Sillanpää M, Witek-Krowiak A (2015) Agricultural waste peels as versatile biomass for water purification–a review. Chem Eng J 270:244–271. https://doi.org/10.1016/j.cej.2015.01.135
61. Park JH, Eom JH, Lee SL, Hwang SW, Kim SH, Kang SW, Yun JJ, Cho JS, Lee YH, Seo DC (2020) Exploration of the potential capacity of fly ash and bottom ash derived from wood pellet-based thermal power plant for heavy metal removal. Sci Total Environ 740:140205. https://doi.org/10.1016/j.scitotenv.2020.140205
62. Sarkar C, Basu JK, Samanta AN (2017) Removal of Ni2+ ion from waste water by Geopolymeric Adsorbent derived from LD slag. J Water Process Eng 17:237–244. https://doi.org/10.1016/j.jwpe.2017.04.012
63. Varjani SJ, Joshi RR, Senthil Kumar P, Srivastava VK, Kumar V, Banerjee C, Praveen Kumar R (2018) Polycyclic aromatic hydrocarbons from petroleum oil industry activities: effect on human health and their biodegradation. Waste Bioremed 185–199
64. Hasani AH, Hosseinzadeh I, Torabifar B (2010) Investigation of using fixed activated sludge system for removing heavy metals (Cr, Ni and Pb) from industrial wastewater
65. Bhat SA, Bashir O, Haq SAU, Amin T, Rafiq A, Ali M, Américo-Pinheiro JHP, Sher F (2022) Phytoremediation of heavy metals in soil and water: an eco-friendly, sustainable and multidisciplinary approach. Chemosphere 303:134788
66. Gadd GM (2010) Metals, minerals and microbes: geomicrobiology and bioremediation. Microbiology 156(3):609–643. https://doi.org/10.1099/mic.0.037143-0
67. Zhou B, Zhang T, Wang F (2023) Microbial-based heavy metal bioremediation: toxicity and eco-friendly approaches to heavy metal decontamination. Appl Sci 13(14):8439
68. Morin E, Kohler A, Baker AR, Foulongne-Oriol M, Lombard V, Nagy LG, Ohm RA, Patyshakuliyeva A, Brun A, Aerts AL, Bailey AM (2013) Erratum: Genome sequence of the button mushroom Agaricus bisporus reveals mechanisms governing adaptation to a humic-rich ecological niche (Proceedings of the National Academy of Sciences of the United States of America. In: Proceedings of the national academy of sciences of the United States of America 109 (17501–17506):110(10)
69. Kapahi M, Sachdeva S (2019) Bioremediation options for heavy metal pollution. J Health Pollut 9(24):191203
70. Memon AR, Schröder P (2009) Implications of metal accumulation mechanisms to phytoremediation. Environ Sci Pollut Res 16:162–175
71. Sharma JK, Kumar N, Singh NP, Santal AR (2023) Phytoremediation technologies and their mechanism for removal of heavy metal from contaminated soil: an approach for a sustainable environment. Front Plant Sci 14:1076876. https://doi.org/10.3389/fpls.2023.1076876
72. Sailesh AR, Riyazuddin S, Kumar KS, Chakraborthy A, Namuduri S (2023) Application of electroremediation coupled with phytoremediation techniques for the removal of trace metals in sewage Sludge. Phytoremediation: management of environmental contaminants, vol 7. Springer International Publishing, Cham, pp 129–144
73. Lemessa G, Gabbiye N, Alemayehu E (2023) Waste to resource: Utilization of waste bagasse as an alternative adsorbent to remove heavy metals from wastewaters in sub-Saharan Africa: a review. Water Pract Technol 18(2):393–407

74. Türkmen D, Bakhshpour M, Akgönüllü S, Aşır S, Denizli A (2022) Heavy metal ions removal from wastewater using cryogels: a review. Front Sustain 3:765592. https://doi.org/10.3389/frsus.2022.765592
75. Abidli A, Huang Y, Rejeb ZB, Zaoui A, Park CB (2022) Sustainable and efficient technologies for removing and recovering toxic and valuable metals from wastewater: recent progress, challenges, and future perspectives. Chemosphere 292:133102. https://doi.org/10.1016/j.chemosphere.2021.133102
76. Shrestha R, Ban S, Devkota S, Sharma S, Joshi R, Tiwari AP, Kim HY, Joshi MK (2021) Technological trends in heavy metals removal from industrial wastewater: a review. J Environ Chem Eng 9(4):105688. https://doi.org/10.1016/j.jece.2021.105688
77. Nedjimi B (2021) Phytoremediation: a sustainable environmental technology for heavy metals decontamination. SN Appl Sci 3(3):286. https://doi.org/10.1007/s42452-021-04301-4
78. Saeed T, Alam MK, Miah MJ, Majed N (2021) Removal of heavy metals in subsurface flow constructed wetlands: application of effluent recirculation. Environ Sustain Indicat 12:100146. https://doi.org/10.1016/j.indic.2021.100146
79. Chang Y, Dang Q, Samo I, Li Y, Li X, Zhang G, Chang Z (2020) Electrochemical heavy metal removal from water using PVC waste-derived NS co-doped carbon materials. RSC Adv 10(7):4064–4070. https://doi.org/10.1039/C9RA09237D
80. Medfu Tarekegn M, Zewdu Salilih F, Ishetu AI (2020) Microbes used as a tool for Bioremediation of heavy metal from the environment. Cogent Food Agricult 6(1):1783174. https://doi.org/10.1080/23311932.2020.1783174
81. Geiger EM, Sarkar D, Datta R (2019) Evaluation of copper-contaminated marginal land for the cultivation of vetiver grass (Chrysopogon zizanioides) as a lignocellulosic feedstock and its impact on downstream bioethanol production. Appl Sci 9(13):2685. https://doi.org/10.3390/app9132685

MIX
Papier aus verantwortungsvollen Quellen
Paper from responsible sources
FSC® C105338

If you have any concerns about our products,
you can contact us on
ProductSafety@springernature.com

In case Publisher is established outside the EU,
the EU authorized representative is:
Springer Nature Customer Service Center GmbH
Europaplatz 3, 69115 Heidelberg, Germany

Printed by Libri Plureos GmbH
in Hamburg, Germany